Springer Series in Optical Sciences Volume 17

Edited by David L. MacAdam

Springer Series in Optical Sciences

1 **Solid-State Laser Engineering**
By W. Koechner

2 **Table of Laser Lines in Gases and Vapors** 2nd Edition
By R. Beck, W. Englisch, and K. Gürs

3 **Tunable Lasers and Applications**
Editors: A. Mooradian, T. Jaeger, and P. Stokseth

4 **Nonlinear Laser Spectroscopy**
By V. S. Letokhov and V. P. Chebotayev

5 **Optics and Lasers** An Engineering Physics Approach
By M. Young

6 **Photoelectron Statistics** With Applications to Spectroscopy and Optical Communication
By B. Saleh

7 **Laser Spectroscopy III**
Editors: J. L. Hall and J. L. Carlsten

8 **Frontiers in Visual Science**
Editors: S. J. Cool and E. J. Smith III

9 **High-Power Lasers and Applications** 2nd Printing
Editors: K.-L. Kompa and H. Walther

10 **Detection of Optical and Infrared Radiation**
By R. H. Kingston

11 **Matrix Theory of Photoelasticity**
By P. S. Theocaris and E. E. Gdoutos

12 **The Monte Carlo Method in Atmospheric Optics**
By G. I. Marchuk, G. A. Mikhailov, M. A. Nazaraliev, R. A. Darbinian, B. A. Kargin, and B. S. Elepov

13 **Physiological Optics**
By Y. Le Grand and S. G. El Hage

14 **Laser Crystals** Physics and Properties
By A. A. Kaminskii

15 **X-Ray Spectroscopy**
By B. K. Agarwal

16 **Holographic Interferometry** From the Scope of Deformation Analysis of Opaque Bodies
By W. Schumann and M. Dubas

17 **Nonlinear Optics of Free Atoms and Molecules**
By D. C. Hanna, M. A. Yuratich, D. Cotter

18 **Holography in Medicine and Biology**
Editor: G. von Bally

19 **Color Theory and Its Application in Art and Design**
By G. A. Agoston

20 **Interferometry by Holography**
By J. Ostrowskij, M. Butussov, G. Ostrowskaja

21 **Laser Spectroscopy IV**
Editors: H. Walther, K. W. Rothe

D. C. Hanna   M. A. Yuratich   D. Cotter

Nonlinear Optics of Free Atoms and Molecules

With 89 Figures

Springer-Verlag Berlin Heidelberg GmbH 1979

DAVID C. HANNA, Ph. D.
MICHAEL A. YURATICH, Ph. D.

Department of Electronics, The University of Southampton
Highfield, Southampton S09 5NH, Great Britain

DAVID COTTER, Ph. D.

Max-Planck-Institut für biophysikalische Chemie
Abt. Laserphysik, D-3400 Göttingen, Fed. Rep. of Germany

ISBN 978-3-662-13493-1 ISBN 978-3-540-34766-8 (eBook)
DOI 10.1007/978-3-540-34766-8

Library of Congress Cataloging in Publication Data. Hanna, David C, 1941-. Nonlinear optics of free atoms and molecules. (Springer series in optical sciences; v. 17) Bibliography: p. Includes index. 1. Nonlinear optics. 2. Atoms. 3. Molecules. I. Yuratich, Michael A., 1952- joint author. II. Cotter, David, 1950- joint author. III. Title. QC 446.2.H36 535'.2 79-19828

Originally published by Springer-Verlag Berlin Heidelberg New York in 1979.
Softcover reprint of the hardcover 1st edition 1979

2153/3130-543210

Preface

Laser physics and nonlinear optics are fields which have been intimately connected from their beginning. Nonlinear optical effects such as second-harmonic generation fulfil vital functions in many laser systems. Conversely advances in laser development quickly lead to progress in nonlinear optics. Of particular importance has been the development of tunable visible and uv lasers. With the ability to tune the laser frequency into close resonance with atomic transition frequencies, one can produce a large resonance enhancement of the nonlinearity. This permits the observation of a great variety of nonlinear optical processes in dilute media such as atomic vapours. In recent years much of the research effort in nonlinear optics has been directed towards the use of such media, and it is this area which forms the subject of the present book.

We review a wide range of nonlinear optical processes in atomic vapours, molecular gases and cryogenic liquids. At the same time we have tried to treat the subject in sufficient depth to be useful to research workers in the field. To achieve this, a measure of selectivity has been introduced by emphasising those nonlinear processes which are seen to have applications as sources of tunable coherent radiation. Thus we have not discussed in any detail those nonlinear processes whose main applications are in spectroscopy, such as Doppler-free two-photon absorption. However, much of the background is appropriate to just such topics as these and we believe that the book should therefore be of use to people working in a number of fields where lasers are involved.

The book divides broadly into two parts; general theory (Chaps.2 and 3) and experimental aspects (Chaps.4-8). In the theoretical chapters we start by considering the nonlinear optical susceptibilities which form the cornerstone of most discussions of nonlinear optics. Then, other approaches to the theory of the nonlinear polarisation are introduced; for example, the ideas of adiabatic ("dressed") states and multiphoton Bloch equations and vector models. The latter approaches emphasise the resonant nature of the process and are a natural extension of ideas developed originally for resonant interactions in two-level atoms.

The close connection between these apparently diverse descriptions is explored in some detail. The other main task of the theoretical section, dealt with in Chap.3, is a consideration of the propagation behaviour of the interacting fields, thus leading to expressions for experimentally important quantities, such as the intensity of the generated radiation. An underlying aim throughout the theoretical section has been to provide a consistent and full description of the notations employed and to give the various formulae in a form suitable for use in calculations. In keeping with modern conventions we use SI-units throughout; the relations between SI and esu definitions are collected in an appendix.

Some readers may not wish to go deeply into the theory at first, and the experimental chapters have been written with this in mind. Thus the formulae given in the experimental chapters usually take a simpler form than their more general counterparts in the theory section. Chapters 4-6 deal with experimental aspects of atomic vapours as nonlinear media. Chapter 7 deals with molecular gases and also molecular cryogenic liquids, drawing extensively on the principles established in earlier sections. Thus attention is focused on the main differences vis à vis atomic vapours, primarily the much richer energy level structure in molecules, the much higher number densities involved, and the fact that resonances occur in the infrared.

Finally, a number of further topics are covered in Chap.8: multipole processes, laser-induced inelastic collisions and phase conjugation. These topics give a good illustration of the variety of possible nonlinear effects, and while they do not fit readily into the earlier experimental sections, they share with these the same theoretical background.

In writing this book we have benefited greatly from discussions with many colleagues, too numerous to mention individually. However, a special word of thanks should go to Mrs. Sue Meen, who typed the draft manuscript with patience and good humour. MAY held a Ramsay Memorial fellowship during the period of writing the book and wishes to thank the trustees for their support. DCH, who spent a sabbatical year in Munich during the final stage of completion of the book wishes to thank the Alexander von Humboldt foundation for its support, and Professor Herbert Walther for his hospitality.

Finally, we all wish to thank our wives for tolerating our long periods of distractions and for giving us so much encouragement.

October 1979 *David C. Hanna, Michael A. Yuratich, David Cotter*

Contents

1. Introduction

Soon after the invention of the laser it was demonstrated that the intense light flux available could produce nonlinear optical effects in various media. The first such demonstration was by FRANKEN et al. [1.1], in which radiation at the second harmonic of the ruby laser was produced when a ruby laser beam was focussed into a quartz crystal. A phenomenological description of this and other nonlinear effects can be given by expressing the polarisation P induced in the medium as a series in ascending powers of the applied field E,

$$P = \varepsilon_0\chi^{(1)}E + \varepsilon_0\chi^{(2)}E^2 + \varepsilon_0\chi^{(3)}E^3 + \dots \quad . \tag{1.1}$$

The first term, containing the first-order susceptibility, describes the familiar linear optical property of an index of refraction. Franken's experiment confirmed the existence of the second term and showed that with the optical field strengths available from lasers, effects due to $\chi^{(2)}$ could become significant. At about the same time, KAISER and GARRETT [1.2] observed another nonlinear optical effect produced by a laser beam—two-photon absorption. This arises from the $\chi^{(3)}$ term.

Following these experiments, there was a rapid growth of interest in the new field of nonlinear optics. There have been three main motivations. First, there is the possibility of exploiting the nonlinear behaviour in various devices. The most important of these are frequency converters, in which laser radiation at one frequency is converted, for example by harmonic generation, sum-frequency generation or difference-frequency generation, into coherent radiation at a new frequency, often with a high efficiency. Because the converted radiation may be at a frequency that is not directly available from a laser source, these frequency-conversion techniques provide an important means of extending the spectral range covered by coherent sources. A second reason for studying nonlinear optical processes is that they set a limit to the light flux that can be passed through a medium. For example, two-photon absorption, stimulated Raman and Brillouin scattering can lead to depletion of the incident light, and self-focussing leads to distortion of the incident-beam profile. Too high an intensity may even lead to irreversible changes in the medium (e.g., damage in the case of solids). A third interest in nonlinear optical effects lies in their use as a means of obtaining information about the microscopic properties of the atoms or molecules that constitute the nonlinear medium. Examples of this are the use of two-photon absorption and coherent anti-Stokes Raman scattering to study energy levels that are inaccessible by single-photon absorption.

Although much of the early nonlinear optics work was on condensed matter, more recently there has been a shift of interest towards gaseous media. The object of this book is to review the progress made in the nonlinear optics of vapours and gases. The theoretical background is presented in a general way, but in discussing uses of these nonlinear effects we place strong emphasis on frequency-conversion applications. We shall also examine a number of other nonlinear processes that compete with and, therefore, impose limitations on the efficiency of the desired frequency-conversion process. First, however, we examine the reasons for the shift of interest from crystals to vapours.

At first crystalline nonlinear media were seen to offer the greatest device potential. The reasons for this view were twofold. First it can be shown by a simple symmetry argument that $\chi^{(2)}$ in (1.1) is zero unless the medium lacks a centre of symmetry. Nonlinear effects in centrosymmetric media (such as atomic vapours and non optically active molecular liquids and gases) would thus depend on higher-order, and therefore presumably smaller, nonlinear terms. Second, it was shown by BASS et al. [1.3], GIORDMAINE [1.4] and MAKER et al. [1.5], that the birefringence of a crystalline medium could be used to match the phase velocities of fundamental and harmonic radiation by compensating the material dispersion. With this "phase-matching", a long interaction length (or coherence length) of up to a few centimetres became possible, and harmonic-conversion efficiencies could then easily reach tens of percent. These promising results were a considerable spur to activity, and at about this time a search began in a number of laboratories for suitable nonlinear crystals. The qualities looked for were many and demanding, including 1) large $\chi^{(2)}$, 2) sufficient birefringence to permit phase matching, 3) good optical quality over a crystal dimension of ~1cm, 4) ability to withstand high optical intensity and 5) good transparency to the incident and generated radiation. The results of this extensive research effort on crystal nonlinear optical devices are covered in a number of review articles

[1.6-10]. Despite the considerable successes of this research, disappointingly few materials have proved capable of satisfying the list of requirements above. Infrared and ultraviolet absorption in crystals are major limitations and these, with other shortcomings, have been the driving force behind attempts to exploit optical nonlinearities in gases and vapours.

A number of significant advantages are offered by gases and vapours. They can be easily prepared, with good optical quality over large dimensions. They do not suffer irreversible (and costly) damage at high intensities, and have good uv and ir transparency. The disadvantages of vapours are their low number densities compared with condensed matter and the fact that because these media have inversion symmetry, the most important nonlinear effects are due to the $\chi^{(3)}$ term. (This remark will need to be qualified when we consider magnetic-dipole and electric-quadrupole contributions to the nonlinear polarisation.) To observe effects due to $\chi^{(3)}$ comparable in magnitude to those due to $\chi^{(2)}$ in the better nonlinear crystals, one of two conditions must be met. Either a very large field strength E is required (for example by using intense mode-locked laser pulses) or the magnitude of $\chi^{(3)}$ must be made resonantly large by choosing the frequencies of the interacting waves to be close to resonance with transition frequencies of the medium.

Pioneering investigations into the nonlinear properties of gases were made by NEW and WARD [1.11], who generated third-harmonic radiation in several atomic and molecular gases (see also [1.12,13]). The conversion efficiencies were very low, however, because the fundamental frequency (provided by a ruby laser) and its third harmonic were well away from any resonance, and also, as the process was not phase-matched, the coherence length was short. HARRIS and MILES [1.14] then made a proposal which overcame these limitations. This involved the use of a metal vapour as the nonlinear medium, with a buffer gas to provide the necessary dispersion for phase matching [1.15]. Metal vapours can provide resonance transitions close to the laser frequencies commonly used; this leads to a much larger susceptibility. Thus, HARRIS and his co-workers went on to demonstrate experimentally [1.16] a value for $\chi^{(3)}$ in rubidium vapour which was 10^6 greater than the $\chi^{(3)}$ measured in He by WARD and NEW [1.12]. This result indicated that high conversion efficiencies should be possible for third-harmonic generation. With a high-power mode-locked Nd:YAG laser as the fundamental source, third-harmonic efficiencies of several percent have now been achieved [1.17,18]. By using harmonics of the Nd:YAG laser output, and then generating harmonics and sum frequencies from these [1.19-21] high powers have been generated at a number of discrete uv and vuv wavelengths.

For many applications, however, tunable uv and vuv sources are required and harmonic generation can provide this if a tunable fundamental source is used. Dye lasers offer the most convenient source of tunable visible radiation, although in their more simple and unsophisticated forms, the peak power available is considerably less than that from a mode-locked Nd:YAG laser. The disadvantage of lower power can, however, be compensated by the tunability, which, besides offering tunability of the generated harmonic, brings with it the possibility of also tuning the dye laser to a frequency at which $\chi^{(3)}$ is resonantly enhanced. For example, using two-photon resonant enhancement (i.e., with the dye laser tuned so that its frequency is half that of an allowed two-photon transition), HODGSON et al. [1.22] obtained significant powers over a wide range in the vuv by sum-frequency generation in Sr vapour of the outputs from two dye lasers pumped by a N_2 laser. Two other reports of the two-photon resonance technique were made around the same time [1.23,24]. BLOOM et al. [1.24] described an experiment in Na vapour in which they showed that with modest laser powers, "up-conversion" from ir to uv photons occurred with an efficiency of 50% (a power-conversion efficiency of ~1600%). Results like these suggest that third-harmonic generation and sum-frequency generation in vapours are likely to play an important role among future sources of coherent uv, vuv and even soft-X-ray radiation [1.25]. We have therefore devoted a whole chapter (Chap.4) to this topic.

In the process of third-harmonic generation (and sum- or difference-frequency generation) the atoms of the nonlinear medium are left, after the scattering process, in their initial state. Such processes are known as "parametric" processes. Characteristically, they need to be phase matched if they are to be efficient. There is another important class of nonlinear processes, which are termed "nonparametric", in which the final state of the atoms after scattering is different from the initial state. Such processes do not involve phase matching. [This distinction between parametric and nonparametric processes is rather loose. It will need qualification later (Chap.3), when we consider some processes that cannot be so clearly categorised.] Two-photon absorption and emission, and stimulated Raman scattering are important examples of nonparametric processes. In stimulated Raman scattering (SRS) a powerful pump wave at frequency ω_p produces a high gain at a Stokes frequency ω_s, where the frequency $\omega_p - \omega_s$ coincides with a Raman-active transition of the medium. Stokes photons generated by spontaneous Raman scattering and travelling in the direction of the pump beam undergo avalanche multiplication, thus creating a Stokes beam that propagates in the same direction as the pump beam, of a power approaching that of the pump beam. SRS is one of the more venerable nonlinear optical processes, being first observed quite unexpectedly by WOODBURY and NG [1.26], as an emission from the nitrobenzene in the Kerr-cell used to Q-switch their ruby laser. In fact, SRS is certainly one of the most easily demonstrated nonlinear optical effects. A number of factors combine to make this so. For example, it can occur in any medium (it is described by a third-order susceptibility $\chi^{(3)}$); furthermore, there is no phase-matching requirement. This means that very long interaction lengths can be achieved. By using a liquid or a high-pressure gas, a high number density is ensured and, despite involving a third-order process, the susceptibility per molecule is high. This is because the pump and Stokes frequencies are in two-photon resonance with the initial and final levels [i.e., $\hbar(\omega_p - \omega_s) = E_f - E_g$; see Fig.1.1]. Thus resonance is automatically achieved without the need, as in two-photon absorption, for a coincidence between twice the incident laser frequency and a transition frequency of the medium.

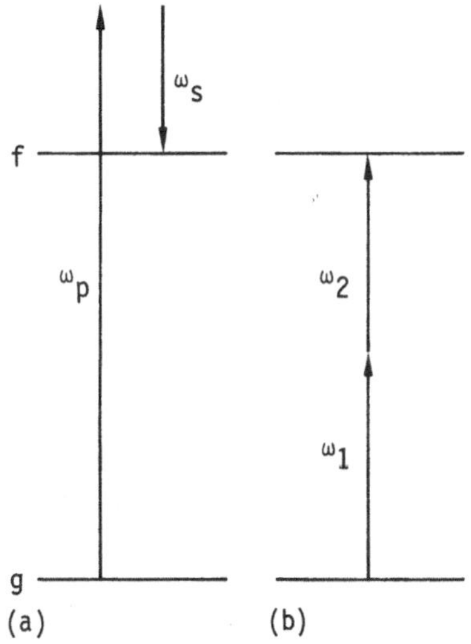

Fig.1.1a,b. Two-photon resonance. (a) Raman scattering $\hbar(\omega_p - \omega_s) = E_f - E_g$ (b) Two-photon absorption $\hbar(\omega_1 + \omega_2) = E_f - E_g$

Before the discovery of tunable dye lasers, SRS from ruby-laser radiation provided the nearest thing to tunable coherent radiation. A large number of organic liquids, and some high-pressure gases, each with different vibrational Stokes shifts, from a few hundred to a few thousand wave numbers, were found capable of SRS, thus providing high-power sources at many wavelengths in the near infrared region. With the arrival of dye lasers, this use of SRS declined. However, even before dye lasers became available the wavelengths

---

$\hbar$ is Planck's constant divided by 2π

thus obtained yielded some close coincidences with allowed transitions in alkali atoms and opened the way to observation of a wealth of nonlinear effects [1.27-31]. One of these effects was stimulated electronic Raman scattering (SERS) [1.32-34]. The word electronic is added to emphasise that the scattering transition is between the purely electronic levels of an atom rather than between the vibrational or rotational levels of a molecule. SERS in atoms differs from SRS in molecules in two important respects. The Stokes shifts are large ($\sim 10^4 cm^{-1}$ typically) and this means that the Stokes wavelength can be well into the infrared for pump wavelengths in the visible region. Also the pump wavelength can be close to an intermediate level; the Raman cross section is then very large (resonant Raman scattering). This in turn implies that the threshold for SERS is reached with a relatively low pump intensity, well within the capabilities of typical dye lasers. With a dye laser as the pump, its wavelength can be tuned to take full advantage of the resonant enhancement. More importantly, it means that the Stokes wavelength can be tuned. SERS, therefore, provides an extremely simple way of directly converting the tunable visible output of a dye laser to a tunable infrared output [1.35-40]. In this way, a major portion of the range 2-15 μm has been covered; over parts of this range, peak photon-conversion efficiencies as high as 50% have been achieved. Another more recent interest in SERS is in the efficient conversion of the uv output from excimer lasers into the visible region of the spectrum [1.41-43]. The subject of stimulated electronic Raman scattering is reviewed in Chap.5, as well as some examples of other related nonparametric processes such as stimulated anti-Stokes Raman scattering, stimulated hyper-Raman scattering, and two-photon emission.

The processes dealt with in Chaps.4,5 are examples of parametric and nonparametric interactions, respectively. There are, however, processes in which both parametric and nonparametric effects are inseparable. The theoretical background for this type of situation is given in Chap.3; in Chap.6 we review the experimental results for a particular example of this hybrid process which has been used for the generation of tunable infrared radiation [1.35]. Briefly, the process may be described as four-wave mixing, i.e., a scheme in which three frequencies ω_1, ω_2, ω_3 are mixed to generate a fourth, ω_4, given in this case by $\omega_4 = \omega_1 - \omega_2 - \omega_3$. The effect is enhanced by arranging that ω_1, ω_2 satisfy a two-photon resonance of the Raman type, i.e., $\omega_1 - \omega_2$ is equal to a Raman transition frequency of the medium (see Fig.1.2). However, instead of supplying ω_2 externally from a laser, it is also generated within the medium by stimulated Raman scattering of the powerful pump wave ω_1. Thus, the overall process can be thought of as a combination of

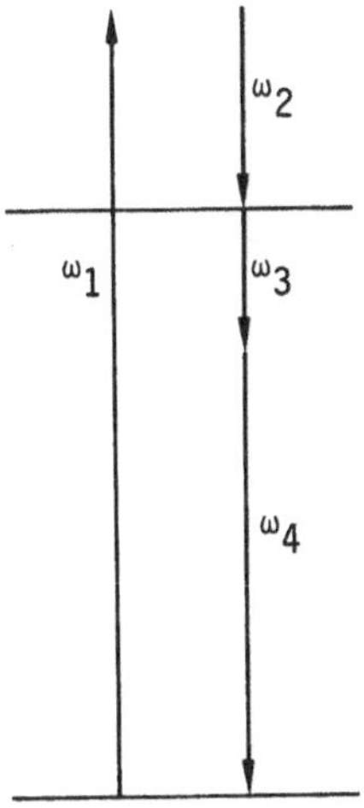

Fig.1.2. Generation of $\omega_4 = \omega_1 - \omega_2 - \omega_3$ by four-wave Raman-resonant difference mixing

nonparametric generation (of ω_2 by SRS) and parametric generation of $\omega_4 = \omega_1 - \omega_2 - \omega_3$ by four-wave mixing. In addition the two-photon absorption of the generated wave has to be considered, because $\omega_3 + \omega_4$ $(= \omega_1 - \omega_2)$ also satisfies two-photon resonance. An interesting result of this type of process is that the efficiency is less critically dependent on phase matching than in a purely parametric process. Although this technique has resulted in generation over wide ranges of the infrared, the observed efficiencies have generally been rather low.

The results reviewed in Chaps.4-6 refer to experiments on atomic-vapour systems. The same ideas of two-photon resonant enhancement can also be applied to molecules; there is growing activity in this field. This is the subject of Chap.7. The main feature of interest is that, owing to their vibrational and rotational motions, molecules have energy-level spacings that correspond to infrared frequencies. Two-photon resonant enhancement can, therefore, be used to achieve efficient nonlinear generation in the ir. For example, BRUECK and KILDAL [1.44,45] used this technique to generate the third harmonic of a CO_2 laser output with an efficiency of ~4%. This is several orders of magnitude more efficient than the best THG conversion efficiency for the CO_2 laser, using the third-order susceptibility of nonlinear crystalline media [1.46]. There is also a revival of interest in SRS in high-pressure molecular gases (such as H_2, D_2 and CH_4) as a means of efficient tunable ir generation [1.47-54], tunable uv generation [1.55,56], efficient Raman shifting of excimer lasers [1.57,58] and high energy pulse compression [1.59,60].

Because the subject of SRS in gases has been extensively covered in the early literature, we shall not discuss this any further in Chap.7. There have also been recent demonstrations of Raman-resonant four-wave mixing in H_2, using a scheme analogous to that described in Chap.6 or variants of it [1.61-65]. Because the results of this work complement the discussion of Chap.6 (and the common theoretical ideas of Sect.3.4) we give a brief appraisal in Sect.7.5. One further topic concludes Chap.7; the extension of SRS generation to the far infrared by the resonance Raman effect in polar molecules [1.61,67].

Most of the necessary theoretical background has been gathered together in Chaps.2,3. Many readers will probably not wish to venture deeply into the theory (the complexity of notation for high-order nonlinearities can be a powerful repellant). An attempt has therefore been made to present the experimental work of Chaps.4-7 in a self-contained way, when possible. Thus, the formulae used in the experimental sections are often simplified particular cases of more general results from Chaps.2,3, for example by assuming all interacting waves to be linearly polarised in the same direction. Nevertheless, we have felt it worthwhile to devote considerable space to the theory for a number of reasons. First, we have set out to establish unambiguously the notation we use for nonlinear susceptibilities, because difficulties have been compounded by different authors using different units and definitions. The most notorious offenders have been factors of two; in discussing second-order susceptibilities ROBINSON [1.68] said of these different conventions that "in particularly fertile ground, these various factors can luxuriate and blossom as factors of 8 in the final answer". Of course, these problems become even greater with higher-order susceptibilities. Expressions for these susceptibilities have been derived by a number of authors using time-dependent perturbation theory [1.69-71]. With some modification of notation to bring it into line with the now more common conventions (including the use of SI units) we adopt BUTCHER's [1.70] expressions for the n'th order susceptibilities as the starting point of Chap.2. (The relations between SI and esu definitions are presented in an appendix.) From these general expressions, we then derive susceptibilities for the particular processes to be discussed in later sections, e.g., third-harmonic generation, sum- and difference-frequency generation, stimulated Raman and hyper-Raman scattering, two-photon absorption, etc.

Roughly half of Chap.2 is given to this task of providing the expressions needed in the subsequent experimental sections. The other main aim we have pursued in Chap.2 is to relate the susceptibility formalism of nonlinear optics to the very different approaches that address themselves to the resonant interaction of radiation with a two-level or three-level atom. In the early days of nonlinear optics, when nonresonant phenomena such as second-harmonic generation in crystals were the major interest, it was natural to express the nonlinear behaviour in terms of susceptibilities, by use of a series expansion, as in (1.1). The more recent interest in resonant nonlinear interactions has seen a growing use of nonperturbative treatments, such as the vector model of FEYNMAN et al. [1.12]. Because the relationship between these two approaches is not always obvious and yet both approaches are widely used, we have given a fairly lengthy discussion of this topic.

An essential feature of nonlinear optics is that the physical dimensions of the medium are usually large compared to the optical wavelength. Wave-propagation effects are therefore important; this forms the subject of Chap.3. Because the emphasis of this review is on device applications, the main aim in Chap.3 has been to produce expressions for the power generated in various nonlinear processes. The analysis is restricted to a plane-wave treatment in Chap.3, although the fact that the fields are confined to a beam can entail important modifications to the results of the plane-wave analysis. For example, in third-harmonic generation the use of focussed gaussian beams modifies the phase-matching condition. In stimulated Raman scattering, a tightly focussed pump beam can lead to a large diffraction loss for the Stokes wave; the threshold condition is then greatly modified. Rather than treat these topics in Chap.3, we have left them to the experimental sections (Chaps.4,5, respectively) where their significance can be made more apparent.

In evaluating possible applications of a nonlinear optical process it is obviously important to know of limiting effects to which the process might be subject. These limiting effects are common to many processes; we have, for convenience, concentrated our discussion of them in Sects.4.6, 5.4. Where possible, we give rough estimates of their magnitudes. These estimates are necessarily rough, because the conditions appropriate to efficient nonlinear generation involve high vapour pressures (high by the standards of typical spectroscopic experiments) and high incident intensities of resonant laser radiation. Under these conditions, a multitude of processes manifest themselves simultaneously, such as multiphoton ionisation, amplified spontaneous emission, various sum- and difference-frequency mixing processes, radiation trapping, level shifts, and so on. Considerable further work is needed to unravel this complex behaviour.

Finally, in Chap.8, some less familiar nonlinear optical effects are briefly described. They provide further illustration of the wide variety of phenomena that may be observed under resonance conditions.

2. Theory of the Nonlinear Optical Susceptibility

2.1 Definition

A central aim of the theory of nonlinear optical interactions is to calculate the polarisation induced in the medium by the applied fields. This forms the subject of this chapter, in which the applied fields are taken to be prescribed. Later, in Chap.3, we will treat the self-consistent generation of these fields by the radiating polarisation, using Maxwell's equations. This leads to a description of the propagation of the waves through an extended medium. It should, however, be appreciated that in many cases, for example scattering of light by a single atom, the polarisation induced by prescribed fields directly determines the radiated field.

The general plan of this chapter is first to define more carefully the phenomenological series indicated in the introduction (Sects.2.1-3), and to compare this briefly with other approaches. We then give formulae for the nonlinear susceptibilities (Sects.2.4,5). At this point, the background is sufficient for understanding the later sections (Chap.3 onwards). However, in Sects.2.6,7, a number of other approaches and their relation to the susceptibility formalism are discussed in more detail. In Sect.2.8 symmetry aspects of the nonlinear susceptibilities are discussed.

We begin by expanding the electric field in the frequency domain,

$$\underline{E}(\underline{r}t) = \int_{-\infty}^{\infty} e^{-i\omega t}\underline{E}(\underline{r}\omega)d\omega \quad ; \quad \underline{E}(\underline{r}\omega)^* = \underline{E}(\underline{r}\ -\omega) \qquad (2.1)$$

and similarly for $\underline{P}(\underline{r}t)$, and then introduce the nonlinear optical susceptibility tensors $\chi^{(n)}(-\omega_\sigma;\omega_1\dots\omega_n)$ by use of their defining relation [1.70][1],

---

[1]The placing of ε_0 in (2.2) is conventional for the linear polarisation, '$P = \varepsilon_0\chi E$', when SI units are employed; retaining it for the nonlinear terms is therefore a logical step. Moreover, $\chi^{(n)}$ then has the simple dimensions of $(\mathrm{Vm}^{-1})^{1-n}$, i.e. of (electric field)$^{1-n}$. The Appendix summarises important relationships between esu and SI units.

$$P_{\mu}^{(n)}(\underline{r}\omega_{\sigma}) = \varepsilon_0 \sum_{\alpha_1\ldots\alpha_n} \int \chi^{(n)}_{\mu\alpha_1\ldots\alpha_n}(-\omega_\sigma;\omega_1\ldots\omega_n)E_{\alpha_1}(\underline{r}\omega_1)\ldots E_{\alpha_n}(\underline{r}\omega_n)$$

$$\times\ \delta(\omega_\sigma - \omega_1\ldots - \omega_n)d\omega_1\ldots d\omega_n \qquad (2.2a)$$

$$\underline{\chi}^{(n)}(-\omega_\sigma;\omega_1\ldots\omega_n)^* = \underline{\chi}^{(n)}(\omega_\sigma;-\omega_1\ldots-\omega_n) \quad . \qquad (2.2b)$$

In (2.2a), $\mu\alpha_1\ldots\alpha_n$ are cartesian subscripts (x,y or z). The Fourier component of the polarisation at frequency ω_σ has been expanded as a power series in the fields, so that

$$\underline{P}(\underline{r}\omega) = \sum_{n=1}^{\infty} \underline{P}^{(n)}(\underline{r}\omega) \quad . \qquad (2.3)$$

(The static polarisation found for example in some crystals can be represented by an n = 0 term, but because it vanishes for isotropic media, it is omitted here.) Equation (2.2b) is the reality condition that ensures that $\underline{P}(\underline{r}t)$ is real. When the frequencies are complex (see [1.70]), condition (2.2b) applies, but with the ω_i in the RHS replaced by ω_i^*.

Because (2.1-3) are at the heart of most discussions of nonlinear optics, and in particular will be studied at length in this chapter, let us give a brief first survey of its range of application.

Equation (2.2) assumes a local spatial reponse of the medium to the applied fields; for the great majority of nonlinear processes in gases, this is a valid approximation. However, in recent years a number of interesting experiments have been reported that make use of what can be viewed as nonlocal response. For example, we will see later that, for free atoms and molecules, the definition (2.2) corresponds to the use of electric-dipole interactions; if higher multipole interactions are to be considered (Sect.8.1) it is necessary to account for the spatial variation of the fields. Another example is the idea of nonlinear processes that involve two or more atoms and molecules, in which, for example, intermolecular forces play a role (Sect.8.2).

A systematic treatment of all such processes is possible by use of spatially dependent nonlinear susceptibilities; a formal definition has been given by FLYTZANIS [2.1], by whom they are used to treat spatially dispersive phenomena in condensed media. The susceptibility is defined in $\underline{k}$-space (spatial Fourier transform) so that in (2.2) $\omega_i \to \underline{k}_i\omega_i$, $d\omega_i \to d\underline{k}_i d\omega_i$ etc., and the local fields etc., are then, in effect, incorporated correct to nth order in $\underline{\chi}^{(n)}$. From what has been said, it is clear that except in a very few special cases, the effects of nonlocal response can be neglected in atomic and molecular gases and vapours.

Accepting then, for the present, that a treatment based on local response is adequate, (2.2) represents a perfectly general expansion of $\underline{P}(\underline{r}t)$ in terms of the applied fields. Thus it contains any arbitrary time dependence of the fields and of the induced atomic-state changes (of populations, etc.), and may be used as a phenomenological series when propagation problems are to be solved (Chap.3). When a microscopic theory of the interaction of the fields with the medium is required, these nonlinear susceptibilities must be expressed in terms of atomic parameters; such expressions will be shown in Sect.2.4, to follow from a straightforward application of time-dependent perturbation theory.

Notwithstanding this general framework, in many cases the expansion in powers of the electric field and/or the use of the frequency domain is not the most suitable approach. As will become clear in Sects.2.6,7, successive orders of nonlinearity take into account such effects as level shifts (optical Stark effect) and power broadening by means of the series expansion. On the other hand, for a given order of nonlinearity, the frequency integrals account for the time response of the medium to the (possibly pulsed) radiation.

Thus, for intense radiation, the effect on the medium may be such as to make the contribution from several orders of nonlinearity comparable in magnitude. Then a different method of calculation may yield a single-term approximation to the polarisation, which is more manageable than the series expansion. One way of doing this is to sum the dominant terms directly—this has been done in studies of multiphoton ionisation, by GONTIER et al. [2.2]. Alternatively, perturbation methods may be devised that automatically take account of, for example, level shifts. These methods generally rely on the simplification produced by restricting attention to the dominant resonant transitions. Examples of this include one- and two-photon vector models (e.g., [1.72,2.3]) and the methods of averages and multiple time scales, and other closely associated approaches (e.g. [2.4-12]).

The results are frequently found to have the same form as the basic susceptibilities, but with field-dependent corrections; this gives rise to the idea of field-dependent ("renormalised") susceptibilities (Sect.2.6).

Turning to time dependence, the frequency integrals in (2.2) can become extremely complicated when short pulses and/or resonant interactions are involved; therefore, a formulation in the time domain can become the more practical approach. Because intense pulses are usually of short duration, these examples must frequently be considered together; in extreme cases they fall into the general category of coherent propagation phenomena.

For these reasons, the main use of the formalism (2.2) (as opposed to its general application) has been to examine a nonlinear process in one order of nonlinearity, and for electric fields consisting of a superposition of (quasi-) monochromatic waves. This use forms the background of Chaps.4-8; we concentrate on it in Sects.2.2,3. This use is not particularly restrictive, because the basic properties of any nonlinear process are reflected in the first term of (2.2) in which it appears; under the so-called adiabatic conditions (Sect.2.6), it is often possible to let the time dependence of the fields enter into monochromatic-wave formulae as a parameter. In any case, for an enormous range of experimental conditions, this basic use is entirely appropriate.

2.2 Permutation Symmetry

The nonlinear susceptibility tensors $\underset{=}{\chi}^{(n)}(-\omega_\sigma;\omega_1\ldots\omega_n)$ possess one rigorous permutation symmetry, intrinsic permutation symmetry; frequently, a second such symmetry occurs, to a good approximation, called "overall permutation symmetry" [1.70]. In this section, these symmetry operations are defined, and used to obtain from (2.2) simple expressions for the polarisation induced by an electric field that consists of a superposition of monochromatic waves.

BUTCHER [1.70] has shown that, as a consequence of the invariance of physical laws under time translations, the nonlinear optical susceptibility in (2.2) is invariant under all permutations of the pairs $\alpha_1\omega_1\ldots,\alpha_n\omega_n$. This is the intrinsic permutation symmetry mentioned above and is seen to be derived from a very general principle; in this context it is, therefore, rigorous.

If the medium is lossless, which here means essentially that none of the frequencies ω_σ, $\omega_1\ldots\omega_n$ or combinations thereof are resonant with transition frequencies in the medium, then overall permutation symmetry holds true. This means that the nonlinear susceptibility tensor components are invariant under all permutations of the pairs $\mu{-}\omega_\sigma;\alpha_1\omega_1\ldots\alpha_n\omega_n$. (Note that the pair $\mu{-}\omega_\sigma$, and not $\mu\omega_\sigma$, is involved in overall permutation symmetry; hence the convention of writing $\underset{=}{\chi}^{(n)}(-\omega_\sigma;\omega_1\ldots\omega_n)$ rather than $\underset{=}{\chi}^{(n)}(\omega_\sigma;\omega_1\ldots\omega_n)$.) It will be seen later that for certain classes of processes, such as Raman scattering, subsets of the $\alpha_i\omega_i$, which may include μ-ω_σ, exhibit permutation symmetry, despite the presence of losses.

To see an important application of intrinsic permutation symmetry, consider the case when the electric field $\underline{E}(\underline{r}t)$ consists of a superposition of monochromatic waves. Then (2.1) becomes

$$\underline{E}(\underline{r}t) = \frac{1}{2}\sum_{\omega\geq 0} (\underline{E}_\omega e^{-i\omega t} + \underline{E}_{-\omega} e^{i\omega t}) \quad ; \quad \underline{E}_{-\omega} = \underline{E}^*_\omega \ . \tag{2.4}$$ [2]

The spatial dependence of $\underline{E}_\omega$ is not displayed explicitly here, as it is not required until Chap.3. An analogous expression may be written for $\underline{P}(\underline{r}t)$.

Substituting (2.4) into (2.1,2) leads to terms for $\underline{P}_\omega^{(n)}$ of the form

$$\sum_{\alpha_1\ldots\alpha_n} \chi^{(n)}_{\mu\alpha_1\ldots\alpha_n}(-\omega_\sigma;\omega_1\omega_2\ldots\omega_n)(E_{\omega_1})_{\alpha_1}(E_{\omega_2})_{\alpha_2}\ldots(E_{\omega_n})_{\alpha_n} \tag{2.5a}$$

$$\sum_{\alpha_1\ldots\alpha_n} \chi^{(n)}_{\mu\alpha_1\ldots\alpha_n}(-\omega_\sigma;\omega_2\omega_1\ldots\omega_n)(E_{\omega_2})_{\alpha_1}(E_{\omega_1})_{\alpha_2}\ldots(E_{\omega_n})_{\alpha_n} \tag{2.5b}$$

and so on, where

$$\omega_\sigma = \omega_1 + \ldots + \omega_n \ . \tag{2.6}$$

In general, there may be several distinct sets of $\omega_1\ldots\omega_n$ that satisfy (2.6). Consider one of these. If in the sample terms (2.5), ω_1 equals ω_2, then only one such term will be found to occur. If, however, $\omega_1 \neq \omega_2$, then by relabelling the dummy subscripts (α_1, α_2) as (α_2, α_1) in (2.5b), and then invoking intrinsic permutation symmetry, it is seen that the two terms (2.5a,b) are the same. The general conclusion is that for a given set of $\omega_1\ldots\omega_n$, only one term need be written, such as (2.5a). The number of times it occurs is equal to the number of distinguishable permutations of the $\omega_1\ldots\omega_n$. (A frequency and its negative are considered distinguishable.) Furthermore, from definition (2.4), a factor of $\frac{1}{2}$ will be associated with each $\underline{E}_{\omega_i}$ if $\omega_i \neq 0$; similarly for $\underline{P}^{(n)}_{-\omega_\sigma}$. .

---

[2] The dimensions of the field's Fourier transform $\underline{E}(\omega)$ and a monochromatic component $\underline{E}_\omega$ are different, as can be seen by comparison of (2.1,4). To avoid confusion, we shall adhere to the convention that the frequency variable ω will be appended as a *subscript* on monochromatic waves (and similarly for the polarisation). Thus $\underline{E}(\omega)$ has units of $Vm^{-1}s$, but $\underline{E}_\omega$ has units of Vm^{-1}. In practice, many equations apply equally well to either quantity, for example (3.3-8a).

Taking these numerical factors into account, we see that for monochromatic waves, the application of intrinsic permutation symmetry to (2.2) yields the simple expression

$$\left(P^{(n)}_{\omega_\sigma}\right)_\mu = \varepsilon_0 \sum_{\alpha_1\ldots\alpha_n}{}' K(-\omega_\sigma;\omega_1\ldots\omega_n)\chi^{(n)}_{\mu\alpha_1\ldots\alpha_n}(-\omega_\sigma;\omega_1\ldots\omega_n)$$

$$(E_{\omega_1})_{\alpha_1}\ldots(E_{\omega_n})_{\alpha_n} \quad . \tag{2.7}$$

The ' in (2.7) reminds us to sum over all of the distinct sets of $\omega_1\ldots\omega_n$; because attention is usually focussed on one such set, it will be dispensed with [bar (2.9)] in the interest of lightening the notation. $K(-\omega_\sigma;\omega_1\ldots\omega_n)$ is the numerical factor described above; for example if ω_σ, $\omega_1\ldots\omega_n$ are all nonzero, and $\omega_1\ldots\omega_n$ are all different, then $K = 2^{1-n}n!$. Equation (2.7) has been written with the frequencies $\omega_1\ldots\omega_n$ in a given order; obviously, in view of intrinsic permutation symmetry we can write them in any order; however, as will become clear later, it is sometimes a conceptual aid to choose one specific order for a given process.

The numerical factor K, as constructed here, is seen to be a logical consequence of the general expressions (2.1,2). It is, however, by no means a universal convention in the literature to display it as in (2.7), because it can be absorbed into a newly defined susceptibility, $\underline{\chi}^{(n)\prime}$ say. This has the disadvantage that when different frequencies are allowed to become equal, or to go to zero, the nonlinear susceptibility $\underline{\chi}^{(n)\prime}$ jumps discontinously.

Much of the literature starts from (2.4), but with the factors of 1/2 omitted. As an example of the possible confusion this may cause, notice that for running waves, (2.4) gives a cycle-averaged intensity for each frequency ω of

$$I_\omega = \frac{1}{2}\varepsilon_0 c n_\omega |\underline{E}_\omega|^2 \quad , \quad (\mathrm{Wm}^{-2}) \tag{2.8}$$

where n_ω is the refractive index at frequency ω, whereas omitting the 1/2's in (2.4) would mean replacing the 1/2 in (2.8) by 2.

It is clear there are many possibilities for the placing of the various numerical factors, and the reader is cautioned that several of them are being used in the literature. We shall adopt (2.7) as our standard notation.

Leaving the vexing matter of numerical factors, we now dispense with the explicit tensor notation in (2.7), by writing

$$\underline{E}_{\omega_i} = \underline{\varepsilon}_i E_i$$

where $\underline{\varepsilon}_i$ and E_i are the (possibly complex) unit polarisation vector and scalar amplitude, respectively. The polarisation $\underline{P}^{(n)}_{-\omega_\sigma}$ is taken to have unit polarisation vector $\underline{\varepsilon}_\sigma$. Defining the scalar nonlinear susceptibility by

$$\chi^{(n)}(-\omega_\sigma;\omega_1\ldots\omega_n) = \sum_{\mu\alpha_1\ldots\alpha_n} \chi^{(n)}_{\mu\alpha_1\ldots\alpha_n}(-\omega_\sigma;\omega_1\ldots\omega_n)(\varepsilon^*_\sigma)_\mu(\varepsilon_1)_{\alpha_1}\ldots(\varepsilon_n)_{\alpha_n} \quad , \tag{2.9a}$$

then (2.7) becomes

$$P^{(n)}_\sigma = \varepsilon_0 \sum{}' K(-\omega_\sigma;\omega_1\ldots\omega_n)\chi^{(n)}(-\omega_\sigma;\omega_1\ldots\omega_n)E_1\ldots E_n \quad , \tag{2.9b}$$

which is the main result of this section. When particular processes are described, the actual frequencies in (2.9) may be negative. The pairs $\underline{\varepsilon}_i\,|\omega_i|$, $\underline{\varepsilon}^*_i\ -|\omega_i|$ are termed positive and negative frequency parts, respectively.

2.3 Nonlinear Optical Processes

By use of (2.9), it is a simple matter to write down the polarisation at a given frequency induced by the electric field. Thus, for third-harmonic generation, $P^{(3)}_{3\omega} = (1/4)\ \varepsilon_0\chi^{(3)}(-3\omega;\omega\omega\omega)E^3_\omega$ is the lowest-order term. In Table 2.1 are listed some nonlinear processes that are of importance in free atoms and molecules, along with the K factors and references, when appropriate, to later sections of the book where they are examined in detail.

One of the most important of the third-order processes is stimulated Raman scattering (SRS). This will be discussed later (Sect.3.3,Chap.5); it suffices here to indicate how the process is described in terms of a susceptibility. Because we are dealing here with classical fields, we cannot expect to give a proper description of spontaneous Raman scattering (although a device that can be used to overcome this limitation is described in Sect.2.7). The stimulated process involves the generation of a polarisation at the Stokes frequency ω_S by the simultaneous action of the fields at the pump (ω_p) and Stokes frequencies. The term "stimulated" arises from the fact that the induced polarisation at ω_S is proportional to the electric field at the same frequency. We have to go to third order to find a susceptibility that has the

Table 2.1. Some processes of importance in free atoms and molecules ($\omega_\sigma = \omega_1 + \ldots + \omega_n$). See Tables 2.2 and 2.3 for resonant susceptibilities

Process	Order, n	$-\omega_\sigma;\omega_1\omega_2\omega_3\omega_4\omega_5$	K[b]	
Linear response	1	$-\omega;\omega$	1	Real part refractive index; imaginary part absorption/emission. Also connected with optical Stark effect (Sect.2.6).
Self-induced intensity-dependent refractive index	3	$-\omega;\omega-\omega\omega$	3/4	First nonlinear correction to linear susceptibility. Can lead to self-focussing/defocussing (Sect.2.6), and generation of time-inverted wavefronts (Sect.8.4).
Stimulated Raman scattering (SRS), optical Kerr effect	3	$-\omega_s;\omega_p-\omega_p\omega_s$	3/2	Generation of Stokes ω_s from pump ω_p. SRS gain coefficient or spontaneous scattered intensity depends on imaginary part (Chap.5). Pump-induced change of refractive index at ω_s depends on Real part; this is the optical Kerr effect. (Sects. 3.3,4.6)
DC Kerr effect	3	$-\omega;\ 0\,0\,\omega$	3/4	DC field-induced refractive index change at frequency ω.
Two-photon emission/absorption/ionisation (TPE/A/I)[a]	3	$-\omega_1;-\omega_2\omega_2\omega_1$	3/2	TPI is a special case of TPA where the final level in the absorption process is in the continuum. (Sect.2.6, Chap.5) The real part is again a contribution to the optical Kerr effect.
Third-harmonic generation	3	$-3\omega;\omega\omega\omega$	1/4	(Chap.4)
Coherent anti-stokes Raman scattering (CARS)	3	$-\omega_{as};\omega_p\omega_p-\omega_s$	3/4	(Sects.3.4,6.1)
General four-wave mixing (4WM)	3	$-\omega_\sigma;\omega_1\omega_2\omega_3$	3/2	(Chaps.3,4,6,7)
Sum mixing	2	$-\omega_\sigma;\omega_1\omega_2$	1	Can occur in free atoms only if multipole transitions are used (Sect.8.2).

Stimulated hyper-Raman scattering (SHRS)	5	$-\omega_s;\omega_{p_1}\omega_{p_2}-\omega_{p_2}-\omega_{p_1}\omega_s$	15/2	(Two pumps ω_{p_1}, ω_{p_2}) (Sects.3.3,5.5)
		$-\omega;\omega_p\omega_p-\omega_p-\omega_p\omega_s$	15/8	(Single pump ω_p) (Sects.3.3,5.5)
Self-induced intensity-dependent refractive index	5	$-\omega;\omega-\omega\omega-\omega\omega$	5/8	Second nonlinear correction to linear susceptibility (Sect.2.6).
Three-Photon E/A/I	5	$-\omega_1-\omega_2-\omega_3\omega_3\omega_2\omega_1$	15/2	(Chaps.4,5,6)
m-photon E/A/I of a single frequency	2m-1	$-\omega;\ldots -\omega\ \omega\ldots\omega$	$4^{1-m}\binom{2m-1}{m}$	
nth harmonic generation	n	$-n\omega;\ \omega\ldots\ \ldots\omega$	2^{1-n}	

[a]Susceptibilities are in fact the same for SRS and TPA, but the frequency arguments are displayed in a different order. This is a matter of convenience. The underlying reason, which will become clear in Sect.2.5 is that for two-photon absorption the medium is resonant with $\omega_1 + \omega_2$, whereas for Raman scattering it is resonant with $\omega_p - \omega_s$. Similarly for SHRS and three-photon absorption, etc.

[b]Applies if differently labelled frequencies are different in value. Thus for TPA of a single frequency K = 3/4; for two different frequencies K = 3/2.

appropriate frequency arguments; the result is

$$P_S^{(3)} = \frac{3}{2}\varepsilon_0\chi^{(3)}(-\omega_S;\omega_p-\omega_p\omega_S)|E_p|^2E_S \quad . \tag{2.10}$$

It will be shown in Sect.3.3 that this polarisation at the Stokes frequency leads to an exponential growth of the Stokes field as it propagates through the pumped medium. Each generated Stokes photon results from the conversion of a single pump photon; the medium absorbs the excess energy $\hbar(\omega_p - \omega_S)$ (see Fig.1.1a). Recalling that the rate for single-photon absorption depends on $\mathrm{Im}\chi^{(1)}(-\omega;\omega)$, we are not surprised to find that the imaginary part $\mathrm{Im}\chi_R$ of the Raman susceptibility $\chi_R = \chi^{(3)}(-\omega_S;\omega_p-\omega_p\omega_S)$ determines the rate at which ω_p is absorbed (and ω_S emitted); i.e., the Stokes gain is proportional to $\mathrm{Im}\chi_R$. Similarly, $\mathrm{Re}\chi_R$ produces a change of refractive index at the Stokes frequency; this refractive index change is known as the optical Kerr effect. [Two-photon absorption is also described by (2.10), but the frequency arguments in the susceptibility have a different relationship to the energy levels in the medium (Fig.1.1b, Sect.3.3)].

Terms such as $\chi^{(3)}(-\omega;\omega-\omega\omega)$ will also lead to intensity-dependent refractive index changes. Indeed, in this class, a "corrected" refractive index could be written:

$$\chi(\omega;E_\omega) = \chi^{(1)}(-\omega;\omega) + \frac{3}{4}\chi^{(3)}(-\omega;\omega-\omega\omega)|E_\omega|^2 + \frac{5}{8}\chi^{(5)}(-\omega;\omega-\omega\omega-\omega\omega)|E_\omega|^4 + \dots \tag{2.11}$$

This equation is simply a geometric progression; it is pleasing to find that it is easy, in some circumstances, to sum the series. The interpretation of this procedure will be left until Sect.2.6; it suffices to say here that $\chi(\omega;E_\omega)$ is essentially the same as $\chi^{(1)}(-\omega;\omega)$ but with the transition frequencies shifted (optical Stark effect) and the resonances broadened (power broadening); cf. Sect.2.1.

2.4 Explicit Formulae for the Nonlinear Atomic Susceptibilities

The first general approach to nonlinear optics is due to ARMSTRONG et al. [1.69]; in that paper, they used time-dependent perturbation theory to calculate the first few orders of nonlinear polarisation, for monochromatic waves. Since then, derivations of the nonlinear susceptibilities for particular processes have repeatedly appeared in the literature (and continue to do so). In this section, we discuss the compact formula due to BUTCHER [1.70], which was the result of an elegant general analysis and contains the majority of other published formulae as special cases. It can, therefore, be used as a basis for all discussions of nonlinear suceptibilities. One further generalisation of Butcher's formula has recently been published by BABIKER [2.13] in which a relativistic quantum-electrodynamic treatment is employed. This is at present perhaps of only academic interest; we refer to it again in Sect.8.2, when multipole interactions are examined. In a number of circumstances, it is possible to use classical forced harmonic and anharmonic oscillator models to describe the nonlinear polarisation (e.g., [2.1,14-17]); examples of this, and the relation to the formulae of this section, are given in Sects.2.7,7.2. BUTCHER, in his derivation of the general nonlinear susceptibility formula, made use of the density-matrix approach. DUCUING [1.71] has since obtained the same formula by means of straightforward time-dependent perturbation theory for the perturbed atomic states. The electric-dipole interaction V between matter and radiation is

$$V = -\underline{Q} \cdot \underline{E}(\underline{r}t) \quad , \tag{2.12}$$

in which $\underline{Q}$ is the electric-dipole-moment operator for the atom or molecule. An atom in an unperturbed stationary state $|g\rangle$ evolves into the state $|g\rangle'$, say; the induced dipole moment $\underline{\mu}(t)$ is then calculated from the expectation value

$$\underline{\mu}(t) = {}'\langle g|\underline{Q}|g\rangle' \quad . \tag{2.13}$$

This expectation value appears as a power series in V, and thus of the field; the terms are then merely sorted according to the form (2.2). The macroscopic polarisation $\underline{P}^{(n)}$ is obtained from $\underline{\mu}^{(n)}$ by use of

$$\underline{P}^{(n)}(\underline{r}t) = \overline{\mathcal{N}\underline{\mu}^{(n)}} \quad , \tag{2.14}$$

where $\mathcal{N}$ is the number density of atoms and the bar denotes an orientation average.

The result (for stationary atoms) is

$$\chi^{(n)}(-\omega_\sigma;\omega_1\ldots\omega_n)$$

$$= \frac{\mathcal{N}}{n!\hbar^n \varepsilon_0} \mathcal{S}_T \sum_{gb_1 \ldots b_n} \rho(g) \frac{\underline{\varepsilon}_\sigma^* \cdot \underline{Q}_{gb_1} \underline{\varepsilon}_1 \cdot \underline{Q}_{b_1 b_2} \cdots \underline{\varepsilon}_n \cdot \underline{Q}_{b_n g}}{(\Omega_{b_1 g} - \omega_1 - \ldots - \omega_n)(\Omega_{b_2 g} - \omega_2 - \ldots - \omega_n) \ldots (\Omega_{b_n g} - \omega_n)} \ . \tag{2.15}$$

Apart from changes of notation, the use of the scalar form (2.9a), and conversion to SI units (Sect.2.1) this formula is identical to that derived by BUTCHER [1.70]. In (2.15), the transition frequencies Ω_{ij} are defined in terms of atomic-energy levels E_i, E_j by $\Omega_{ij} = (E_i - E_j)/\hbar$. The $\underline{Q}_{ij}$ are matrix elements $\langle i|\underline{Q}|j\rangle$ between stationary states $|i\rangle$, $|j\rangle$ of the unperturbed atom[3]. The summations $\sum_{gb_1 \ldots b_n}$ are over *all* of the atomic states, which include those for which $\Omega_{b_i g} = 0$, and also include integrations over the continuum states.

A weighting over the unperturbed equilibrium distribution of initial states g has been included by means of the factor $\rho(g)$, which is the unperturbed diagonal density-matrix element for state g. For atoms and rotating molecules, $\sum_g \rho(g)$ is

$$\sum_{M_g} \rho(\gamma_g J_g M_g) = \overline{\rho(\gamma_g J_g)}(2J_g + 1)^{-1} \sum_{M_g} \ , \tag{2.16}$$

where the states are labelled $|\gamma JM\rangle$ in the usual fashion. The factor $\overline{\rho(\gamma_g J_g)}$ represents the probability for an unperturbed atom to be in a state of energy $\hbar\Omega_{\gamma_g J_g M_g}$, of which there are $2J_g + 1$ due to M_g degeneracy. The remainder of the RHS of (2.16) may be shown to impose isotropic symmetry on (2.15); it is, therefore, equivalent to an orientation average, so that the bar in (2.14) is unnecessary here (see Sect.2.8).

The last item to be defined in (2.15) is $\mathcal{S}_T$, the overall permutation operator [1.70]. This requires that the expression following it be summed over all permutation of the pairs $\underline{\varepsilon}^*$-$\omega_\sigma$, $\underline{\varepsilon}_1\omega_1 \ldots \underline{\varepsilon}_n\omega_n$. There are, therefore, $(n + 1)!$ expanded terms in (2.15), although, of course, if some of the frequencies are equal, the number of distinct terms may be greatly reduced. It

---

[3]It is frequently necessary to refer to the z component of this matrix element; we shall denote this by $\mu_{ij} = (Q_z)_{ij}$, for brevity.

is obvious that $\mathscr{S}_T$ imposes overall (and therefore intrinsic) permutation symmetry on (2.15)[4].

With a little practice it is not difficult to write down from (2.15) the susceptibility formulae for particular processes; we will give a number of such special cases later. In general, diagrammatic representations afford the easiest means of keeping track of the terms in (2.15) generated by $\mathscr{S}_T$. Diagrammatic perturbation theory is familiar from a wide range of disciplines; for example, transition rates for multiphoton processes have been thus described by WALLACE [2.18] and LOUDON [2.19]. WARD [2.20] has given one such scheme for calculating the nonlinear polarisation.

Besides their obvious pictorial advantages as a notation, diagrammatic techniques lend themselves to an interpretation of the denominators in (2.15). YURATICH [2.21] has given an algorithm directly based on this expression, which we now summarise.

Adapting the photon creation and destruction operator notions from quantised field theory, we can regard the polarisation vectors $\underline{\varepsilon}_i$, $\underline{\varepsilon}_i^*$ [or the field components $\underline{E}(\omega_i)$, $\underline{E}(\omega_i)^*$] as commuting creation and destruction operators, since, in this book, classical fields are employed. To be more precise, the positive frequency pair $\underline{\varepsilon}_i|\omega_i|$ is associated with the destruction of a photon, and the negative frequency pair $\underline{\varepsilon}_i$ $-|\omega_i|$ is associated with creation of a photon. Therefore, in (2.15) the denominator $\hbar(\Omega_{b_na} - \omega_n)$ represents the net change of energy in the atom-field system as the atom undergoes the transition $a \rightarrow b_n$ and the field loses a photon of frequency ω_n through absorption; the associated matrix element is $\underline{\varepsilon}_n \cdot \underline{Q}_{b_na}$. In diagrammatic form, the RHS of (2.15) becomes

$$\frac{\mathscr{N}}{n!\hbar^n\varepsilon_0}\sum_g \rho(g)\,\mathscr{S}_T$$ 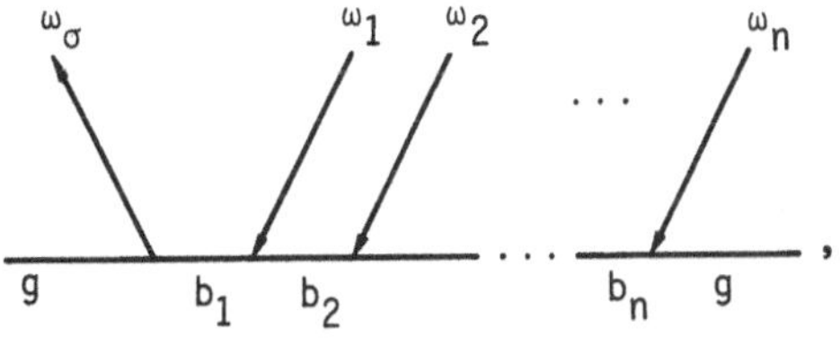

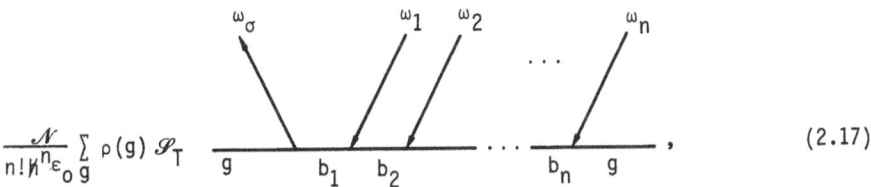

, (2.17)

where summation over the intermediate states $b_1 \ldots b_n$ is implied. A downward pointing arrow (absorption) is associated with the positive-frequency pairs

[4] If the K factor (Sect.2.2) was included in (2.15), then it could be dispensed with (apart from the factors of 1/2) by redefining $\mathscr{S}_T$ to $\mathscr{S}_T'$, say, where $\mathscr{S}_T'$ generates only distinguishable permutations; this has the disadvantage mentioned in Sect.2.2.

and an upward pointing arrow (emission) with negative-frequency pairs. Although in (2.17) all of the arrows (excepting ω_σ) have been drawn pointing downwards, if in a particular process any of the frequencies are negative, then the corresponding arrows may be reversed and labelled with the associated positive-frequency pair. In (2.17), $\mathscr{S}_T$ generates the sum of all (n + 1)! diagrams that result from permutations of the arrows in the given diagram. These diagrams may easily be transcribed into the corresponding parts of (2.15), and enumerate the varying but equivalent routes of absorption and emission of the field frequencies that return the atom to state $|g\rangle$. By noting the resonant states in a particular atom, it is possible to write down the dominant parts of (2.15) directly, by use of (2.17). It is not possible to pursue this approach in more detail here, although the foregoing description is complete insofar as transcription of (2.15) is concerned.

We now consider a simple case of (2.15), that of third-harmonic generation (THG). Because three of the frequency arguments in $\chi^{(3)}(-3\omega;\omega\omega\omega)$ are identical, $\mathscr{S}_T$ generates four distinct terms, each of which occurs six times (see Fig.2.1b for the diagrammatic representation). The four distinguishable permutations are readily determined, to yield

$$\chi^{(3)}(-3\omega;\omega\omega\omega) = \frac{\mathscr{N}}{\hbar^3\varepsilon_0}\sum_{gabc}\rho(g)$$

$$\times\left(\frac{\underline{\varepsilon}_{3\omega}^{*}\cdot\underline{Q}_{ga}\underline{\varepsilon}_{\omega}\cdot\underline{Q}_{ab}\underline{\varepsilon}_{\omega}\cdot\underline{Q}_{bc}\underline{\varepsilon}_{\omega}\cdot\underline{Q}_{cg}}{(\Omega_{ag}-3\omega)(\Omega_{bg}-2\omega)(\Omega_{cg}-\omega)}\right.$$

$$+\frac{\underline{\varepsilon}_{\omega}\cdot\underline{Q}_{ga}\underline{\varepsilon}_{3\omega}^{*}\cdot\underline{Q}_{ab}\underline{\varepsilon}_{\omega}\cdot\underline{Q}_{bc}\underline{\varepsilon}_{\omega}\cdot\underline{Q}_{cg}}{(\Omega_{ag}+\omega)(\Omega_{bg}-2\omega)(\Omega_{cg}-\omega)}$$

$$+\frac{\underline{\varepsilon}_{\omega}\cdot\underline{Q}_{ga}\underline{\varepsilon}_{\omega}\cdot\underline{Q}_{ab}\underline{\varepsilon}_{3\omega}^{*}\cdot\underline{Q}_{bc}\underline{\varepsilon}_{\omega}\cdot\underline{Q}_{cg}}{(\Omega_{ag}+\omega)(\Omega_{bg}+2\omega)(\Omega_{cg}-\omega)}$$

$$\left.+\frac{\underline{\varepsilon}_{\omega}\cdot\underline{Q}_{ga}\underline{\varepsilon}_{\omega}\cdot\underline{Q}_{ab}\underline{\varepsilon}_{\omega}\cdot\underline{Q}_{bc}\underline{\varepsilon}_{3\omega}^{*}\cdot\underline{Q}_{cg}}{(\Omega_{ag}+\omega)(\Omega_{bg}+2\omega)(\Omega_{cg}+3\omega)}\right). \tag{2.18}$$

This formula is at the heart of most of Chap.4. Because for atoms the states may all be labelled by parity, and $\underline{Q}$ is an operator of odd parity, the states $|b\rangle$ and $|g\rangle$ in (2.18) must have the same parity. Hence, with ref-

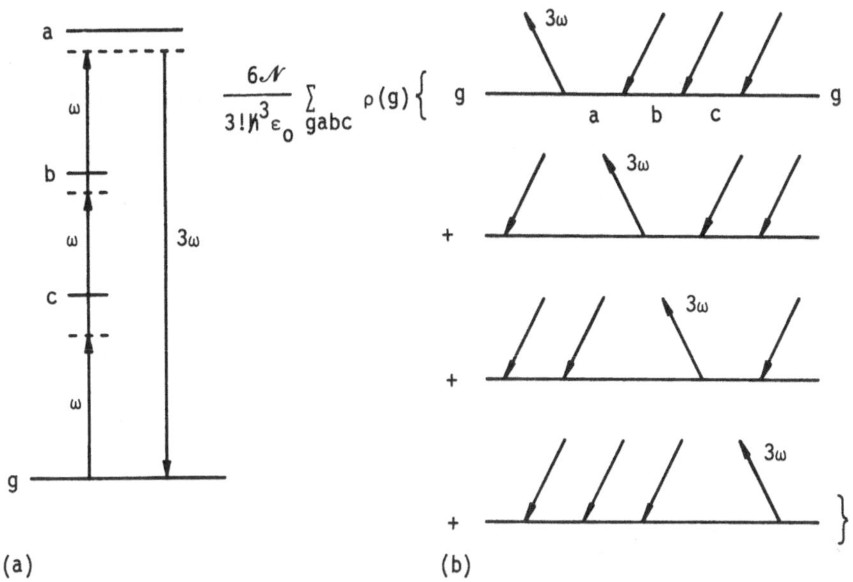

Fig.2.1a,b. Third-harmonic generation (THG). (a) Resonances occur when the frequencies ω, 2ω, 3ω are nearly coincident with transitions in the medium. With energy levels and frequencies as depicted, the first term in (2.18) is large: it is usually dominant. (b) Diagrammatic representation of (2.18). The unmarked arrows are for the fundamental photon (ω); the state labelling matches (2.18), although a, b, c are dummy indices. The factor of six arises from indistinguishable permutations of the ω arrows; the first subdiagram corresponds to (a)

erence to the first two terms in (2.18) the susceptibility may be enhanced by tuning 2ω close to some transition Ω_{bg}, without the limitation of strong single-photon absorption that would occur if ω were tuned to some Ω_{cg} (Fig.2.1a). This "two-photon resonance enhancement" (and its equivalents in other processes) is for this reason an essential ingredient in most efficient nonlinear-optical experiments in atomic vapours. Of course, a certain amount of single-photon enhancement of the fundamental or harmonic photons can be employed. It is also possible to make use of resonances of the form $\Omega_{bg} + 2\omega \simeq 0$, where $\Omega_{bg} < 0$. For the majority of experiments in atoms, however, g is the ground state, and such resonances cannot occur. On the other hand, the atoms can be prepared in excited states, for example by optical pumping (Sect.5.2). In molecules, there is usually a significant population in excited rotational levels (Sect.7.6); in such cases use can be made of these resonances.

The experimental aspects of resonance enhancement will be discussed at length in later sections. Here we have to consider what happens as a resonant denominator in (2.15) approaches zero—i.e., a singularity occurs. This may happen in two ways, classified as resonant or secular by ORR and WARD [2.4]. Resonance is typified by the example of $\Omega_{bg} - 2\omega \to 0$ in (2.16); a combination of frequencies is resonant with some nonzero atomic transition frequency. The simple device for handling (2.15) in this case, and its implementation, is detailed in the next section.

The second type of singularity in (2.15), secularity, would arise if a denominator were zero because both the transition frequency *and* the combination of field frequencies that occurred in it were zero. There really is no singularity, as will be demonstrated in Sect.2.6.5.

2.5 Resonant Susceptibilities

The divergence due to resonance is clearly unphysical, and ultimately has its origin in the neglect of higher-order terms. When these are taken into account, the large changes in the atoms implied by the resonance introduce level shifts into the denominator; therefore, a finite polarisation results. However, for many purposes, damping mechanisms play a more important role than these field-dependent processes; it is then sufficient to retain only one order of nonlinearity, and to introduce appropriate damping parameters. It is common practice to model the varied and usually complex damping mechanisms, such as forms of collisional dephasing, by inserting into resonant denominators an empirically chosen parameter Γ. The actual choice of Γ is a rather delicate matter, which is hardly surprising considering the demands made of it, and is left to later sections; here we take it as given. It is of interest here to point out that if quantised fields were employed, then the idea of spontaneous emission would lead to the choice of Γ as the natural lifetime (in the absence of collisions, etc.). This can be regarded as the power broadening induced by the field in the limit of vanishing classical amplitude (i.e., the vacuum field); in view of the foregoing remarks, it is satisfying to find that the usual expression for Γ can be obtained by direct summation of higher-order terms [2.22].

In the most general case, the denominators in (2.15) take on the form (denominator factor) $\to$ (denominator factor $\pm i\Gamma$). Thus $\Omega_{bg} - 2\omega$ in the first part of (2.18) becomes $\Omega_{bg} - 2\omega - i\Gamma$. Unlike its magnitude, the choice of the sign with which Γ enters a denominator is fixed, and may be determined from causality. It is discussed at the end of this section. For the present, the reader is asked to accept the choices presented; in any case, the signs may usually be deduced by direct physical reasoning when final expressions are obtained. For example, if we take ground-state atoms, then in SRS the requirement that gain be positive fixes the sign, as does the need for absorption to be positive in single-photon processes.

It is convenient at this point to introduce the first-order transition hyperpolarisability (Sect.2.7), $\alpha_{fg}(\omega_1;\omega_2)$, defined as

$$\alpha_{fg}(\omega_1;\omega_2) = \alpha_{fg}(\omega_2;\omega_1)$$

$$= \frac{1}{\hbar}\sum_i \left(\frac{\underline{\varepsilon}_1\cdot\underline{Q}_{fi}\,\underline{\varepsilon}_2\cdot\underline{Q}_{ig}}{\Omega_{ig}-\omega_2} + \frac{\underline{\varepsilon}_2\cdot\underline{Q}_{fi}\,\underline{\varepsilon}_1\cdot\underline{Q}_{ig}}{\Omega_{ig}-\omega_1}\right) , \qquad (2.19)$$

where, for simplicity, damping parameters have been omitted from the frequency denominators. [Permutation of ω_1 and ω_2 in $\alpha_{fg}(\omega_1;\omega_2)$ implies that the pairs $\underline{\varepsilon}_1\omega_1$ and $\underline{\varepsilon}_2\omega_2$ are to be interchanged in the RHS of (2.19).]

Then (2.15) can be shown to factorise into a product of two α_{fg}'s for all two-photon resonant third-order nonlinear susceptibilities. For example, (2.18) becomes

$$\chi^{(3)}(-3\omega;\omega\omega\omega) = \frac{\mathcal{N}}{2\hbar\varepsilon_0}\,\frac{1}{\Omega_{fg}-2\omega-i\Gamma}\sum_{\substack{\text{degeneracy}\\ \text{of } f,g}}\left(\rho(g) - \rho(f)\right)\alpha_{fg}(3\omega;-\omega)^{*}\alpha_{fg}(\omega;\omega) \quad . \qquad (2.20)$$

In (2.20), the resonant transition is $|g\rangle \rightarrow |f\rangle$; therefore, the general sum over $\rho(g)$ in (2.17) is restricted to $\rho(g)$ and $\rho(f)$, giving a net factor of $\rho(g) - \rho(f)$. Thus, the resonant susceptibility is proportional to the equilibrium population difference $\mathcal{N}(\overline{\rho(g)} - \overline{\rho(f)})$ of the transition.

As already mentioned, two-photon resonance is one of the most important features in the nonlinear optics of atomic vapours. Any two-photon resonant third-order nonlinear susceptibility may be put into one of four general classes. This classification is set out in Fig.2.2; to help explain it, we first remind the reader of the usual classical treatment of vibrational Raman scattering. In this, there is considered to be a natural polarisability that oscillates at the vibrational frequency Ω_{fg}; an incident wave ω_p is, therefore, modulated, with the result that sidebands at the frequencies $\omega_p \pm \Omega_{fg}$ are generated — the Stokes and anti-Stokes frequencies.

Raman resonant   TPA resonant

Difference
generation

$\omega_3 \simeq \Omega_{fg} - \omega_4$

$\omega_4 \simeq \Omega_{fg} - \omega_3$

(a) $\chi^{(3)}(-\omega_4;-\omega_3-\omega_2\omega_1)$ (c) $\chi^{(3)}(-\omega_4;-\omega_3\omega_2\omega_1)$

Sum generation

$\omega_4 \simeq \Omega_{fg} + \omega_3$

Difference
generation

$\omega_3 \simeq \omega_4 - \Omega_{fg}$

(b) $\chi^{(3)}(-\omega_4;\omega_3-\omega_2\omega_1)$ (d) $\chi^{(3)}(-\omega_4;\omega_3\omega_2\omega_1)$

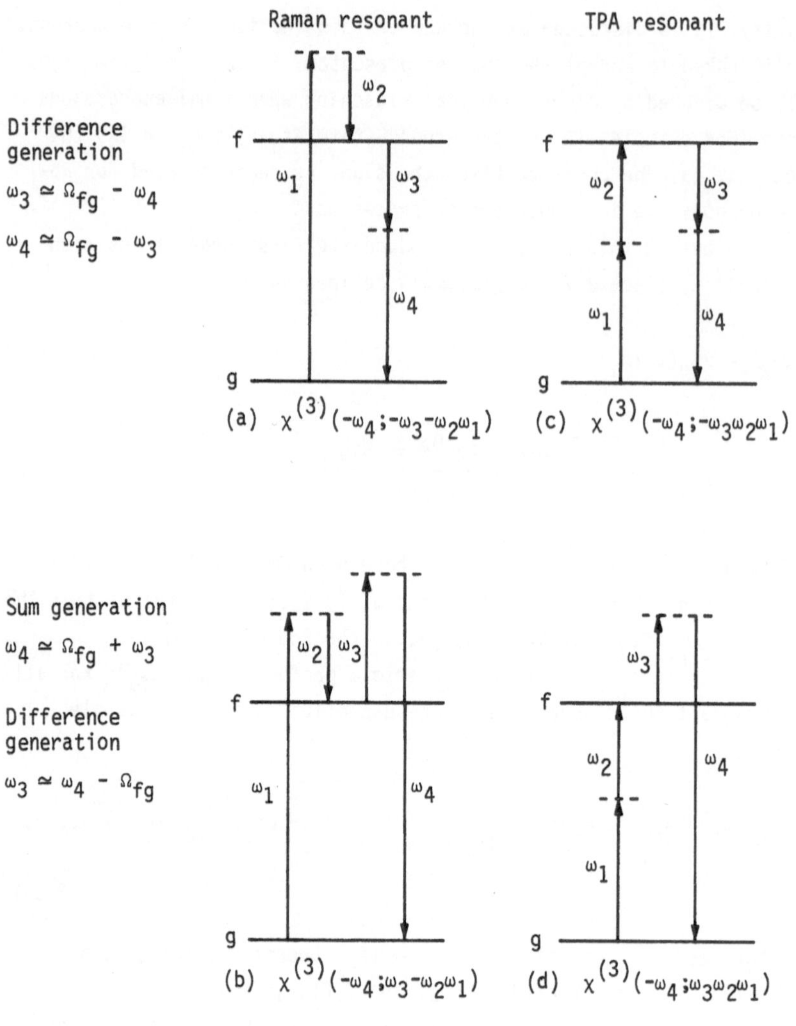

Fig.2.2a-d. Classification of two-photon resonant third-order nonlinear sus-
ceptibilities. Fields at ω_1, ω_2 act through two-photon absorption (TPA) or
Raman-type resonances in conjunction with ω_3 to generate ω_4, or conversely,
with ω_4, to generate ω_3. (a) Raman-resonant difference generation of ω_3 or
ω_4 (Chap.6). (b) Raman-resonant sum generation of ω_4, i.e., coherent anti-
Stokes Raman scattering (CARS). (Frequently with $\omega_1 = \omega_3$. ω_4 is the anti-
Stokes frequency) (Chap.6) / Raman-resonant difference generation of ω_3, i.e.,
coherent Raman mixing or biharmonic pumping (Sects.3.4,7.5)/ Stimulated
Raman scattering is the degenerate case $\omega_1 = \omega_4 = \omega_p$, $\omega_2 = \omega_3 = \omega_s$ (Sects.
2.7,3.3,Chap.5). (c) TPA-resonant difference generation of ω_3 or ω_4 (Sect.3.4;
as a competitor to SHRS in Chap.5) / TPA is the degenerate case $\omega_1 = \omega_4$,
$\omega_2 = \omega_3$ (Sects.2.6,3.3). (d) TPA-resonant sum generation of ω_4, i.e., up-
conversion, of which THG is a special case (Chap.4) / TPA-resonant difference generation of ω_3

Turning to Fig.2.2a, consider the generation of frequency ω_4 by the incident waves ω_1, ω_2 and ω_3. The action of ω_1 and ω_2 is to set up, via a Raman-type resonance, an excitation of the medium, which oscillates at the transition frequency Ω_{fg}, i.e., the medium is put into a superposition of the resonant states g,f. This superposition state produces a macroscopic *polarisability* of the medium. As explained, the third wave ω_3, passing through the medium, generates a *polarisation* at the sideband frequencies $\Omega_{fg} - \omega_3$ (Fig.2.2a) and $\Omega_{fg} + \omega_3$ (Fig.2.2b). We term these processes Raman-resonant difference (Chap.6) and sum-frequency generation of ω_4, respectively. For difference mixing there is a symmetry between ω_3 and ω_4; this is not so for the sum process. If Fig.2.2b referred to the generation of ω_3 (with ω_4 the incident wave), the process could equally well be described as a type of Raman-resonant difference-frequency generation, with $\omega_3 = \omega_4 - \Omega_{fg}$. In fact, the processes shown in Fig.2.2b may be distinguished by their specific names of coherent anti-Stokes Raman scattering (CARS) if ω_4 is generated and biharmonic pumping or coherent Raman mixing (Sects.3.4,7.5) if ω_3 is generated.

Similarly, in Fig.2.2c,d the polarisability is set up by a TPA type of resonance, $\omega_1 + \omega_2 \simeq \Omega_{fg}$, and modulates the third wave ω_3 or ω_4. Thus, the classification is completed with TPA-resonant sum and difference mixing.

This description has been based on the idea that the generated wave is produced by one of the input waves interacting with the polarisability induced by the other two. Of course, when the generated wave increases, the reverse process occurs, bringing with it the possibility of complete reversal, or a limiting of the process, if a stable balance is reached. Such matters are the concern of large-signal analyses, an example of which is discussed in Sect.3.4. However, this does serve to illustrate the point that the direction of any arrow in diagrams such as those in Fig.2.2 does not always imply that the process occurs in that direction; it is safest to regard the arrows as a form of vector diagram that must form a closed loop, to demonstrate the relation $\omega_\sigma - \omega_1 - \omega_2 - \omega_3 = 0$. Furthermore, the idea of absorption and emission of photons, as discussed in connection with the diagrammatic representation of (2.15), is seen to be fraught with difficulties of interpretation if taken too literally; this matter cannot, unfortunately, be pursued here.

We now return to the susceptibility description of these two-photon resonant processes; the connection with the polarisability picture will be given in Sect.2.7. All of the susceptibilities for these processes fall into the form

$$\chi^{(3)}(-\omega_\sigma;\omega_1\omega_2\omega_3)_{res} = \frac{f\mathcal{N}/\hbar\varepsilon_0}{\Delta \pm i\Gamma} \sum_{\deg f,g} [(\rho(g) - \rho(f)]\alpha_{fg}(1)^* \alpha_{fg}(2) \ , \qquad (2.21)$$

where $\Delta \pm i\Gamma$ is the resonant denominator, α_{fg} is defined in (2.19), 1,2 refer to pairs of frequencies chosen from $-\omega_\sigma \ldots \omega_3$, and f is a simple numerical factor. As mentioned in Sect.2.2, it is convenient to order the arguments in $\chi^{(3)}$ so as to match the pairs 1 and 2; this is shown in Table 2.2, which also enumerates the parameters of (2.21) for the most important cases. MAKER and TERHUNE [2.23] have given an early discussion of the two-photon resonant susceptibility.

Obviously single-photon resonance enhancement can also be employed. (Strong single-photon resonance enhancement in Raman scattering is known as "resonance Raman scattering".) Often, the presence of a single photon resonance determines which of the Raman or TPA resonances dominate; the dotted lines in Fig.2.2 illustrate possible positions for the dominant intermediate states.

Similarly, any resonant nonlinear susceptibility will factorise (YURATICH [2.21] gives general formulae). For three-photon resonant fifth-order nonlinear susceptibilities the general form is the same as (2.21). However, Δ is now a three-photon resonant denominator, and α_{fg}(1 or 2) is replaced by the second-order transition polarisability β_{fg} (1 or 2); 1,2 refer to triplets of frequencies chosen from $-\omega_\sigma;\ \omega_1 \ldots \omega_5$, and β_{fg} is given by

$$\beta_{fg}(\omega_1;\omega_2\omega_3) = \frac{1}{\hbar}\sum_{ij}$$

$$\times \left\{ \frac{\varepsilon_1\cdot\mathcal{Q}_{fi}\,\varepsilon_2\cdot\mathcal{Q}_{ij}\,\varepsilon_3\cdot\mathcal{Q}_{jg}}{(\Omega_{ig}-\omega_2-\omega_3)(\Omega_{jg}-\omega_3)} + \frac{\varepsilon_1\cdot\mathcal{Q}_{fi}\,\varepsilon_3\cdot\mathcal{Q}_{ij}\,\varepsilon_2\cdot\mathcal{Q}_{jg}}{(\Omega_{ig}-\omega_2-\omega_3)(\Omega_{jg}-\omega_2)} \right.$$

$$+ \frac{\varepsilon_2\cdot\mathcal{Q}_{fi}\,\varepsilon_1\cdot\mathcal{Q}_{ij}\,\varepsilon_3\cdot\mathcal{Q}_{jg}}{(\Omega_{ig}-\omega_1-\omega_3)(\Omega_{jg}-\omega_3)} + \frac{\varepsilon_3\cdot\mathcal{Q}_{fi}\,\varepsilon_1\cdot\mathcal{Q}_{ij}\,\varepsilon_2\cdot\mathcal{Q}_{jg}}{(\Omega_{ig}-\omega_1-\omega_2)(\Omega_{jg}-\omega_2)}$$

$$\left. + \frac{\varepsilon_2\cdot\mathcal{Q}_{fi}\,\varepsilon_3\cdot\mathcal{Q}_{ij}\,\varepsilon_1\cdot\mathcal{Q}_{jg}}{(\Omega_{ig}-\omega_1-\omega_3)(\Omega_{jg}-\omega_1)} + \frac{\varepsilon_3\cdot\mathcal{Q}_{fi}\,\varepsilon_2\cdot\mathcal{Q}_{ij}\,\varepsilon_1\cdot\mathcal{Q}_{jg}}{(\Omega_{ig}-\omega_1-\omega_2)(\Omega_{jg}-\omega_1)} \right\} . \qquad (2.22)$$

Notice that because damping is omitted from (2.22), β_{fg} exhibits overall permutation symmetry of its arguments, cf. (2.19). The most important three-photon resonant fifth-order processes are stimulated hyper-Raman scattering (SHRS) and three-photon absorption (3PA) or ionisation (3PI). These are

Table 2.2. Two-photon resonant third-order nonlinear susceptibilities. The pairs of frequencies and the numerical factor f are to be inserted in (2.21)

Process	Reference to Fig.2.2	Susceptibility	Resonance $\simeq 0$	f	Pair 1	Pair 2
Raman-resonant difference mixing	a	$\chi^{(3)}(-\omega_4;-\omega_3-\omega_2\omega_1)$	$\Omega_{fg}+\omega_1-\omega_2-i\Gamma$	1/6	$\omega_3;\omega_4$	$-\omega_2;\omega_1$
CARS	b	$\chi^{(3)}(-\omega_4;\omega_3-\omega_2\omega_1)$	$\Omega_{fg}+\omega_1-\omega_2-i\Gamma$	1/6	$-\omega_3;\omega_4$	$-\omega_2;\omega_1$
		$\chi^{(3)}(-\omega_{as};\omega_p\omega_p-\omega_s)$	$\Omega_{fg}+\omega_p-\omega_s-i\Gamma$	1/3	$-\omega_p;\omega_{as}$	$-\omega_s;\omega_p$
Biharmonic pumping	b	$\chi^{(3)}(-\omega_3;\omega_4-\omega_1\omega_2)$	$\Omega_{fg}+\omega_2-\omega_1+i\Gamma$	1/6	$-\omega_4;\omega_3$	$-\omega_2;\omega_1$
SRS	b	$\chi^{(3)}(-\omega_s;\omega_p-\omega_p\omega_s)$	$\Omega_{fg}+\omega_p-\omega_s+i\Gamma$	1/6	$-\omega_s;\omega_p$	$-\omega_s;\omega_p$
TPA-resonant difference mixing	c	$\chi^{(3)}(-\omega_4;-\omega_3\omega_2\omega_1)$	$\Omega_{fg}-\omega_1-\omega_2-i\Gamma$	1/6	$\omega_3;\omega_4$	$\omega_2;\omega_1$
TPA	c	$\chi^{(3)}(-\omega_1;-\omega_2\omega_2\omega_1)$	$\Omega_{fg}-\omega_1-\omega_2-i\Gamma$	1/6	$\omega_2;\omega_1$	$\omega_2;\omega_1$
		$\chi^{(3)}(-\omega_1;-\omega_1\omega_1\omega_1)$	$\Omega_{fg}-2\omega_1-i\Gamma$	1/6	$\omega_1;\omega_1$	$\omega_1;\omega_1$
Up-conversion	d	$\chi^{(3)}(-\omega_4;\omega_3\omega_2\omega_1)$	$\Omega_{fg}-\omega_1-\omega_2-i\Gamma$	1/6	$-\omega_3;\omega_4$	$\omega_2;\omega_1$
		$\chi^{(3)}(-\omega_4;\omega_3\omega_1\omega_1)$	$\Omega_{fg}-2\omega_1-i\Gamma$	1/6	$-\omega_3;\omega_4$	$\omega_1;\omega_1$
THG	d	$\chi^{(3)}(-3\omega;\omega\omega\omega)$	$\Omega_{fg}-2\omega-i\Gamma$	1/2	$-\omega;3\omega$	$\omega;\omega$
TPA-resonant difference mixing	d	$\chi^{(3)}(-\omega_3;\omega_4-\omega_2-\omega_1)$	$\Omega_{fg}-\omega_1-\omega_2+i\Gamma$	1/6	$\omega_1;\omega_2$	$-\omega_3;\omega_4$

shown in Fig.2.3; in Table 2.3, the parameters of (2.22) are set out analogously to Fig.2.2 and Table 2.2.

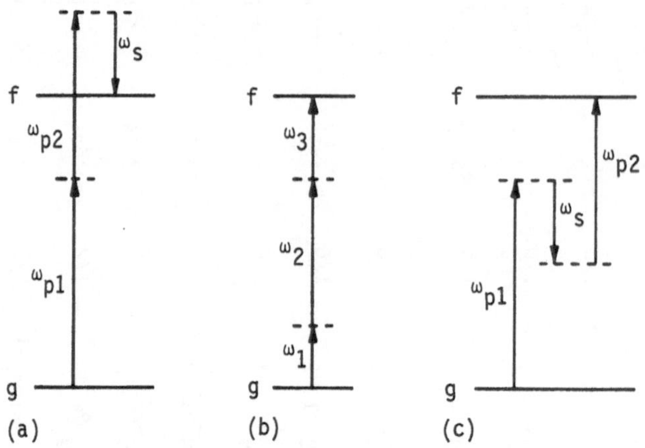

Fig.2.3a-c. Three-photon resonant fifth-order nonlinear processes. (a) Stimulated hyper-Raman scattering (SHRS) showing conventional resonances. (b) Three-photon emission/absorption/ionisation (3PE/A/I). (c) SHRS-type process showing resonances exploited by GRYNBERG et al. [2.24]

To simplify the notation, we shall generally dispense with $\sum_{\text{deg } f,g}$ in writing out resonant susceptibilities, or use a bar to indicate the orientation average *and* the sum over degeneracies of the final level. The factor $\mathcal{N}[\overline{\rho(g)} - \overline{\rho(f)}]$ remains, which is the effective number density for the process. We will give this the symbol N, although it should be noted that in many cases the use of $\mathcal{N}$ and N is interchangeable—for example, if the atoms are in their ground state.

To conclude this section on resonant susceptibilities, we tackle the question of the choice of sign of the damping factor in resonant denominators. This choice has been discussed from the point of view of causality by BUTCHER [1.70]; the guiding result is that the poles of the nonlinear susceptibility should be in the upper-half complex plane. This requirement can be seen in a direct way, by noting that the denominators in (2.15) arise from time integrals such as $\int_{-\infty}^{t} \exp[i(\Omega_{b_n g} - \omega_n)t]\,dt$ (e.g. [1.71]). To ensure convergence at the lower limit of the integral, it is convenient to let $\omega_n \to \omega_n + i\varepsilon$, where $\varepsilon > 0$ is a real infinitesimal quantity, commonly known as a switching parameter. Physically, this guarantees that the field vanishes in the distant past. Then, in successive orders of perturbation theory, the perturbed initial state, $|g\rangle'$, is found to be of

Table 2.3. Three-photon resonant fifth-order nonlinear susceptibilities. The triplets of frequencies and numerical factor f are to be inserted into a formula analogous to (2.21), see text

Process	Reference to Figure 2.3	Susceptibility	Resonance $\simeq 0$	f	Triple 1	Triple 2
SHRS	a,c	$\chi^{(5)}(-\omega_s;\omega_{p1}\omega_{p2}-\omega_{p2}-\omega_{p1}\omega_s)$	$\Omega_{fg}+\omega_s-\omega_{p1}-\omega_{p2}+i\Gamma$	1/120	$-\omega_s;\omega_{p1}\omega_{p2}$	$-\omega_s;\omega_{p1}\omega_{p2}$
	a,c	$\chi^{(5)}(-\omega_s;\omega_p\omega_p-\omega_p-\omega_p\omega_s)$	$\Omega_{fg}+\omega_s-2\omega_p+i\Gamma$	1/120	$-\omega_s;\omega_p\omega_p$	$-\omega_s;\omega_p\omega_p$
3PA	b	$\chi^{(5)}(-\omega_1;-\omega_2-\omega_3\omega_3\omega_2\omega_1)$	$\Omega_{fg}-\omega_1-\omega_2-\omega_3-i\Gamma$	1/120	$\omega_1;\omega_2\omega_3$	$\omega_1;\omega_2\omega_3$
		$\chi^{(5)}(-\omega_1;-\omega_1-\omega_1\omega_1\omega_1\omega_1)$	$\Omega_{fg}-3\omega-i\Gamma$	1/120	$\omega_1;\omega_1\omega_1$	$\omega_1;\omega_1\omega_1$
Three-photon resonant 5th-harmonic generation	-	$\chi^{(5)}(-5\omega;\omega\omega\omega\omega\omega)$	$\Omega_{fg}-3\omega-i\Gamma$	1/12	$5\omega;-\omega-\omega$	$\omega;\omega\omega$

the schematic form $|g\rangle' \sim [(\ldots-3i\varepsilon)(\Omega_{ig}-\omega'-2i\varepsilon)(\Omega_{jg}-\omega-i\varepsilon)]^{-1}|g\rangle$, i.e., the denominators contain $-i\varepsilon$, $-2i\varepsilon$, $-3i\varepsilon\ldots$ successively, towards the left. Because the polarisation is defined by the expectation value (2.13), denominators to the left of matrix elements of $\underline{\varepsilon}_\sigma^*\cdot\underline{Q}$ in (2.15) contain $+mi\varepsilon$, and those to the right contain $-mi\varepsilon$. Therefore, (2.15) must be rewritten as $n+1$ terms, corresponding to the $n+1$ permutations of $\underline{\varepsilon}_\sigma^* - \omega_\sigma$ with (itself, $\underline{\varepsilon}_1\omega_1\ldots\underline{\varepsilon}_n\omega_n$), into which the parameter ε may be inserted according to the foregoing prescription. Each of these terms retains intrinsic permutation symmetry, defined by the generator $\mathcal{S}$ of all permutations of the pairs $\underline{\varepsilon}_1\omega_1\ldots\underline{\varepsilon}_n\omega_n$; but, now, overall permutation symmetry is violated; this is in accord with the statements of Sect.2.2.

As an example, the third-order susceptibility becomes

$$\chi^{(3)}(-\omega_\sigma;\omega_1\omega_2\omega_3) = \frac{\mathcal{N}}{6\hbar^3\varepsilon_0}\mathcal{S}\sum_{glmn}\rho(g)$$

$$\times\left\{\frac{\underline{\varepsilon}_\sigma^*\cdot\underline{Q}_{gl}\underline{\varepsilon}_1\cdot\underline{Q}_{lm}\underline{\varepsilon}_2\cdot\underline{Q}_{mn}\underline{\varepsilon}_3\cdot\underline{Q}_{ng}}{(\Omega_{lg}-\omega_\sigma-3i\varepsilon)(\Omega_{mg}-\omega_2-\omega_3-2i\varepsilon)(\Omega_{ng}-\omega_3-i\varepsilon)}\right.$$

$$+\frac{\underline{\varepsilon}_1\cdot\underline{Q}_{gl}\underline{\varepsilon}_\sigma^*\cdot\underline{Q}_{lm}\underline{\varepsilon}_2\cdot\underline{Q}_{mn}\underline{\varepsilon}_3\cdot\underline{Q}_{ng}}{(\Omega_{lg}+\omega_\sigma-\omega_2-\omega_3+i\varepsilon)(\Omega_{mg}-\omega_2-\omega_3-2i\varepsilon)(\Omega_{ng}-\omega_3-i\varepsilon)}$$

$$+\frac{\underline{\varepsilon}_1\cdot\underline{Q}_{gl}\underline{\varepsilon}_2\cdot\underline{Q}_{lm}\underline{\varepsilon}_\sigma^*\cdot\underline{Q}_{mn}\underline{\varepsilon}_3\cdot\underline{Q}_{ng}}{(\Omega_{lg}+\omega_\sigma-\omega_2-\omega_3+i\varepsilon)(\Omega_{mg}+\omega_\sigma-\omega_3+2i\varepsilon)(\Omega_{ng}-\omega_3-i\varepsilon)} \tag{2.23}$$

$$\left.+\frac{\underline{\varepsilon}_1\cdot\underline{Q}_{gl}\underline{\varepsilon}_2\cdot\underline{Q}_{lm}\underline{\varepsilon}_3\cdot\underline{Q}_{mn}\underline{\varepsilon}_\sigma^*\cdot\underline{Q}_{ng}}{(\Omega_{lg}+\omega_\sigma-\omega_2-\omega_3+i\varepsilon)(\Omega_{mg}+\omega_\sigma-\omega_3+2i\varepsilon)(\Omega_{ng}+\omega_\sigma+3i\varepsilon)}\right\}.$$

If none of the denominators is small due to resonance, ε may safely be taken as zero, and (2.15) recovered.

The use of ε is rigorous for limiting analyses, such as in the discussion of secularity (Sect.2.6.5). For the treatment of damping, outlined earlier, $i\Gamma$ is simply assumed to enter resonant denominators with the sign of ε. This ad hoc approach cannot be expected to work when large population changes are induced by the fields, for example by multiple close resonances, because no account is taken of real decay into and out of the various states: Γ is a dephasing parameter appropriate to weak collisions. Population changes are best handled directly, by use of density-matrix equations with damping. FLYTZANIS [2.1] has derived susceptibilities to third order in this way. Terms appear that have their origins in the transfer of real population to intermediate states (Sect.2.6.2), from which further processes can occur: there are thus resonant denominators that contain $\Omega_{bb'}$(b,b' excited states) rather than just Ω_{bg}. Intrinsic permutation symmetry of course remains; if the damping is "turned off", the lossless formulae are recovered. In Sect.2.6.8, damping in resonant processes is further reviewed.

With these points in mind, we will continue to use the ad hoc method as a basis for discussion, but restricted to a single weakly damped resonance. For example in the two-photon resonant processes of Fig.2.2, only the $\pm 2i\varepsilon$ factors of (2.23) would be retained. Consider SRS, where the resonant denominator is $\Omega_{fg} + \omega_s - \omega_p$, and take the atom to be in its initial state g. From (2.23), this denominator must arise from $(\Omega_{fg} - \omega_2 - \omega_3 - i\Gamma)$ or $(\Omega_{fg} + \omega_\sigma - \omega_3 + i\Gamma)$, where $\omega_\sigma \ldots \omega_3$ are to be selected from the arguments of $\chi^{(3)}(-\omega_s;\omega_p{-}\omega_p\omega_s)$. Clearly, the second denominator is required, i.e., the last two parts of (2.23). Similarly, the first two parts of (2.23) lead to TPA. Such considerations lead immediately to the entries of $\pm i\Gamma$ in Tables 2.2 and 2.3. (The selection of $\pm i\Gamma$ becomes a trivial exercise if diagrammatic techniques are used; we need only write down a typical term of the susceptibility; if the resonant denominator appears to the right of the ω_σ arrow, take $-i\Gamma$; if it is to the left, take $+i\Gamma$; moreover, the typical term can be written down directly from the susceptibility arguments.)

2.6 Resonant Processes

As has been mentioned in Sect.2.1, although the susceptibility formalism is perfectly general, for time-dependent resonant processes that involve intense radiation it is not always the most practical approach. A further example of this occurred in Sect.2.5, when the notion of a material excitation was introduced in qualitative terms; that such an excitation exists was not obvious from the theory presented up to that point. In this and the following section (Sect.2.7) we attempt to redress the balance, paying particular attention to the relation of the susceptibility approach to the various methods designed specifically to handle time-dependent processes.

Particular points that are examined include the question of how to deal with the secular terms in the susceptibility, the theory of the intensity-dependent refractive index, the optical Stark effect, and criteria for when time dependence becomes important. In the course of this discussion, use will be made of the unifying idea of adiabatic states; these are first introduced in Sect.2.6.2. In later chapters, these results will prove useful when limiting mechanisms for various nonlinear processes are discussed. Therefore, our account of resonant processes will be limited to the topics needed later, and to the connection with susceptibility theory; we do not consider transient coherent optical phenomena, such as photon echoes and self-induced transparency (e.g., [2.25-27,117]).

2.6.1 Nonlinear Polarisation for Quasi-Monochromatic Radiation

We first examine in more detail the linear polarisation induced by a pulsed field, and then discuss the extension to nonlinear effects. The main aim is to give the conditions under which the transient nature of the field becomes important. The results are obtained rather simply, and are equivalent to those of PUELL and VIDAL [2.28]. These authors calculated the polarisation from first principles, using the density matrix; in the course of their calculations they expanded the results in terms of time derivatives and integrals of the field; they obtained the coefficients of these expansions as complicated functions of atomic matrix elements and transition frequencies. We will show how to deduce these coefficients in terms of derivatives with respect to frequency of the nonlinear susceptibilities; the latter can be written down from (2.15). Thus the present method may be readily extended to the case of a field consisting of pulses of differing carrier frequencies.

Consider then, the case where the electric field is a quasi-monochromatic wave, i.e., is of the form

$$E(t) = \frac{1}{2} E_{\omega_0}(t)\, e^{-i\omega_0 t} + \text{c.c.} \tag{2.24a}$$

with $E_{\omega_0}(t)$ a slowly varying envelope function that modulates the carrier wave of frequency ω_0. A similar expansion is used for the linear polarisation $P^{(1)}(t)$, where the envelope function is written $P^{(1)}_{\omega_0}(t)$. The Fourier transform (2.1) of (2.24a) yields

$$E(\omega) = \frac{1}{2} \bar{E}_{\omega_0}(\omega - \omega_0) + \frac{1}{2} \bar{\bar{E}}_{\omega_0}(\omega + \omega_0) \quad , \tag{2.24b}$$

where the bar denotes the transform of the envelope functions. Because the radiation is in the form of a pulse, $\bar{E}_{\omega_0}(\Omega)$ is sharply peaked around $\Omega = 0$. Using this fact, and the rigorous relation [from (2.2)]

$$P^{(1)}(t) = \varepsilon_0 \int_{-\infty}^{\infty} \chi^{(1)}(-\omega;\omega)E(\omega)\ e^{-i\omega t}d\omega \tag{2.25}$$

we see that

$$P^{(1)}_{\omega_0}(t) = \varepsilon_0 \int_{-\infty}^{\infty} \chi^{(1)}(-\omega;\omega)\bar{E}_{\omega_0}(\omega - \omega_0)\ e^{-i(\omega-\omega_0)t}d\omega\ . \tag{2.26}$$

From (2.15), we can write an expression for the linear susceptibility [(2.33) below]; therefore, in principle, the polarisation is known. Examples of where (2.26) has been used with particular field envelopes are given by BRILLOUIN [2.29], GARRETT and McCUMBER [2.30], and CRISP [2.31,32]. We consider here two limiting cases of interest, in which the pulse shapes are arbitrary.

First take ω_0 to be in a transparency region of the medium, so that $\chi^{(1)}(-\omega;\omega)$ is a slowly varying function of frequency when $\omega \sim \omega_0$ (this is unnecessarily restrictive, as will become clear later). The susceptibility may then be expanded in a Taylor series about ω_0, so that

$$\begin{aligned} P^{(1)}_{\omega_0}(t) &= \varepsilon_0\chi^{(1)}(-\omega_0;\omega_0) \int_{-\infty}^{\infty} \bar{E}_{\omega_0}(\omega - \omega_0)\ e^{-i(\omega-\omega_0)t}d\omega \\ &+ \varepsilon_0 \left.\frac{\partial\chi^{(1)}(-\omega;\omega)}{\partial\omega}\right|_{\omega_0} \int_{-\infty}^{\infty} (\omega - \omega_0)\bar{E}_{\omega_0}(\omega - \omega_0)\ e^{-i(\omega-\omega_0)t}d\omega \\ &+ \dots\ . \end{aligned} \tag{2.27}$$

The first integral is simply $E_{\omega_0}(t)$; the second is $idE_{\omega_0}(t)/dt$. Thus

$$P^{(1)}_{\omega_0}(t) = \varepsilon_0\chi^{(1)}(-\omega_0;\omega_0)E_{\omega_0}(t) + i\varepsilon_0 \left.\frac{\partial\chi^{(1)}}{\partial\omega}\right|_{\omega_0} \frac{dE_{\omega_0}(t)}{dt} + \dots\ , \tag{2.28}$$

where the series continues with higher-order derivatives of $\chi^{(1)}$ and $E_{\omega_0}(t)$. We can thus define a quantity $\bar{\chi}^{(1)}(\omega_0)$ (which is *not* a susceptibility) by

$$\bar{\chi}^{(1)}(\omega_0) \equiv \chi^{(1)}(-\omega_0;\omega_0) + i \left.\frac{\partial\chi^{(1)}}{\partial\omega}\right|_{\omega_0} \frac{1}{E_{\omega_0}(t)} \frac{dE_{\omega_0}(t)}{dt} + \dots\ , \tag{2.29}$$

so that

$$P^{(1)}_{\omega_0}(t) = \varepsilon_0 \bar{\chi}^{(1)}(\omega_0) E_{\omega_0}(t) \quad . \tag{2.30}$$

The relation (2.30) is, however, misleading, because it obscures the fact that, in general, the polarisation at time t is not determined simply by the instantaneous field $E_{\omega_0}(t)$, but rather that it depends on the history of the field. As we will see, this is implicit in (2.25) and (2.2), and is the origin of the derivative terms in (2.28,29). Thus we must resist the temptation to think of (2.30) as a simple generalisation of the rigorous relation for monochromatic waves,

$$P^{(1)}_{\omega} = \varepsilon_0 \chi^{(1)}(-\omega;\omega) E_{\omega} \quad , \tag{2.31}$$

to general time-dependent problems. However, in a sufficiently non-dispersive medium or for a sufficiently slowly varying field envelope, the derivative terms in (2.29) may be dropped; whence $\bar{\chi}^{(1)}(\omega_0) = \chi^{(1)}(-\omega_0;\omega_0)$, i.e., the polarisation induced in the medium at time t then depends only upon the instantaneous field. This is referred to as the adiabatic limit, in which time may be allowed to enter monochromatic-wave formulae such as (2.31), as a parameter. As the field frequency ω_0 approaches a resonance line in the medium, or for transient pulses, the second term in (2.29) can become important; the condition that it may be ignored is

$$\left| \left(\frac{1}{\chi^{(1)}(-\omega_0;\omega_0)} \left. \frac{\partial \chi^{(1)}}{\partial \omega} \right|_{\omega_0} \right) \left(\frac{1}{E_{\omega_0}(t)} \frac{dE_{\omega_0}(t)}{dt} \right) \right| \ll 1 \quad . \tag{2.32}$$

As is well-known, or from (2.15),

$$\chi^{(1)}(-\omega;\omega) = \frac{\mathcal{N}}{\hbar\varepsilon_0} \sum_n \overline{|\mu_{ng}|^2} \left\{ \frac{1}{\Omega_{ng}-\omega-i\Gamma} + \frac{1}{\Omega_{ng}+\omega+i\Gamma} \right\} \quad , \tag{2.33}$$

where the atoms are assumed to be in their ground state; a damping parameter has been added, in the usual fashion. Thus

$$\left. \frac{\partial \chi^{(1)}}{\partial \omega} \right|_{\omega_0} = \frac{\mathcal{N}}{\hbar\varepsilon_0} \sum_n \overline{|\mu_{ng}|^2} \left\{ \left(\frac{1}{\Omega_{ng}-\omega_0-i\Gamma} \right)^2 - \left(\frac{1}{\Omega_{ng}+\omega_0+i\Gamma} \right)^2 \right\} \tag{2.34}$$

{[2.28], equation (25)}. As ω_0 approaches a transition frequency Ω_{ng}, (2.32) and (2.34) give

$$\left|\frac{1}{\Delta - i\Gamma}\,\frac{1}{E_{\omega_0}}\,\frac{dE_{\omega_0}}{dt}\right| \ll 1 \quad , \tag{2.35}$$

where Δ is the detuning $\Delta = \Omega_{ng} - \omega_0$. {CRISP [2.32] has used the on-resonance limit of (2.35).} If we denote the characteristic rate of change of the field envelope by τ_c, then (2.35) becomes $|\Delta - i\Gamma|\tau_c \gg 1$. For smooth pulses, τ_c might be the pulse length, or rise time. Frequently, the pulses used have much structure, and a more realistic estimate of τ_c is obtained from $\tau_c \sim 1/\Delta\omega$, where $\Delta\omega$ is the laser linewidth. Equation (2.35) thus states that the frequency spread of the pulse should not overlap the atomic-transition frequency. A final way of expressing the adiabatic condition, which is intuitively obvious, is to require the time scale of changes in the field, τ_c, to be much longer than the response time of the medium, which for radiation of frequency ω_0 is characterised by $|\Delta - i\Gamma|^{-1}$.

As an example from nonlinear optics, consider third-harmonic generation. The algebra is only a little more complicated than the foregoing; it starts from (2.2). It is readily found that, to first order in $dE_{\omega_0}(t)/dt$,

$$P^{(3)}_{3\omega_0}(t) = \frac{1}{4}\,\varepsilon_0 \left\{\chi^{(3)}(-3\omega_0;\omega_0\omega_0\omega_0) + 3i\,\frac{\partial\chi^{(3)}}{\partial\omega}\bigg|_{\omega_0} \frac{1}{E_{\omega_0}}\,\frac{dE_{\omega_0}}{dt}\right\}[E_{\omega_0}(t)]^3 \quad , \tag{2.36}$$

where

$$\frac{\partial\chi^{(3)}}{\partial\omega}\bigg|_{\omega_0} \equiv \frac{\partial}{\partial\omega}\,\chi^{(3)}(-(2\omega_0+\omega);\omega_0\omega_0\omega)\bigg|_{\omega_0} \quad , \tag{2.37}$$

which may be compared with (2.28-30). The susceptibility in (2.37) may be written from (2.15); when the derivative is evaluated, the condition (2.35) is again found, where $\Delta - i\Gamma$ is now the smallest of any of the denominators in the third-harmonic susceptibility (2.17). For example, for a two-photon resonant susceptibility, we have $|\Omega_{fg} - 2\omega - i\Gamma|\tau_c \gg 1$. When this condition is satisfied, we can let the monochromatic-wave formula for the third-harmonic polarisation become time dependent, i.e., retain only the first part of (2.36). It is obvious that a similar condition applies for all processes.

In practice, the adiabatic condition is fulfilled for an extremely wide range of off- and near-resonant processes, particularly where nanosecond pulses are concerned. A further obvious requirement is that any generated waves should satisfy the adiabatic condition; this can pose a more stringent limit. For example, in Sect.5.3.2 it is shown that in stimulated Raman scattering, the pump induces changes in the Stokes wave on a time scale that may be much shorter than its own characteristic time, so that although the pump may satisfy the adiabatic (or quasi-steady-state) condition, the process as a whole need not. This case is in the realm of transient Raman scattering. Another factor to be borne in mind is the effect of the fields on the atoms, as manifested by higher-order nonlinearities. This will be taken up more fully later; here we note that the adiabatic-following regime of coherent interactions is described by the adiabatic terms [e.g., the first parts of (2.28,36)], in each order of nonlinearity; corrections may be obtained by including derivative terms. By jumping ahead to the wave-propagation equation (3.8b) the reader can see that the contributions to the polarisation from the first time derivative of the field have the effect of changing the pulse velocity, (see e.g., [2.32]).

Now consider what happens to the linear polarisation in the limiting case when ω_0 is tuned into resonance with a transition $g \rightarrow n$ of the medium, and where perhaps the product $\Gamma\tau_c$ violates (2.35). Although it is possible to continue to work in the frequency domain, it is easier and more instructive to transform to the time domain. Thus, (2.25) may be rewritten as

$$P^{(1)}(t) = \varepsilon_0 \int_{-\infty}^{\infty} R(t - r)E(r)dr \quad , \qquad (2.38)$$

where R(t-r) is the response function [1.70;2.1,33]. As required by causality, and easily proved from (2.25,33,38), R(t - r) vanishes if r > t, i.e., the integral in (2.38), in effect, ranges over only $r = -\infty$ to t. (The notion of response tensors can form the basis of the susceptibility formalism (2.2) [1.70]).

In a nondispersive medium, $\chi^{(1)}$ is a constant; it can then be shown that $R(t - r) = \chi^{(1)}\delta(t - r)$, so that from (2.38), $P^{(1)}(t) = \varepsilon_0\chi^{(1)}E(t)$ – the medium exhibits instantaneous response, as discussed earlier.

It is not difficult to show that from (2.25,33,38)

$$R(t - r) = \frac{\mathcal{N}}{\hbar\varepsilon_0} \sum_n \overline{|\mu_{ng}|^2}\, e^{-(t-r)\Gamma}\, e^{i\Omega_{ng}(t-r)} \quad . \quad (t \leq r) \qquad (2.39)$$

Therefore, the usual exponential decay is associated with a lorentzian line shape. Inserting (2.39) into (2.38) gives

$$P^{(1)}_{\omega_0}(t) = \frac{\mathcal{N}}{\hbar}\overline{|\mu_{ng}|^2}\int_{-\infty}^{t} E_{\omega_0}(r)\, e^{-(i\Delta+\Gamma)(t-r)}dr \quad , \tag{2.40}$$

where the sum over n has been omitted because we wish to consider the case of resonance; $\Delta = \Omega_{ng} - \omega_0$ as before. A series expansion of the integral may be obtained by integration by parts, so that

$$P^{(1)}_{\omega_0}(t) = \frac{\mathcal{N}}{\hbar}\overline{|\mu_{ng}|^2}\left\{\int_{-\infty}^{t} E_{\omega_0}(r)dr + i(\Delta - i\Gamma)\int_{-\infty}^{t} dr \int_{-\infty}^{r} E_{\omega_0}(r')dr' + \dots\right\} \quad , \tag{2.41}$$

which shows clearly the importance of the history of the field. The first integral is called the field area. Such results are familiar in the theory of coherent interactions, for example, in self-induced transparency (e.g., [2.26]).

Both (2.28,41) are exact expansions of the polarisation, but are complementary, in that the first term in (2.41) is a good approximation only if $|\Delta - i\Gamma|\tau_c \ll 1$, i.e., the converse of (2.35). (This follows from noting that the field area is proportioned to $\tau_c E_{\omega_0}$.) As an extreme example, in the absence of damping, and for exact resonance, the adiabatic condition can never be fulfilled, whereas (2.41) terminates at the first term. Generally speaking, (2.41) is appropriate for ultra-short pulsed radiation, and is of most interest in the study of coherent pulse propagation.

2.6.2 Adiabatic States

Having discussed the idea of adiabaticity from the point of view of the applied field, we now turn to the associated effect on the atoms. To simplify the description, we use the ubiquitous two-level atom model.

Fortunately, the results thereby obtained are of wide applicability, and lead also to an elegant formulation of the theory of two-photon resonant processes in multilevel atoms. In Sect.2.6.6, the vector model is introduced, but until then attention is focussed on the atomic states themselves. (For spectroscopic applications, mainly in the rf and microwave spectrum, see [2.34-37]).

Consider then the two-level atom, Fig.2.4a. A linearly polarised field is applied, of the form (2.24a), so that the electric-dipole interaction is

$$V = -\frac{1}{2}\mu E_\omega(t)\left(e^{-i\omega t} + e^{i\omega t}\right) \quad . \tag{2.42}$$

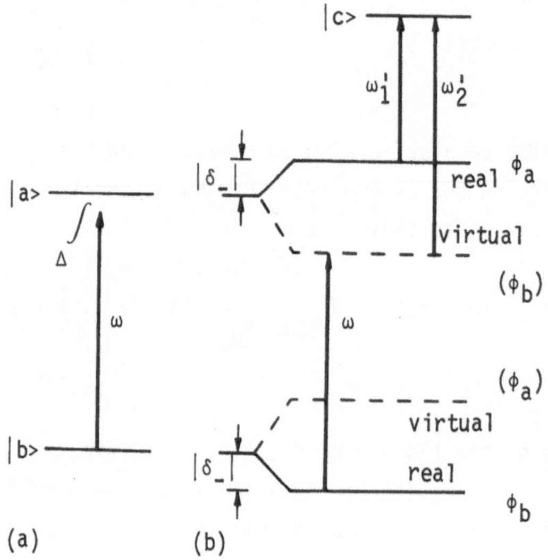

Fig.2.4.a+b. Adiabatic states in a two-level atom. (a) Unperturbed atom. (b) Perturbed atom showing real and virtual (dashed) amplitudes. ϕ_b is the adiabatic ground state and is composed of the real amplitude $|b\rangle$ and the virtual amplitude $|a\rangle$; these components are joined by the ω-arrow. ω_1' and ω_2' are probe frequencies, as described in the text

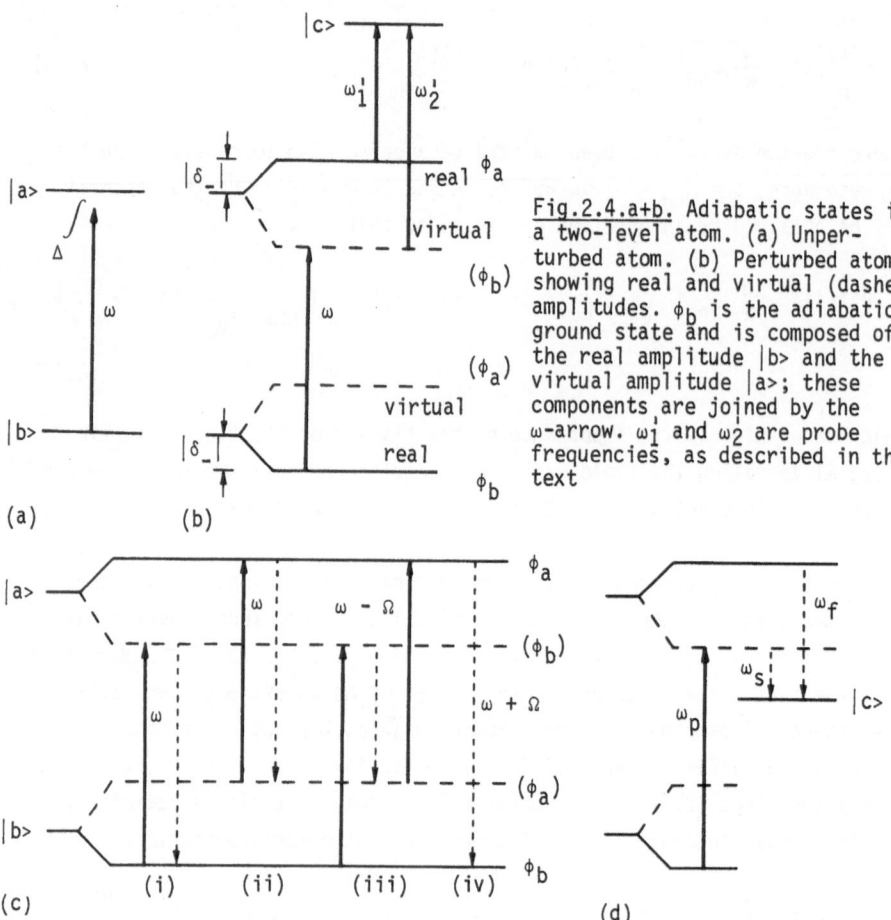

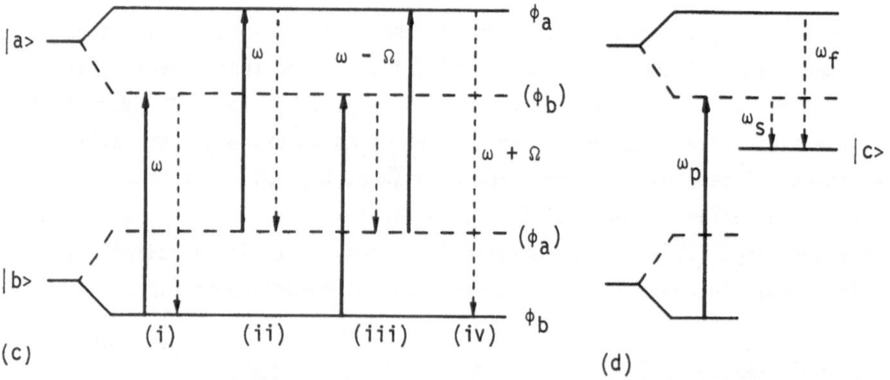

Fig.2.4c+d. (c) Spontaneous-emission processes that lead to three-frequency spectrum in resonance fluorescence. (d) Resonance Raman scattering; ω_s is scattering from the virtual component of ϕ_b, and ω_f is fluorescence from the real component of ϕ_a

For simplicity, the field envelope function is taken to be positive and real. The atom has two states of energies $\hbar\Omega_a$ and $\hbar\Omega_b$ with $|a\rangle$ the upper state; ω is detuned from resonance by

$$\Delta = \Omega_{ab} - \omega \quad . \tag{2.43}$$

Unless otherwise stated, the field envelope will be assumed to be sufficiently slowly varying that the adiabatic condition (2.35) is satisfied. Because damping is ignored until Sect.2.6.8, this means that exact resonance,

$\Delta = 0$, is prohibited. The states $|a\rangle$, $|b\rangle$ may be chosen so that (in the absence of a magnetic field), the matrix element μ_{ab} is positive and real; this is assumed.

In general, the atom will be in superposition state

$$\psi = a\, e^{-i\Omega_a t}|a\rangle + b\, e^{-i\Omega_b t}|b\rangle \quad . \tag{2.44a}$$

Knowledge of the wave function ψ is all that is required for a complete description of the response of the atom to the applied field; for example, $|a|^2$ is the probability that the atom is in the upper state. However, quantities such as $|a|^2$ are often composed of terms of quite different physical character; this suggests that an expansion in a different set of basis states, i.e.,

$$\psi = A\, e^{-i\Omega_a' t}\phi_a + B\, e^{-i\Omega_b' t}\phi_b \quad , \tag{2.44b}$$

may separate out these parts in a meaningful way. This is indeed the case if the so-called adiabatic states are used as the new basis ϕ_a, ϕ_b. Our procedure is to start with (2.44a) and then to show how the expansion (2.44b) arises quite naturally in the course of the calculations.

Using (2.42,44a) in the Schrödinger equation leads in the usual way to

$$i\dot{b} = -\frac{1}{2}\beta e^{-i\Delta t}a \quad ; \quad i\dot{a} = -\frac{1}{2}\beta e^{i\Delta t}b \quad , \tag{2.45}$$

where the dots denote time derivatives and β is the (real) on-resonance Rabi frequency (or as we shall refer to it, simply the Rabi frequency)

$$\beta = \mu_{ab}E_\omega(t)/\hbar \quad . \tag{2.46}$$

Notice that β is time dependent; use will be made of the adiabatic condition, so that terms in $\dot{\beta}$ may be dropped. Equations (2.45) have been approximated by ignoring terms that oscillate at the frequency $\Omega_{ab} + \omega$ (the rotating-wave approximation); the justification for this is that otherwise when (2.45) is integrated expressions with denominators Δ, $\Omega_{ab} + \omega$ are obtained, and that the latter denominator is by far the larger in magnitude.

From (2.45),

$$\ddot{b} + i\Delta\dot{b} + \frac{1}{4}\beta^2 b = 0 \quad , \tag{2.47}$$

which has natural frequencies $-i\delta_\pm$, where

$$\delta_\pm = \frac{1}{2}\,[\Delta \pm (\Delta^2 + \beta^2)^{\frac{1}{2}}] \quad . \tag{2.48}$$

For definiteness, the root in (2.48), which we call Ω, is taken to have the sign of Δ, i.e.,

$$\Omega \equiv \mathrm{sgn}(\Delta)\,|\Delta^2 + \beta^2|^{\frac{1}{2}} \quad . \tag{2.49}$$

Thus $\delta_\pm = \Delta - \delta_\mp = \frac{1}{2}\,(\Delta \pm \Omega)$, and δ_+ has the same sign as Δ but δ_- has the opposite sign. The coefficients a and b are given by

$$a = -(2/\beta)(\delta_+ A'\, e^{i\delta_- t} + \delta_- B'\, e^{i\delta_+ t}) \tag{2.50a}$$

$$b = A'\, e^{-i\delta_+ t} + B'\, e^{-i\delta_- t} \quad , \tag{2.50b}$$

where A' and B' are arbitrary constants.

At this stage, it is possible to specify initial conditions, from them calculate A', B' and thus obtain the wave function (2.44a). An example of this is where a monochromatic field is switched on at $t = 0$. This is, of course, not an adiabatically applied field. If the atom is initially in the ground state $|b\rangle$, then the upper-state population is

$$|a|^2 = \frac{1}{4}\,\beta^2 \sin^2(\Omega t/2)/(\Omega/2)^2 \quad ,$$

where $|\Omega|$ is the off-resonance Rabi frequency [2.38]. The calculated polarisation oscillates at the forcing frequency ω and the sidebands $\omega \pm \Omega$; these sidebands are ringing terms that arise from the transient application of the field; in the presence of damping they die out.

Rather than calculate A', B' in this way, we first insert (2.50) into (2.44a), and collect terms in A' and B'. The result is just the wave function (2.44b) with A and B new arbitrary constants, and ϕ_a, ϕ_b the manifestly orthonormal wave functions

$$\phi_a = \cos\frac{\theta}{2}\,|a\rangle - \sin\frac{\theta}{2}\,e^{i\omega t}|b\rangle \tag{2.51a}$$

$$\phi_b = \sin\frac{\theta}{2}\,e^{-i\omega t}|a\rangle + \cos\frac{\theta}{2}\,|b\rangle \quad , \tag{2.51b}$$

where

$$\cos\frac{\theta}{2} = \left(\frac{\delta_+}{\Omega}\right)^{\frac{1}{2}} \quad ; \quad \sin\frac{\theta}{2} = \mathrm{sgn}(\Delta)\left(\frac{-\delta_-}{\Omega}\right)^{\frac{1}{2}} \quad , \tag{2.52a}$$ [5]

so that

$$\sin\theta = \beta/\Omega \quad ; \quad \cos\theta = \Delta/\Omega \quad . \tag{2.52b}$$

The frequencies Ω_a', Ω_b' are

$$\Omega_a' = \Omega_a - \delta_- \; ; \quad \Omega_b' = \Omega_b + \delta_- \quad . \tag{2.53}$$

The definition of the angle in (2.52a) is restricted to $|\theta| < \pi/2$, so that, as the field is turned off, $\theta \to 0$, and as $\beta \to \infty$, $\theta \to \pm\,\pi/2$ depending on the sign of Δ. For this range of θ, ϕ_a and ϕ_b are distinct states.

Some insight into the meaning of the states ϕ_a, ϕ_b can be obtained by expanding to lowest order in β. Thus

$$\phi_b \simeq |b\rangle + \frac{\beta}{2\Delta}\, e^{-i\omega t}|a\rangle \tag{2.54a}$$

and

$$\Omega_b' \simeq \Omega_b - \beta^2/4\Delta \quad . \tag{2.54b}$$

For a static perturbation, $\omega \to 0$ and $V_{DC} = -\mu E_{DC}$, it is readily seen that the substitution $\hbar\beta \to 2V_{DC}$ must be made, in which case (2.54) reduces to

$$\phi_b = |b\rangle + (V_{DC}/\hbar\Omega_{ab})|a\rangle \tag{2.55a}$$

$$\Omega_b' = \Omega_b - V_{DC}^2/\hbar^2\Omega_{ab} \quad . \tag{2.55b}$$

Thus the quantity $\exp(-i\Omega_b' t)\phi_b$ is immediately recognisable as a stationary state (calculated to lowest order) of the system whose hamiltonian is $H_0 + V_{DC}$, where H_0 is the unperturbed hamiltonian that gives rise to the

---

[5] The function sgn(Δ) (sign of Δ) arises from the definition (2.49); if Ω had been defined in the conventional way, as the positive root of $(\Delta^2+\beta^2)$ then sgn(Δ) could be omitted from (2.52a). The choice (2.49) facilitates some of the later discussion, because it allows the cases $\Delta > 0$, $\Delta < 0$ to be treated together; unfortunately neither choice is entirely satisfactory for defining Ω.

states $|a\rangle$, $|b\rangle$. If the static field had been switched on infinitely slowly, then an unperturbed atom initially in the state $|b\rangle$ of energy $\hbar\Omega_b$ would have evolved into the state ϕ_b of dc Stark-effect-shifted energy $\hbar\Omega_b'$. The evolution of $|a\rangle$ into ϕ_a is similar. This is in accord with the adiabatic theorem of perturbation theory ([2.39] Sect.1.8, [2.40] Chap.35), in which, for a sufficiently slowly applied perturbation, a system evolves through a continuous succession of stationary states without any transitions.

The extension to optical fields is straightforward, although no proofs are given here. For field envelopes that satisfy the adiabatic condition of slowness (2.35), the states $|a\rangle$, $|b\rangle$ evolve into the quasi-stationary states ϕ_a,ϕ_b of the perturbed hamiltonian, of perturbed (optical Stark-effect-shifted) energies Ω_a', Ω_b', respectively. These are the so-called *adiabatic* (e.g., [2.41,42]) or *dressed* states (e.g., [2.35-37]).

From what has been said, it is clear that if, at any time, the atom was in state ϕ_b say, then it remains in ϕ_b, i.e., $A = 0$, $B = 1$ in (2.45). The choice $B = 1$ is fixed by normalisation. This condition is supplied in lieu of an initial condition, because the definition of adiabatic slowness renders meaningless the use of an initial time. This is most obvious when the convergence parameter ε (Sect.2.5) is used to switch on Fourier components of the field in the distant past. Thus, for a slow pulse, it is possible to say that the state ϕ_b was well approximated by $|b\rangle$ in the past; nevertheless the atom is strictly always in the adiabatic state ϕ_b.

From (2.44a,50) or (2.44b,51) it is seen that, in general, the wave function ψ has components that oscillate at four frequencies; the corresponding energies are drawn in Fig.2.4b. Therefore, the state $|a\rangle$ enters with two frequencies [see (2.50a)]; these are shown connected to the unperturbed state $|a\rangle$; similarly for $|b\rangle$. The definition of the adiabatic ground state shows that for adiabatic excitation only two of these four frequencies are excited, one from each amplitude a, b; these are shown connected by the ω arrow in the figure. The $|b\rangle$ component of ϕ_b has the energy Ω_b'; it is here termed the *real*-state part of ϕ_b; the $|a\rangle$ component of ϕ_b has the energy $\Omega_b' + \omega$; it will be called the *virtual*-state part of ϕ_b; it is shown dotted in Fig.2.4b. Similarly, for ϕ_a, the $|a\rangle$ and $|b\rangle$ parts are respectively real and virtual.

This idea of energies can be put on a firmer basis by considering the atom-field system. Then the states ϕ_b can be considered to be made up of a mixture of the two atom-field states $|n + 1, b\rangle$ and $|n,a\rangle$, where the first label in each part refers to the number of photons in the radiation mode of the field. Thus, in the atom-field system, ϕ_b (and similarly ϕ_a) is a stationary state that is made up of a superposition of the direct-product basis states [2.35,36,43]; the factors $\exp(\pm i\omega t)$ in our semiclassical wave functions correspond to the difference of photon energies of the two parts.

Because damping has so far been ignored, an atom in its adiabatic ground state ϕ_b has a slowly varying energy $\hbar\Omega_b'$, and remains in that state if (2.35) is satisfied[6]. In the presence of damping, however, even if (2.35) is satisfied, then transitions between ϕ_b and ϕ_a can occur. From the point of view of the field, an irreversible (or more precisely, incoherent) loss of energy to the medium takes place. Thus, we can distinguish between the admixture of $|a\rangle$ into ϕ_b (virtual) and of $|a\rangle$ into ϕ_a (real). The former occurs only while the field is applied, i.e., while ϕ_b is created. The latter is produced by absorption from the lower state ϕ_b, and will persist even after the field is removed, until it is dissipated by damping. A particularly clear example of this may be taken from the work of PUELL and VIDAL [2.28], mentioned earlier (Sect.2.6.1). In our notation, their equation (34) gives for the population of state $|a\rangle$ of an atom initially in state $|b\rangle$ [see (2.44a)], to lowest order in β,

$$|a|^2 = \frac{\Gamma/2}{\Delta^2+\Gamma^2}\int_{-\infty}^{t} \beta^2 dt + \frac{\Delta^2-\Gamma^2}{(\Delta^2+\Gamma^2)^2}\,\frac{\beta^2}{4} \; . \tag{2.56}$$

This adiabatically induced population may be contrasted with the transiently induced expression given earlier. Therefore, as mentioned in the introduction to this subsection, use of the wave function (2.44) can lead to an expression for $|a|^2$ that is composed of quite different terms. The first part of (2.56) depends on the integrated field intensity, and, therefore, is nonzero even after the radiation has left the medium. It is absorptive, and vanishes in the absence of damping; it may be identified with $|A|^2$ in (2.45). The second part of (2.56) exists only in the presence of the field; in the absence of damping it reduces to $\beta^2/4\Delta^2$. This is just the population of the virtual admixture of $|a\rangle$ in the adiabatic ground state ϕ_b, (2.54a).

---

[6] In the present notation (2.35) is then $|\Delta^{-1}\dot\beta/\beta| \ll 1$; it was derived by considering one order of nonlinearity. In (2.51), the exact solution for the wave function allows the more precise statement $|\Delta^{-1}\dot\theta/\theta| \ll 1$ to be made. Indeed, $|\Delta^{-1}\dot\theta/\theta|$ is on the order of the error introduced into the hamiltonian by assuming the adiabatic hypothesis to be exact [2.44]. For a nonadiabatic pulse, reversible transitions between the adiabatic states can occur. From the point of view of the field, a coherent exchange of energy with the medium takes place, with resultant changes of group velocity.

The population induced in ϕ_a by transitions from ϕ_b because of damping has been termed incoherent by GRISCHKOWSKY [2.41]; conversely, the population of the component state $|a\rangle$ of ϕ_b is termed coherent, although as pointed out by COURTENS and SZÖKE [2.44] this phraseology is perhaps not ideal.

Use of adiabatic states can lead to interesting insights into nonlinear processes. Take as an example a three-level system, with two levels and a strong field, as depicted in Fig.2.4a, and a third level $|c\rangle$ lying above $|a\rangle$. A weak probe field ω' connects $|a\rangle$ and $|c\rangle$; further, we assume there to be no matrix element connecting $|c\rangle$ and $|b\rangle$. Then the TPA susceptibility for parallel linearly polarised light is

$$\chi^{(3)}(-\omega';-\omega\omega\omega') = \mathcal{N}\mu_{ca}^2\mu_{ab}^2/6\hbar^3\varepsilon_0\Delta^2(\Omega_{cb} - \omega - \omega' - i\Gamma) \quad ,$$

where $|a\rangle$ is the intermediate state and $|b\rangle$ the initial state (Sect.2.5). Now consider the description of TPA in terms of adiabatic states, Fig.2.4b, where the probe is assumed sufficiently weak that the state $|c\rangle$ is unperturbed (i.e., ϕ_c and $|c\rangle$ coincide). An additional complication here is that the level shifts have been accounted for, but are not necessary for the following argument. To lowest order,

$$e^{-i\Omega_b t}\phi_b = e^{-i\Omega_b' t}|b\rangle + (\beta/2\Delta)\, e^{-i(\Omega_b'+\omega)t}|a\rangle$$

[see (2.54a)]; as discussed in the foregoing, the virtual-state component has energy $\hbar(\Omega_b' + \omega)$. In the absence of the strong field $\phi_a = |a\rangle$; if there were some population in $|a\rangle$, the probe would experience absorption at the frequency $\omega' = \Omega_{ca}$. When the strong field is applied, the state $|a\rangle$ becomes ϕ_a, and the line moves to $\omega_1' = \Omega_c - \Omega_a' = \Omega_{ca} + \delta_-$. Now, however, there is a second line in the spectrum, because ϕ_b contains a (virtual) admixture of $|a\rangle$; it is, therefore, connected to $|c\rangle$. Because the virtual state has energy $\hbar(\Omega_b' + \omega)$, the second line is at $\omega_2' = \Omega_c - \Omega_b' - \omega$, i.e., it satisfies the two-photon resonance condition $\omega + \omega_2' = \Omega_c - \Omega_b'$.

Setting aside the question of level shifts, then $\omega_1' = \Omega_{ca}$ and $\omega + \omega_2' = \Omega_{cb}$. The first line is due to ordinary single-photon absorption; if there were damping of $|a\rangle$, then the population in ϕ_a could have resulted from absorption from the real part $|b\rangle$ of ϕ_b. This would be two-step two-photon absorption. (This illustrates the comments made at the end of Sect.2.5: the ad hoc insertion of damping would not account for this process in the third-order susceptibility.) The second line is just two-photon absorption from the real part of ϕ_b; it is viewed, however, as single-photon absorption from the virtual part of ϕ_b. Therefore, treating the virtual part of ϕ_b as an ordinary state, it has a population $N_a = \mathcal{N}\beta^2/4\Delta^2$; the linear susceptibility is

$$\chi_a^{(1)}(-\omega_2';\omega_2') = N_a\mu_{ca}^2/\hbar\varepsilon_0(\Omega_{cb} - \omega - \omega_2' - i\Gamma) \quad .$$

When this is used to calculate the polarisation at ω_2', the third-order susceptibility $\chi^{(3)}(-\omega_2',-\omega\omega\omega_2')$, given above, results. GRISCHKOWSKY [2.41] and FLUSBERG and HARTMANN [2.42] have discussed these ideas in some detail; in an elegant experiment with rubidium vapour, GRISCHKOWSKY monitored the $|a\rangle$ components of ϕ_a and ϕ_b, directly and separately. The use of the two-photon resonant optical Kerr effect as a polarisation rotator has been demonstrated by LIAO and BJORKLUND [2.45]; a theoretical treatment of this, based on adiabatic states, has been given by GRISCHKOWSKY [2.41] and LIAO and BJORKLUND [2.46].

Some other examples of interest have been discussed by COURTENS and SZÖKE [2.44], SZÖKE and COURTENS [2.47], and CARLSTEN et al. [2.48]. Thus, rather than introducing a weak probe field, spontaneous emission rates can be calculated (by use of Fermi's Golden Rule) from the real and virtual states of a resonantly excited atom (Fig.2.4c,d). In Fig.2.4c, various contributions to resonance fluorescence are shown. Rayleigh scattering of a frequency ω is shown in (i) and (ii). Thus, in (i) spontaneous emission occurs by transition from the virtual state $|a\rangle$ of ϕ_b to the real part $|b\rangle$; in (ii), spontaneous emission occurs by transition from the real part $|a\rangle$ of ϕ_a to the virtual state $|b\rangle$ of ϕ_a. In (iii), a three-photon process occurs, with spontaneous emission of frequency $\Omega_b' + \omega - (\Omega_a' - \omega) = \omega - \Omega$; in (iv) spontaneous emission occurs by transition between the real states $|b\rangle$ and $|a\rangle$, of frequency $\Omega_a' - \Omega_b' = \omega + \Omega$. Thus, three lines in the emission spectrum are expected. This method, by its naive usage of a transition-rate formula, does not take into account the coherence of the various processes; therefore, it cannot account properly for details of the spectrum, although it does predict the existence and correct positions of the three lines [2.35, 37,44]. [These lines are quite different in origin from the transient frequencies mentioned after (2.50).] Finally, the interpretation of resonance Raman scattering is shown in Fig.2.4d. The pump field generates population in the real state $|a\rangle$ (component of ϕ_a) because of damping or violation of (2.35), and the virtual state $|a\rangle$ (component of ϕ_b). Spontaneous emission then occurs with the fluorescence frequency $\omega_f = \Omega_a' - \Omega_c$ from the real state

$|a\rangle$, and with the Stokes frequency $\omega_s = \Omega_b' + \omega_p - \Omega_c$ from the virtual state $|a\rangle$. If the level shifts are ignored, $\omega_f = \Omega_{ac}$ and $\omega_s = \omega_p - \Omega_{ca}$. Therefore, the generation of ω_s is ordinary Raman scattering, whereas ω_f is fluorescence, and may form the starting noise for amplified spontaneous emission [2.49].

To reiterate, a process can be considered either as a multiple absorption-emission sequence from the unperturbed initial state, as in earlier sections, or as shorter sequences between real and virtual levels. In particular, the perturbed ground state $|g\rangle'$ introduced in Sect.2.4 is just the adiabatic ground state of a multilevel atom; all of the unperturbed states that are mixed into it have the significance of virtual states. However, because, in general, several frequencies are applied, the description can become involved. Equation (2.54a) is the lowest-order approximation to ϕ_b. A second-order approximation introduces a coefficient proportional to $\beta^3 \exp(-i\omega t)$, which may be written as $\beta\exp(-i\omega t)[\beta\exp(-i\omega t)]^*\beta\exp(-i\omega t)$. Therefore, it can be interpreted as the population in the virtual state $|a\rangle$ produced by absorption, emission and a further absorption of ω between the real state $|b\rangle$ and itself; this is part of the ladder approximation to be mentioned in Sect.2.6.4. Alternatively, β^3 can be viewed simply as part of the expansion of the virtual-state population $\sin^2\theta/2$ in terms of the unperturbed states $|a\rangle$ and $|b\rangle$. If the rotating-wave approximation had not been made, components $\sim\exp(\pm i3\omega t)$ would also have been produced; in general, the exact wave function can be said to have virtual states at all harmonics of the fundamental frequency [2.50]. Therefore, in the case in which several frequencies are applied, virtual states can be considered to be at all combinations of the applied frequencies, although the number of terms in the combination is limited by the order of the approximation used for the wave function. Moreover, by use of the alternative view explained earlier, a given virtual state is made up of a linear combination of all of the unperturbed states of the atom; this is the origin of the intermediate-state summations in (2.15). Because there is only one possible such state in the two-level atom, this summation is not so obvious. To lowest order, it is the single term $(\beta/2\Delta)\exp(-i\omega t)|a\rangle$ in ϕ_b, say, (2.54a). Because of resonance enhancement, a virtual state may be dominated by one intermediate state, whose energy is close to it. Thus, in third-harmonic generation, Fig.2.1a, it is justified to draw virtual states at the energies ω, 2ω, 3ω above the ground state, whose populations derive mainly from the unperturbed states whose energies are nearest to theirs. (A note of caution: these virtual levels really refer to *amplitudes*, which may interfere if there are several intermediate states; the idea of a population is then not valid for calculations.)

2.6.3 The ac or Optical Stark Effect

We now consider the optical Stark shifts in more detail, noting that a good early review has been given by BONCH-BRUEVICH and KHODOVOI [2.43]. They are of importance in spectroscopy where shifts of several hundred MHz are readily observed [2.51] and in nonlinear optics whenever strong resonances and/or laser intensities are employed—for example, shifts on the order of 10 cm^{-1} have been reported by VREHEN and HIKSPOORS [2.52], in studies of SHRS in caesium vapour using a Nd:YAG pump (Sect.5.5.1).

When monochromatic radiation of frequency ω passes through a medium, it experiences absorption and a change of velocity due to the imaginary and real parts of the refractive index, respectively. The irreversible energy flow W (i.e., the dissipated power density, $Jm^{-3}s^{-1}$) from the field to the medium at any point is given by

$$W = -\langle \underline{E} \cdot \dot{\underline{P}} \rangle = \frac{1}{2}\omega \, \mathrm{Im}(\underline{E}_\omega^* \cdot \underline{P}_\omega) \tag{2.57}$$

as shown in texts on electromagnetic theory; the angle brackets denote a cycle average. Using $P_\omega^{(1)} = \varepsilon_0 \chi^{(1)}(-\omega;\omega)E_\omega$ and $\chi^{(1)}(-\omega;\omega) = \mathcal{N}\alpha(-\omega;\omega)/\varepsilon_0$, where $\alpha(-\omega;\omega)$ is the polarisability for a ground-state atom, we have the familiar result for linear response that

$$W = \frac{1}{2}\mathcal{N}\omega\alpha''(-\omega;\omega)|E_\omega|^2 = \frac{\mathcal{N}\omega}{\varepsilon_0 c}\alpha''(-\omega;\omega)I_\omega, \quad \text{vacuum} \quad . \tag{2.58}$$

This quantity accounts for real transitions from the adiabatic (or if a weak field, unperturbed) ground state.[7]

On the other hand, the real part of the refractive index corresponds to a reversible repartitioning of the energy of the field-atom system. Under the influence of the field, there is an energy of interaction, which may be calculated in many ways; we examine two, and assume ω to be distant from any absorption lines. The refractive index may be approximated by $\eta_\omega \equiv (1 + \chi^{(1)})^{1/2} \simeq 1 + \chi^{(1)}/2$, so that $\eta'_\omega - 1 \simeq \mathcal{N}\alpha'(-\omega;\omega)/2\varepsilon_0$. As a consequence of the refractive index the energy density of the monochromatic

---

[7] α', α'' are the real and imaginary parts of α respectively.

field is increased from its vacuum value of $\varepsilon_0|E_\omega|^2/2$ to $\varepsilon_0 n'_\omega|E_\omega|^2/2$, i.e., an increase of $\mathcal{N}\alpha'(-\omega;\omega)|E_\omega|^2/4$. This energy-density increase is a continuous function of an adiabatically switched field; therefore, in the total atom-field system, in which energy is conserved, there must be a reversible flow of energy from the medium — slowly turning off the field restores the atom to its original state. This energy comes not from irreversible transitions, but from an adiabatic variation of the ground-state energy, i.e., the creation of the adiabatic ground state[8]. Each atom, therefore, experiences a level shift for state $|g\rangle$ of

$$\Delta E_g = -\frac{1}{4}\alpha'(-\omega;\omega)|E_\omega|^2 \quad , \tag{2.59}$$

where α is the polarisability for this state. [$\alpha = \alpha_{gg}(-\omega;\omega)$ in (2.19).] This energy of interaction can be calculated more rigorously as the lowest order approximation to

$$\Delta E_g = -\mathcal{N}^{-1}\langle\int \underline{P}\cdot d\underline{E}\rangle \quad , \tag{2.60}$$

as indicated in LANDAU and LIFSCHITZ [2.53] pp.52-, which yields (2.59) and is $\mathcal{N}^{-1}$ times the difference between the total free-energy density of the atom-field system and the energy density of the field in vacuum. Note, however, that this interpretation of (2.60) assumes the field to be applied adiabatically. Inserting (2.19) into (2.59), we have

$$\Delta E_g = -\frac{1}{4\hbar}\sum_n |\underline{E}_\omega\cdot\underline{\mathcal{D}}_{ng}|^2\left(\frac{1}{\Omega_{ng}-\omega}+\frac{1}{\Omega_{ng}+\omega}\right) = -\frac{1}{2\hbar}\sum_n |\underline{E}_\omega\cdot\underline{\mathcal{D}}_{ng}|^2\frac{\Omega_{ng}}{\Omega_{ng}^2-\omega^2} \quad , \tag{2.61}$$

the standard formula for the (quadratic) ac Stark effect; see, e.g., BONCH-BRUEVICH and KHODOVOI [2.43] and LIAO and BJORKHOLM [2.51]. When $\omega = 0$, this reduces to the quadratic dc Stark effect, when the change $|E_\omega|^2/2 \to E_{DC}^2$ is made. This reinforces the interpretation of ΔE_g as the shift averaged over an optical cycle.

We can compare (2.61) with the ground-state shift δ_- obtained for the two-level atom. Off resonance, corresponding to (2.61), $\delta_- \sim -\beta^2/4\Delta$ which is

---

[8]The extension to the case of population distributed over several states is trivial, and (2.59) et seq. apply equally to all states; see also HEITLER [2.11], pp.172.

precisely (2.61) when $(\Omega_{ng} + \omega)^{-1}$ is dropped, i.e., the rotating-wave approximation is made [9]. On resonance, and/or for very strong fields, (2.61) breaks down. We can either introduce higher-order corrections by use of $\chi^{(3)}$, $\chi^{(5)}$... or solve more accurately with fewer levels—such as in the two-level case, when the shift is δ_-. Therefore, as the field strength increases, $\delta_- \sim -\frac{1}{2}\beta\,\mathrm{sgn}(\Delta)$, which shows the usual turnover from quadratic to linear Stark effects for strong fields. The shift, linear in the field strength $(\sim I_\omega^{1/2})$, has been observed by, for example, LIAO and BJORKHOLM [2.54], MOODY and LAMBROPOULOS [2.55], and GRAY and STROUD [2.56]; see also FENEUILLE [2.37].

In the atom-field system, at exact resonance the states $|n+1,b\rangle$ and $|n,a\rangle$ are degenerate; therefore, it is expected that the perturbation due to the atom-field interaction will induce a linear splitting; this is simply $|2\delta_-| = \beta$, the Rabi frequency. However, because we usually deal directly with the atomic states, we find that the splitting in an atom-field system reduces to linear shifts of the atomic upper and lower states ϕ_a and ϕ_b. Ignoring the complication introduced by damping, take the case of an atom initially in its adiabatic ground state. Away from exact resonance, there will be only one line at the two-photon absorption frequency $\Omega_c - (\Omega_b' + \omega)$, because there is no population in ϕ_a. On the other hand, at exact resonance ($\Delta = 0$) although the excitation is no longer adiabatic, we can see from (2.51) that the states ϕ_a and ϕ_b contain equal proportions of the state $|a\rangle$. Moreover, the energies of the virtual state $|a\rangle$ of ϕ_b and the real part $|a\rangle$ of ϕ_b are symmetrically disposed about the energy $\hbar\Omega_a$. Hence a probe field (as in Sect.2.6.2) is expected to detect two lines of equal strength and of frequencies separated by an amount β. In this case, as might be surmised by the degeneracy in the atom-field system, no distinction can be made between the real and virtual aspects of the state $|a\rangle$. Therefore, we might use (2.44a) rather than (2.44b) and the two components of a in (2.50a) will be equally excited. We see then that the absorption *line* is split when $\Delta = 0$; it is important to distinguish between this splitting in the spectroscopic sense AUTLER and TOWNES [2.50] and the equivalent splitting of the *states* of the atom-field system, a distinction sometimes obscured in the recent literature.

---

[9]Setting $\Omega_{ng} + \omega \simeq 2\Omega_{ng} \to 2\Omega_{ab}$ for the two-level atom gives a non-resonant contribution to the shift of $-\hbar\beta^2/(8\Omega_{ab})$, i.e., $\Omega_{ab} \to \Omega_{ab} + \beta^2/(4\Omega_{ab})$. This is known as the Bloch-Siegert shift and is negligible for optical transitions.

Experimental investigations of this Autler-Townes line splitting, predominantly in the rf and microwave spectral regions, have been reviewed by FENEUILLE [2.37]. There are a number of studies in the optical spectrum, for example in neon [2.57], and in sodium, MOODY and LAMBROPOULOS [2.55], and GRAY and STROUD [2.56]. An interesting feature of the latter experiment is that the probe ω' consists of two parts. The two-level atom was modelled by resonantly exciting the $3^2S_{1/2}$ - $3^2P_{1/2}$ transition. Then the probe was tuned to the electric-quadrupole transition $3^2P_{1/2}$ - $5^2P_{3/2}$, thereby ensuring that it was a weak probe. The atoms were then ionised from the $5^2P_{3/2}$ state by ω or ω'; consequently, the absorption spectrum of ω' may be detected by collecting the electrons, an extremely sensitive technique. BONDAREVA et al. [2.58] describe an experiment in potassium vapour, in which the applied field consisted of ruby-laser radiation and its 1st Stokes component produced by SRS in nitrobenzene. The ruby-laser and Stokes frequencies were resonant with the $4^2P_{3/2}$ - $6^2S_{1/2}$ and $4^2S_{1/2}$ - $4^2P_{3/2}$ transitions, respectively. Including the $4^2P_{1/2}$ state, a total of four states have to be taken into account, which means that the adiabatic ground state consists of the real state $4^2S_{1/2}$ and virtual-state admixtures of $4^2P_{1/2,\,3/2}$ and $6^2S_{1/2}$. Therefore, four energy levels are associated with the adiabatic ground state; for the four adiabatic states, there are a total of sixteen levels (reduced in this special case by degeneracy to twelve, see BONDAREVA et al.). In the spectroscopic sense, each unperturbed line, therefore, splits into four lines. By considering transitions between the various virtual levels BONDAREVA et al. were able to analyse their results qualitatively. This demonstrates the remarks in Sect.2.6.2 that it is possible to analyse multiphoton processes by consideration of the virtual states; however, in complicated systems both methods have their difficulties, and must be regarded as complementary approaches.

To conclude, we consider briefly the effect of damping (see also Section 2.6.8). If in the usual way, damping is inserted into the formulae (2.59-61), then complex level shifts arise. The imaginary parts represent power (i.e., field-dependent) broadening of the state, and are associated with absorption (e.g. [2.59]). In general, the real shift and power broadening are inseparably connected by Kramers-Kronig-type relations for the real and imaginary parts of the susceptibilities. An example of this arose in a theoretical study of two-photon resonant third-harmonic generation in alkali vapours [2.60]. The third-harmonic frequency exceeded the ionisation limit; therefore, when the level shift of the excited two-photon resonant level was computed [by use of (2.61)], it was necessary to include the continuum in the sum over intermediate states. This, of course, introduces an imaginary part into the level shift, which is simply the broadening due to ionisation from the excited level. KNIGHT [2.61] has discussed the Autler-Townes effect in two-photon ionisation.

2.6.4 Intensity-Dependent Refractive Index

In this subsection, we derive an expression for the polarisation induced by $E_\omega(t)$ in the two-level atom, as an illustration of the series (2.11). The field is taken to be adiabatic, so the atom is in state ϕ_b. The induced dipole moment is then

$$\mu(t) = \langle\phi_b|\mu|\phi_b\rangle = \mu_{ba} \sin\frac{\theta}{2} \cos\frac{\theta}{2} (e^{-i\omega t} + \text{c.c.})$$

using (2.51b), so that the component of the macroscopic polarisation at ω is

$$P_\omega(t) = \mathcal{N}\mu_{ba} \sin\theta = \mathcal{N}\mu_{ba}\beta/\Omega \quad , \tag{2.62}$$

yielding a power-dependent "susceptibility"

$$\varepsilon_0\chi\Big(\omega;E_\omega(t)\Big) = \mathcal{N}\mu_{ba}^2/[(\hbar\Delta)^2 + \mu_{ba}^2 E_\omega(t)^2]^{\frac{1}{2}} \quad . \tag{2.63}$$

The RHS of (2.63) is to be taken to have the sign of Δ, as follows from (2.49). This expression appears to have been given first by GRISCHKOWSKY [2.62] (who derived it by use of the vector model), in connection with a study of self-defocussing of laser light in potassium vapour. GRISCHKOWSKY and ARMSTRONG [2.63] report a careful experimental and theoretical study of self-focussing in rubidium vapour, based on (2.62). The susceptibility for an atom in its upper adiabatic state ϕ_a is the negative of (2.63).

A sketch of the susceptibility (2.63) is given in Fig.2.5 for zero field, a field such that $\beta = 1$ (arbitrary units), and for comparison, the real part of the linear susceptibility $\chi^{(1)}(-\omega;\omega)$, with a damping parameter $\Gamma = 1$. Near exact resoncance, the effect of the field is to remove the pole at $\Delta = 0$, and in effect to broaden the dispersion curve. However, in general it is difficult to distinguish in an obvious way between power broadening and level shifts near a resonance. Thus, when off resonance, where $\Delta^2 \gg \beta^2$, the denominator in (2.63) may be approximated, to give

$$\varepsilon_0\chi\Big(\omega;E_\omega(t)\Big) \simeq \mathcal{N}\mu_{ba}^2\Big/\Big(\hbar\Delta + \frac{1}{2}\mu_{ba}^2 E_\omega(t)^2/\hbar\Delta\Big) \quad , \tag{2.64}$$

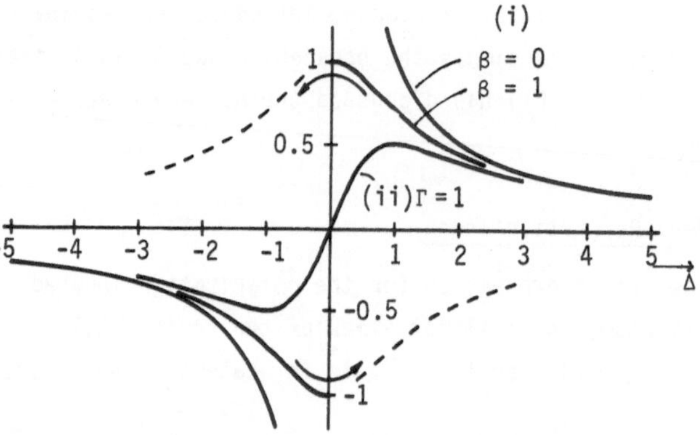

Fig.2.5. Intensity-dependent refractive index in a two-level atom. $\chi(\omega;E_\omega) \sim 1/\Omega$, equation (2.63), as a function of detuning. (a) Solid curves show (i) Ω^{-1}, for zero field ($\beta = 0$), $\beta = 1$ (arbitrary units); for comparison, (ii) $\Delta/(\Delta^2 + \Gamma^2)$, which is proportional to the real part of the linear refractive index, with a damping parameter $\Gamma = 1$. These are for an atom in its adiabatic ground state ϕ_b. The dotted curve shows the case $\beta = 1$ (no damping) for an atom in its upper adiabatic state ϕ_a. If the detuning is swept through the resonance, the atom passes from ϕ_b to ϕ_a or vice versa, as indicated by arrows (see text)

which has the appearance of a level-shifted denominator. Moreover, the shift in (2.64) is just what would be calculated from the shifted detuning $\Omega_a' - \Omega_b' - \omega \simeq \Omega_{ab} + \beta^2/4\Delta - (-\beta^2/4\Delta) - \omega = (\Omega_{ab} + \beta^2/2\Delta) - \omega$. Hence, off resonance the denominator in (2.63) may be termed level shifted. (If damping were introduced into the shift (Sect.2.6.3), the curve would be flattened slightly; this may be termed power broadening of the denominator.) As resonance is approached, the shift in (2.64) becomes a rapid function of Δ; finally we must revert to (2.63); looked at in this way, and because there is no damping, it is reasonable to persist in terming the denominator (2.63) level shifted. This is finally confirmed by writing the denominator as $\Omega = \Delta - (\Delta - \Omega) = \Delta - 2\delta_-$. The quantity $2\delta_-$ is just the level shift for the transition, which becomes linear in the field near resonance. When $\Delta = 0$ the shift is $\pm\beta$; i.e., the field can never be on resonance with the shifted transition frequency; therefore, there is no pole in the susceptibility.

Referring again to Fig.2.5, the discontinuities at $\Delta = 0$ require further discussion. In a classical model the pole at $\Delta = 0$ is simply removed by the addition of damping. However, if we consider the two-level atom, an interesting interpretation arises. As described in Sects.2.6.2-3, when $\Delta = 0$ the adiabatic states ϕ_a and ϕ_b become degenerate in the atom-field system. From (2.51) we see that

$$\phi_a = \left(|a\rangle - \mathrm{sgn}(\Delta)e^{i\omega t}|b\rangle\right)/\sqrt{2} \tag{2.65a}$$

$$\phi_b = \left(\mathrm{sgn}(\Delta)e^{-i\omega t}\right)\left(|a\rangle + \mathrm{sgn}(\Delta)e^{i\omega t}|b\rangle\right)/\sqrt{2} \ . \tag{2.65b}$$

The function $\mathrm{sgn}(\Delta)$ takes account of the direction from which Δ approaches zero. The first part of (2.65b) is an unimportant phase factor, which we will ignore. Therefore, from the left of Fig.2.5, $\Delta \ll 0$, towards resonance, the state $\phi_b \sim |a\rangle - \exp(i\omega t)|b\rangle$, whereas from the right, $\phi_b \sim |a\rangle + \exp(i\omega t)|b\rangle$. On the other hand, if the atom was initially in the state ϕ_a, the susceptibility would be the negative of (2.63). This is shown dotted in Fig.2.5a. When resonance is approached from the right, the state $\phi_a \sim |a\rangle - \exp(i\omega t)|b\rangle$, and when resonance is approached from the left $\phi_a \sim |a\rangle + \exp(i\omega t)|b\rangle$. Thus, it is seen that as Δ passes through zero, the state ϕ_a changes smoothly into ϕ_b or vice versa (indicated by the arrows in Fig.2.5a) — this will be referred to in Sect.2.6.5 as adiabatic rapid passage, and has no classical counterpart. For the present, we see that the apparently nonphysical jump of susceptibility at $\Delta = 0$ is a consequence of the naïve assumption that an atom in one state for Δ positive or negative remains in that state as it is swept *through* resonance. In fact, the macroscopic susceptibility for the atom should be written more properly as $\mathcal{N}[\overline{\rho(b)} - \overline{\rho(a)}]\mu_{ba}^2/\hbar\varepsilon_0\Omega$ (Sect.2.5). For an atom in state ϕ_b, the occupation probability $\overline{\rho(b)} = 1$, and $\overline{\rho(a)} = 0$. As the detuning is swept through resonance, the atom changes state; therefore, the effective density N changes sign along with Ω, resulting in no change of sign of the macroscopic susceptibility. The polarisation thus varies smoothly as Δ is swept through resonance, accompanied by the change of state of the atom.

Continuing the discussion of (2.64), for $\Delta^2 \gg \beta^2$ we see that (2.63) may be expanded as a power series, to yield

$$\varepsilon_0\chi\left(\omega;E_\omega(t)\right) = \frac{\mathcal{N}\mu_{ba}^2}{\hbar\Delta} - \frac{\mathcal{N}\mu_{ba}^4E_\omega(t)^2}{2\hbar^3\Delta^3} + \dots \ . \tag{2.66a}$$

Clearly this is an example of the series (2.11), where the identifications

$$\chi^{(1)}(-\omega;\omega) = \mathcal{N}\mu_{ba}^2/\hbar\varepsilon_0\Delta \tag{2.66b}$$

$$\chi^{(3)}(-\omega;\omega-\omega\omega) = -2\mathcal{N}\mu_{ba}^4/3\varepsilon_0\hbar^3\Delta^3 \tag{2.66c}$$

must be made[10]. A rigorous proof of (2.66c) will be given in Sect.2.6.5. Equation (2.66) shows that the Stark shifts in the atom are quite properly accounted for by the susceptibility series. Equation (2.66c) is usually known as the susceptibility for intensity-dependent refractive index; therefore, again it is seen that, as in Sect.2.6.3, the refractive index as seen by the field and the Stark shift as seen by the atom are two aspects of the same phenomenon. Equation (2.60) enables us to calculate these shifts to any order of approximation. Actually, wherever repeated frequency arguments of the form $(\omega,-\omega)$ appear in a susceptibility, there will be terms in the full susceptibility expression that represent contributions to the shift (or refractive index). This may be familiar to the reader if the series (2.11)/(2.66) is illustrated schematically,

These are the so-called ladder terms of perturbation theory (e.g., [2.64]).

If (2.62) is inserted into the level-shift formula (2.60), the correct result $\Omega_b' = \Omega_b + \delta_-$ is obtained, which is a satisfying demonstration of the self-consistency of the various equations.

In conclusion, it is interesting to observe that BLOEMBERGEN [2.15] and FLYTZANIS [2.1] have discussed the use of (2.60) to *define* the nonlinear susceptibilities, i.e., the reverse procedure of that adopted here.

2.6.5 Secularity

We now turn to the problem of secular divergences in the nonlinear susceptibility formula (2.15), mentioned in Sect.2.5. Secularity is easily discussed now that adiabatic states have been introduced. This subsection is otherwise self-contained, so the reader may pass on to Sect.2.6.6 with no loss of continuity.

The total wave function for an atom in its adiabatic ground state has been seen to be of the form $\psi = \exp(-i\Omega_g' t)\phi_g$, where both Ω_g' and ϕ_g may be expressed in terms of powers of the field. For the distant past, this state approaches $\psi \sim \exp(-i\Omega_g t)|g\rangle$, i.e., the atom is initially in its unperturbed

---

[10]Note the use of the K factor in (2.66c), cf. (2.11).

state. When the polarisation $\langle\psi|\mathcal{Q}|\psi\rangle$ is calculated, the phase factors cancel out, to give $\langle\phi_g|\mathcal{Q}|\phi_g\rangle$. On the other hand, if the wave function is expanded in the interaction picture, as in (2.44a), and the resulting equations for the coefficients, such as (2.45), are evaluated by time-dependent perturbation theory, then the calculated wave function is $\psi = |g\rangle'$, of the form $\sum_n a_n \exp(-i\Omega_n t)|n\rangle$. This wave function leads to the polarisation '$\langle g|\mathcal{Q}|g\rangle$' introduced in Sect.2.4 (2.13). Clearly, these two forms for the wave function and polarisation must be identical. We therefore see that $|g\rangle'$ must incorporate the phase factor $\exp(-i\Omega_g' t)$ expanded in powers of the field. Because the phase factor cancels out when the polarisation is evaluated, the nonlinear susceptibility (2.15), calculated from (2.13), must include a large number of redundant terms, which correspond to that cancellation. This is the clue to the secular divergences, for they arise when denominators of the form $\Omega_{gg} - \omega' = 0$ occur (ω' is a vanishing sum of field frequencies). In terms of the diagrammatic interpretation (Sect.2.4), such denominators correspond to transitions that leave the atom in its initial state. Therefore they are self-energy terms that correspond to a shift of energy $\Omega_g' - \Omega_g$ of the ground state, and hence are associated with the expanded phase factor.

Before considering the removal of redundant terms from (2.15), we mention the rather simple modification of perturbation theory needed in order to obtain ψ in an adiabatic form. This is lucidly described by SOBEL'MAN ([2.65], pp.272ff) and in great detail for higher order by LANGHOFF et al. [2.66]. Thus taking (2.44a), we may extract the coefficient b as a factor, and write ψ as

$$\psi = e^{-i\int_{-\infty}^{t} \Omega_b(t)dt} (|b\rangle + a'|a\rangle) \quad , \tag{2.67}$$

where $a' = [\exp(i\Omega_{ba}t)]a/b$ and the phase-shift integral $\int_{-\infty}^{t}\Omega_b(t)dt$ is defined as $i\ln b$. We can insert (2.67) into the Schrödinger equation and obtain equations for $\Omega_b(t)$ and a', rather than for a and b, (2.45). Obviously, the resultant wave function is the same, but now it is of the required form. We see that the real part of the phase shift corresponds to a level shift and the imaginary part to the normalisation constant, i.e., it accounts for the loss of amplitude from the state $|b\rangle$ caused by the admixture of $|a\rangle$. Thus, for the exact solution (2.51b), we can see that

$$\int_{-\infty}^{t} \Omega_b(t)dt \equiv \Omega_b' t + i\ln(\cos\tfrac{\theta}{2}) \quad ,$$

although for a general multilevel atom the phase shift can be given only as a perturbation expansion.

ORR and WARD [2.4] applied the method of averages to the calculation of the nonlinear polarisation, one of their aims being to handle the secular terms. As they show, this method makes use of the fact that resonance and off-resonance terms vary on quite different time scales, and such terms can be singled out for separate treatment. For example, WARD and SMITH [2.67] later derived the two-photon resonant field-dependent denominator (Section 2.6.8) by these means. However, in the actual calculations of ORR and WARD [2.4], only minor use was made of the method; it was reduced to looking for a wave function in the form (2.67) (extended, or course, to a multilevel system) with $\Omega_b(t)$ real, i.e., just a level shift. This was because no strong resonances were allowed, or at least only those that could be adequately handled by damping parameters (cf. Sect.2.5). Moreover, for very slowly varying fields, or for a superposition of monochromatic waves, the method further reduces to looking for a phase shift $\Omega_b' t$ with Ω_b' constant; this is a simple matter for perturbation theory. For reference, we quote Orr and Ward's result for the third-order susceptibility (ignoring damping, $|g\rangle$ non-degenerate, and with some changes of notation)

$$\chi^{(3)}(-\omega_\sigma;\omega_1\omega_2\omega_3) = \frac{\mathcal{N}}{6\hbar^3\varepsilon_0}\mathcal{S}_T\sum_g \rho(g)$$
$$\times\left\{\sum_{lmn}{}' \frac{\underline{\varepsilon}_\sigma^*\cdot\underline{\mathcal{Q}}_{gl}\underline{\varepsilon}_1\cdot(\underline{\mathcal{Q}}_{lm}-\underline{\mathcal{Q}}_{gg})\underline{\varepsilon}_2\cdot(\underline{\mathcal{Q}}_{mn}-\underline{\mathcal{Q}}_{gg})\underline{\varepsilon}_3\cdot\underline{\mathcal{Q}}_{ng}}{(\Omega_{lg}-\omega_\sigma)(\Omega_{mg}-\omega_2-\omega_3)(\Omega_{ng}-\omega_3)}\right.$$
$$\left.-\sum_{mn}{}' \frac{\underline{\varepsilon}_\sigma^*\cdot\underline{\mathcal{Q}}_{gm}\underline{\varepsilon}_1\cdot\underline{\mathcal{Q}}_{mg}\underline{\varepsilon}_2\cdot\underline{\mathcal{Q}}_{gn}\underline{\varepsilon}_3\cdot\underline{\mathcal{Q}}_{ng}}{(\Omega_{mg}-\omega_\sigma)(\Omega_{ng}-\omega_3)(\Omega_{ng}+\omega_2)}\right\}, \quad (2.68)$$

(Primes indicate that g is to be omitted in the summations over intermediate states.)

YURATICH (unpublished, 1977) showed that the third-order susceptibility taken from (2.15), and (2.68) are identical, being merely rearrangements of one another[11]. This is, of course, as expected in the light of the

---

[11]The somewhat tedious but not particularly long method used was to single out from (2.15) those terms in the summations for which g is an intermediate state. These were then manipulated by means of permutation symmetry and occasional use of partial-fraction expansions until (2.68) was obtained. In the course of this, the secular terms cancelled out. The $\chi^{(2)}$ of ORR and WARD was also proved equivalent to (2.15) in this way; $\chi^{(1)}$ does not possess any secular terms in (2.15) or ORR and WARD, and both formulae are trivially identical.

above discussion. The form (2.68) has the advantage that it is manifestly nonsingular. However, secure in the knowledge that (2.15) is also nonsingular, we can make use of (2.15) with its greater advantage of simplicity of form, an example of which is its straightforward diagrammatic representation.

It should be emphasised that (2.15,68) and most other methods can still present difficulties if $|g\rangle$ is degenerate; this arises from an unsuitable choice of initial basis states and is familiar in time-independent perturbation theory.

From (2.68), the third-order contribution to the intensity-dependent refractive index is

$$\chi^{(3)}(-\omega;\omega-\omega\omega) \simeq \frac{2\mathcal{N}}{3\hbar^3\varepsilon_0}\left[\sum_{lmn}{}' \frac{\underline{\varepsilon}^*\cdot\mathcal{Q}_{gl}\underline{\varepsilon}^*\cdot\mathcal{Q}_{lm}\underline{\varepsilon}\cdot\mathcal{Q}_{mn}\underline{\varepsilon}\cdot\mathcal{Q}_{ng}}{(\Omega_{lg}-\omega)(\Omega_{mg}-2\omega)(\Omega_{ng}-\omega)} - \left(\sum_m{}' \frac{|\underline{\varepsilon}\cdot\mathcal{Q}_{mg}|^2}{(\Omega_{mg}-\omega)}\right)\left(\sum_n{}' \frac{|\underline{\varepsilon}\cdot\mathcal{Q}_{ng}|^2}{(\Omega_{ng}-\omega)^2}\right)\right] \quad , \qquad (2.69)$$

where only one- and two-photon resonant terms have been retained, and where the atom is taken to be in its ground state. For a two-level atom, (2.69) reduces to (2.66c), as it should. LEHMBERG et al. [2.68][who derived (2.69) from first principles] have used (2.69) to describe two-photon resonant self-focussing in rubidium and cesium vapours, where (2.69) makes the lossless adiabatic contribution to the refractive index (Sect.4.6.2). It is of interest to note that the removal of secular terms from (2.15) does not amount simply to deleting the initial state from intermediate-state summations, because in that case, the second part of (2.69), clearly important in this example, would be lost.

To conclude this subsection, we indicate an alternative approach to secularity. This makes use of the switching parameter mentioned in Sect.2.5, and is particularly useful when the susceptibility (2.15) has been written for a given process. As an illustration, consider the problem of computing $\chi^{(3)}(0;000)$ from (2.15). This gives a mathematically undefined expression that involves products of Ω_{gg}^{-1}, where $\Omega_{gg} \equiv 0$. However, from (2.23) we obtain denominators such as $[(\Omega' - 3i\varepsilon)(\Omega - 2i\varepsilon)(-i\varepsilon)]^{-1}$, and so on, where $\varepsilon \to 0_+$. It is not correct to approximate $\Omega - 2i\varepsilon$ by Ω, etc., which would give the singular $(-i\varepsilon\Omega'\Omega)^{-1}$, because the denominator has a finite real part. By carefully collecting real and imaginary parts, we find that only the imaginary terms are singular, i.e., of the form $\sim i/\varepsilon$, and these cancel out. (The numerical coefficients of ε are important.) The resultant real susceptibility is just what would be obtained from (2.68). Hence, it would be desirable to have a general method, based on the limiting procedure $\varepsilon \to 0_+$, of removing secularities from (2.15); a clue is perhaps given by the work of HEITLER [2.11]. In this, certain secular divergences in the perturbation theory of free-electron scattering are removed by making use of properties of the singular zeta function. See also GELL'MAN and LOW [2.69].

2.6.6 The Vector Model

This account of adiabatic states would not be complete without a brief description of the vector model, introduced by FEYNMAN et al. [1.72] to handle two-level atom problems, and closely connected to the Bloch equations of magnetic resonance, RABI et al. [2.70]. For the two-level atom, the density matrix elements in the interaction picture are given by $\rho_{aa} = |a|^2$, $\rho_{bb} = |b|^2$, $\rho_{ba} = b^*a$ where a, b are the coefficients in (2.44a). Strictly speaking, ensemble averages of these bilinear products are to be used, but this is of no account until dephasing terms are introduced. When (2.45) or the Liouville equation for ρ are used directly, and $\sigma_{ab} = \rho_{ab} \exp(i\omega t)$ and the real quantities

$$u = \sigma_{ab} + \sigma_{ba}$$

$$v = i(\sigma_{ab} - \sigma_{ba})$$

$$w = \rho_{aa} - \rho_{bb} \tag{2.70}$$

are introduced, it follows that

$$\dot{u} = \Delta v$$

$$\dot{v} = -\Delta u - \beta w$$

$$\dot{w} = \beta v \quad . \tag{2.71a}$$

It is an elementary matter to prove from (2.71a) that

$$u^2 + v^2 + w^2 = 1 \tag{2.71b}$$

is a constant of the motion; the constant is 1 because of the normalisation condition $|a|^2 + |b|^2 = 1$ in (2.44), and is equivalent to $\mathrm{Tr}\{\rho\} = 1$.

The polarisation is given by

$$P(t) = \mathcal{N}\mathrm{Tr}(\rho\mu) = \mathcal{N}(e^{i\Omega_{ab}t}\rho_{ab}\mu_{ba} + \mathrm{c.c.}) = \mathcal{N}\mu_{ba}(\sigma_{ba}e^{i\omega t} + \mathrm{c.c.}) \quad , \tag{2.72a}$$

so that

$$P_\omega(t) = 2\mathcal{N}\mu_{ba}\sigma_{ba} = \mathcal{N}\mu_{ba}(u + iv) \quad . \tag{2.72b}$$

The quantities u and v are therefore proportional to the in-phase and quadrature components, respectively, of the polarisation.

For an adiabatically applied field, a solution of (2.71) is immediately obtained. Set $\dot{u} = \dot{v} = \dot{w} = 0$; then

$$u = -\beta w/\Delta \quad , \quad v = 0 \quad . \qquad (2.73)^{12}$$

When (2.71b) is used in (2.73),

$$u = \mp\beta/\Omega \quad , \quad v = 0 \quad , \quad w = \pm\Delta/\Omega \tag{2.74}$$

is obtained.

Comparison of (2.72b) and the solution (2.74), with (2.62) shows that the lower sign in (2.74) corresponds to an atom in its ground state. Moreover, from (2.52) w is then simply $w = -\cos\theta = \sin^2\theta/2 - \cos^2\theta/2$, i.e., it is the inversion of the virtual-state component $|a\rangle$ over the real component $|b\rangle$ in the adiabatic ground state ϕ_b. For weak fields, $w \to -1$ and $\phi_b \to |b\rangle$, as expected. It is apparent, therefore, that the solution (2.74) is equivalent to the adiabatic-states picture presented earlier, where the lower sign corresponds to the atom in state ϕ_b and the upper sign to ϕ_a.

Returning to (2.71), if a unit real cartesian vector $\underline{r} = (u, v, w)$ is defined, then we soon find that

$$\dot{\underline{r}} = \underline{\Omega} \times \underline{r} \quad , \tag{2.75}$$

---

[12] A first correction to (2.73) is to calculate v from the first of (2.71a) but by use of the relation for u given in (2.73). Thus, $v = -\Delta^{-2}d(\beta w)/dt$, $= \mp\Delta^{-1}d(\beta/\Omega)/dt$ from (2.71). A systematic method of obtaining corrections has been given by CRISP [2.71], including the effects of damping. See also Sect.2.6.1 and GRISCHKOWSKY et al. [2.3], GRISCHKOWSKY [2.62], GRISCHKOWSKY and BREWER [2.118].

where the *torque vector* $\underline{\Omega}$ is $(\beta, 0, -\Delta)$. Equation (2.75) describes the precession of $\underline{r}$ around $\underline{\Omega}$; this is conveniently illustrated by the vector model (Fig.2.6a). The torque vector $\underline{\Omega}$ is in the $\hat{\underline{u}}$, $\hat{\underline{w}}$, plane (carets denote orthogonal unit vectors) and is of length $|\Omega| = (\Delta^2 + \beta^2)^{1/2}$, the off-resonance Rabi frequency. The vector $\underline{r}$ is shown as precessing around $\underline{\Omega}$; in the general case, such a picture depends on their relative rates of precession. An advantage of the vector model is that u, v, w are readily visualised, because they are given by the projection of $\underline{r}$ onto the appropriate axes.

The adiabatic solution for a ground-state atom, $\underline{r} = \underline{r}_b = (\beta/\Omega, 0, -\Delta/\Omega)$ makes an angle θ with the negative $\underline{w}$ axis, where $\tan\theta = \beta/\Delta$, and lies in the $\hat{\underline{u}} - \hat{\underline{w}}$ plane. Thus $\underline{r}_b$ lies along $\underline{\Omega}$, and, as shown in Fig.2.6b, points in the same direction if $\Delta > 0$ and in the opposite direction if $\Delta < 0$. The upper-state solution $\underline{r} = \underline{r}_a$ is just $\underline{r}_a \equiv -\underline{r}_b$; it is also shown in Fig.2.6b. So long as the pulse remains adiabatic, $\underline{r}_a$ and $\underline{r}_b$ will follow the torque vector; hence this solution is termed *adiabatic-following*.

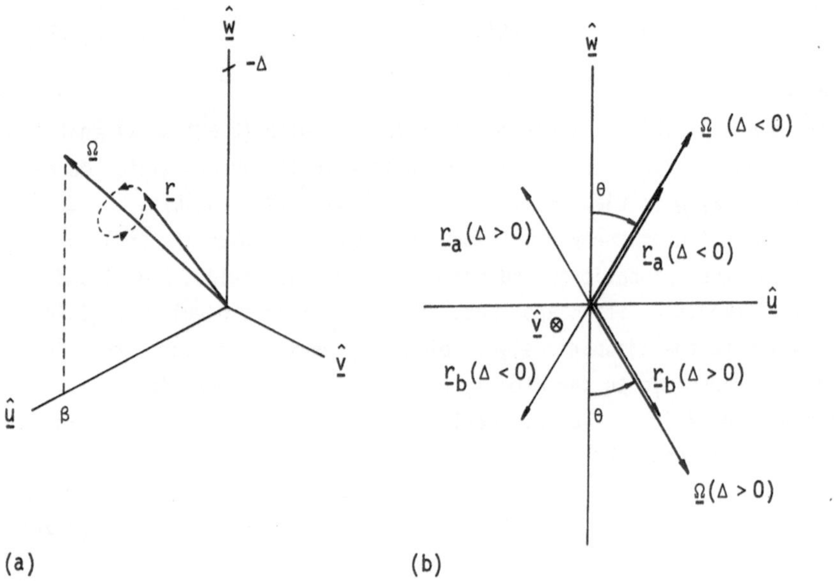

Fig.2.6. (a) Vector model. The tip of the $\underline{r}$ vector moves on the surface of the unit sphere, and precesses around the torque vector $\underline{\Omega}$. (b) Adiabatic-following solution. The $\underline{r}$ vector is aligned parallel or anti-parallel to $\underline{\Omega}$, depending on the detuning and whether the atom is in the adiabatic ground state (ϕ_b; $\underline{r} = \underline{r}_b$) or upper state ($\phi_a$; $\underline{r} = \underline{r}_a$). In adiabatic rapid passage the $\underline{r}$ vector moves smoothly from $\underline{r}_b$ to $\underline{r}_a$ or vice versa, as the detuning is swept through resonance

To see what happens when the detuning Δ is swept through resonance, take the atom to be initially in state ϕ_b, and $\Delta > 0$, for definiteness. Then, for a given field (i.e., β), as Δ is decreased, $\underline{r}_b$ moves so as to line up, from below, with the positive $\hat{\underline{u}}$ axis. Similarly, an atom in state ϕ_a, with $\Delta < 0$, moves so as to line up, from above, with the positive $\hat{\underline{u}}$ axis; hence, as Δ is swept from $\Delta > 0$ through resonance, the state ϕ_b changes into ϕ_a (cf. Sect.2.6.4) and the $\underline{r}$ vector changes from $\underline{r}_b$ to $\underline{r}_a$; this is shown on Fig.2.6b. This is known as *adiabatic rapid passage*, in which the atomic transition is completely inverted simply by sweeping Δ through resonance. The effect is well known in magnetic resonance, and has been demonstrated in the optical region by LOY [2.72], who used a dc Stark field to shift the transition frequencies and thereby sweep Δ through resonance.

It will be remembered that the definition of θ in Sect.2.6.2 was restricted to $|\theta| \leq \pi/2$; the reason for this can now be seen. When $|\theta| \leq \pi/2$, the vector $\underline{r}_b$ is constrained to lie below the $\hat{\underline{u}}$-$\hat{\underline{v}}$ plane, and $\underline{r}_a$ above it, as the states ϕ_b and ϕ_a are uniquely defined. If θ is allowed to range over $|\theta| \leq \pi$, then the states are no longer unique - they change over at $\theta = \pm\pi/2$. The adiabatic states themselves have rarely been considered in discussions of such phenomena as adiabatic rapid passage; the analyses are usually entirely in terms of the vector model. In this case, the vector $\underline{r}$ moves smoothly with the detuning, from $\theta = 0$ to $\pm\pi$ (depending on the initial sign of Δ); as it crosses the $\hat{\underline{u}}$ - $\hat{\underline{v}}$ plane, the system is manifestly inverted, because w changes sign. This is suggestive of a jump of the state of the system, whereas we have seen that the state actually varies smoothly.

An elegant pictorial link between the adiabatic states ϕ_a, ϕ_b and the vector model may be made if we use a result due to VENKATESH and SARKAR [2.73].[13] They considered the stereographic projection of the general $\underline{r}$ vector onto the $\hat{\underline{u}}$ - $\hat{\underline{v}}$ plane. This is defined here as the complex number $z = \tilde{u} + i\tilde{v}$ obtained from the point of intersection $(\tilde{u}, \tilde{v}, 0)$ of a line drawn from the point (0, 0, 1) through the tip of the $\underline{r}$ vector and onto the $\hat{\underline{u}}$ - $\hat{\underline{v}}$ plane. Elementary geometry reveals that $z = (u + iv)/(1 - w)$. This reduces, after use of (2.70), to $z = a/b$, where a and b are the coefficients of the wave function ψ, (2.44a). Thus as $\underline{r}$ varies, z traces out the mixture of $|a\rangle$ and $|b\rangle$ in ψ. For the adiabatic-following solution $\underline{r} = \underline{r}_b$, Fig.2.7 shows that $z = u/(1 - w) = \tan\theta/2$. Now the adiabatic wave function is just

---

[13] We further adapt their result to the situation at hand. See also [2.74].

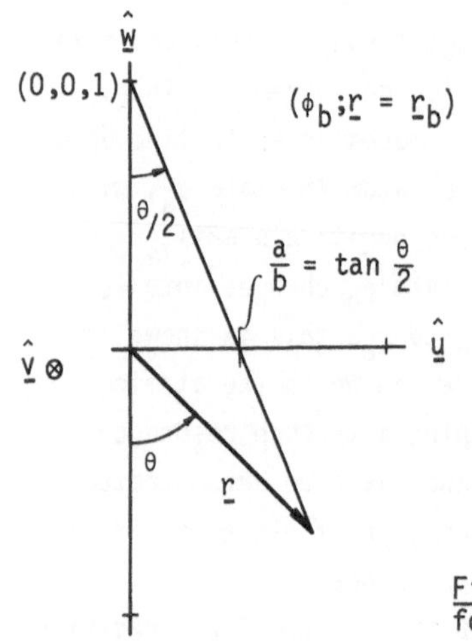

Fig.2.7. Stereographic projection of adiabatic-following solution

$\psi = \exp(-i\Omega_b't)\phi_b$, which can be written [see (2.51b)] as $e^{-i\delta_- t}(\cos(\theta/2)\exp(-i\Omega_b t)|b\rangle + \sin(\theta/2)\exp(-i\Omega_a t)|a\rangle)$; i.e., $a/b = \tan(\theta/2)$. Hence z traces out the state ϕ_b. If the angle θ is allowed to increase beyond π/2, when we call it θ', then, as Δ passes from Δ > 0 through resonance, the stereographic projection moves smoothly if $\underline{r}$ is changed from $\underline{r}_b$ to $\underline{r}_a$. The angle θ in ϕ_a is measured from the positive $\hat{\underline{w}}$ axis towards the negative $\hat{\underline{u}}$ axis (Figure 2.6b), therefore, $\theta' = \pi + \theta$; whence $z = \tan(\theta/2) = \cot(\theta/2)$, in agreement with $z = a/b$ as calculated from (2.51a). Thus ϕ_a is also traced out. The reader may have noticed that the form of ϕ_a, ϕ_b in (2.51) is that of a rotation of the basis |a>, |b> to ϕ_a, ϕ_b under the spin ($D^{(\frac{1}{2})}$) representation of SU(2). This is the covering group of the group of proper rotations in three dimensions, $R^+(3)$; the vector $\underline{r}$ rotates according to the $D^{(1)}$ representation of that group. The stereographic projection is a means of relating these representations, via their basis vectors (ϕ_a, ϕ_b) and $\underline{r} = (u, v, w)$. If the vector $\underline{r}$ is rotated from pointing downwards, $\theta = 0(\underline{r} = \underline{r}_b)$, through $\theta = \pi(\underline{r} = \underline{r}_a)$, and on to $\theta = 2\pi$, then $\underline{r}$ is unchanged, but the state $\phi_b \rightarrow -\phi_b$. A physical realisation of this sign change is not possible in adiabatic following; but in self-induced transparency, where the $\underline{r}$ vector traces out complete circuits on the sphere, the passage of a pulse can result in a sign change of the wave function (ELGIN, private communication).

It can readily be seen from $\underline{r} = \underline{r}_b$ in Fig.2.6b that the polarisation (2.72b) is identical to that of (2.62). Of more interest, however, is that as the angle of tilt, θ, increases from zero with increasing field strength (i.e., increasing β), it corresponds to the need to include higher-order susceptibilities in the calculation of the polarisation. This provides a pictorial representation of the expansion (2.2). Moreover, for small tilts ($\beta \to 0$), we have $\theta \simeq \beta/\Delta$, i.e., the expansion parameter β is more properly replaced by $|\beta/\Delta|$ for resonant processes, and this gives a more accurate guide as to the need to include higher-order terms. Thus, there are two expansion parameters, $|\beta/\Delta|$ for the field strength, and $|\dot{\beta}/\beta\Delta|$ for the deviation from adiabaticity of the field.

2.6.7 The Two-Photon Vector Model

The vector model may be extended into the realm of two-photon resonant nonlinear optics, as reviewed by GRISCHKOWSKY et al. [2.3]. Building, in particular, on their earlier work and that of KHRONOPULO [2.75] and TAKATSUJI [2.9,76,77], they used a sequence of unitary transformations in order to produce a representation in which the problem reduces to an effective interaction between the two-photon resonant adiabatic states. (Such methods are familiar in time-independent perturbation theory. See, e.g., HEITLER [2.11], ZIMAN [2.78].) The fields are assumed to be applied adiabatically with respect to any other states of the atom. The new $\underline{r}$ vector, $\underline{r}'$, which is related to the polarisation through a sequence of transformations, precesses about a torque vector $\underline{\Omega}'$, which has components $(\Delta', 0, \beta')$.

Here Δ' is the two-photon resonant detuning, such as a denominator in Table 2.2, shifted by the quadratic optical Stark effect for each level. For example, consider a Raman amplifier where, ignoring damping, we have $\Delta' = \Omega_{fg} + \omega_s - \omega_p + \delta_f - \delta_g$. The shifts $\delta_{f,g}$ may be calculated from (2.63). [The changes to Δ' and β' (in the sequel) for a general resonance are obvious.] Notice that because, in principle, a number of intermediate states are to be considered in (2.63), then in general $\delta_f \neq -\delta_g$, whereas for the two-level atom, $\delta_b = -\delta_a = \delta_-$. Also, the shifts $\delta_{f,g}$ must each incorporate the effects of both I_s and I_p, the Stokes and pump intensities. Frequently, (see e.g., Sects.4.5,5.4), ω_p will be resonant with a transition Ω_{ig}, so that δ_g will be dominated by the contribution from I_p; similarly, ω_s will be resonant with Ω_{if}, so that δ_f is dominated by I_s. It is important to recognise that Δ' does *not* represent the total shifted detuning of the resonance, in the sense of the single-photon case (see p.70).

The parameter β' is the two-photon Rabi frequency; in this example it is given by

$$\beta' = \frac{1}{2\hbar}\alpha_{fg}(-\omega_s;\omega_p)E_sE_p \quad , \tag{2.76}$$

where α_{fg}, E_s, and E_p are taken to be real and positive. To compare β' with the single-photon Rabi frequency β, we can write $\beta' = \mu_{fg}E_s/2\hbar$, where $\mu_{fg} = \alpha_{fg}(-\omega_s;\omega_p)E_p$ is termed the induced Stokes transition-dipole moment (Sect.2.7.1).

The two-photon vector model may be solved as for the two-level-atom model, where in (2.71) all quantities become primed. We shall see in Sect.2.7, where the new equations are derived in an heuristic fashion, that among the contributions to the polarisation (again taking the example of a Raman amplifier) is a quantity equivalent to $P_s(t) = \mathcal{N}\mu_{fg}(u' + iv')$. This is analogous to (2.72). Although the individual fields must be adiabatic with respect to single-photon transitions, the product E_sE_p associated with the two-photon resonance need not be; in general $\underline{r}'$ can have any of the various solutions of the single-photon vector model (e.g., [2.26]). Of particular interest, however, is the adiabatic-following solution. The criterion for this to be valid is just $|\dot{\beta}'/\beta'\Delta'| \ll 1$, which may be derived along the lines of Sect.2.6.1. Hence, similarly to Sects.2.6.4 and 2.6.6, and as shown by GRISCHKOWSKY et al. [2.3], a general two-photon resonant third-order susceptibility must have its denominator Δ (such as in Table 2.2) replaced by Δ', and then

$$\Delta' \rightarrow \mathrm{sgn}\{\Delta'\}|\Delta'^2 + \beta'^2|^{\frac{1}{2}} \equiv \Omega' \quad , \tag{2.77}$$

[cf. (2.63) where Δ^{-1} of the linear susceptibility was replaced by Ω^{-1}], where we continue to use Ω' as a signed quantity. GRISCHKOWSKY et al. term the quantity $|\Delta/\Omega'|$ ($|\Omega'| \equiv \gamma$ in their notation) a "power factor"; we shall discuss its meaning below. The considerations of Sect.2.6.4 require the susceptibility of an atom in its adiabatic ground state (or upper state) to change sign with Δ', although the polarisation itself (or the macroscopic susceptibility) does not change sign, if Δ' is swept through resonance; in this respect, the vector model obviously gives the polarisation correctly.

In Fig.2.8, we have drawn the states involved in the construction of the two-photon vector model, for the resonance $\Delta \equiv \Omega_{fg} - \omega_1 - \omega_2 \simeq 0$. Thus, Fig.2.8a shows the relation of the frequencies ω_1 and ω_2 to the unperturbed initial and final states $|g\rangle$, $|f\rangle$, and the various intermediate states $|i\rangle$. The unitary transformations referred to earlier recast the problem in terms of the adiabatic states ϕ_g and ϕ_f. These states are calculated by taking into account the coupling, via ω_1 and ω_2, of $|g\rangle$ to $|i\rangle$, to produce ϕ_g, and $|f\rangle$ to $|i\rangle$, to produce ϕ_f. The requirement that ω_1 and ω_2 be adiabatic with respect to the intermediate states allows ϕ_g and ϕ_f to be well approximated

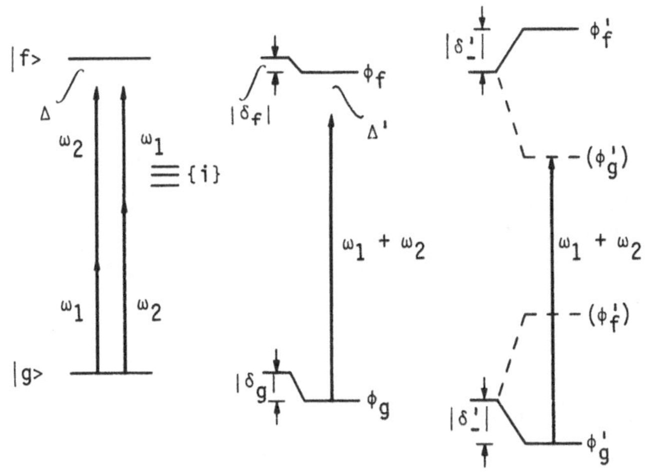

(a) (b) (c)

Fig.2.8a-c. Two-photon resonant atom. (a) Unperturbed atom. The sum $\omega_1 + \omega_2$ is resonant with the transition g - f, with a two-photon coupling of ω_1 and ω_2 via the intermediate states {i}. (b) Perturbed atom. The coupling of g to i and i to f is taken into account to produce adiabatic states ϕ_g, ϕ_f. These states in turn act as unperturbed basis states of a two-level "atom", with an interaction of frequency $\omega_1 + \omega_2$ (cf. Fig.2.4a). (c) Adiabatic solution for "atom", analogous to Fig.2.4b, showing level shifts $|\delta'_-|$

by lowest-order perturbation theory, i.e., they are of the form (2.54a), with quadratic shifts (2.54b). [Equation (2.54) must be extended in an obvious way to incorporate a sum over all of the intermediate states, and over both frequencies.] This results in Fig.2.8b, in which the new basis states are ϕ_g, ϕ_f shifted in energy by δ_g, δ_f respectively, giving a new detuning $\Delta' = \Delta + \delta_f - \delta_g$. The states are coupled by an effective two-photon interaction β', indicated by the arrow $\omega_1 + \omega_2$, and we see that the analogy to the two-level atom system, Fig.2.4a, is complete. The new system may thus be described in terms of the two-photon vector model. For the adiabatic-following solution, we see that a new set of states ϕ'_f and ϕ'_g, evolves. These are just the adiabatic states that would arise in a two-level system with unperturbed states ϕ_f, ϕ_g. As shown in Fig.2.8c, and by analogy with Fig.2.4b, we again have real and virtual states, with ϕ'_f, ϕ'_g shifted by

$$\delta'_- = \frac{1}{2}(\Delta' - \Omega') \quad . \qquad (2.78)$$

Thus it is seen that Δ' refers to the shift of the two-photon resonance when only the separate couplings of $|g\rangle$ and $|f\rangle$ to the intermediate states are taken into account. The quantity $\Omega' = \Delta' - 2\delta_-' = \Delta + (\delta_f - \delta_g - 2\delta_-')$ gives the shift of the two-photon resonance Δ when the coupling of $|f\rangle$ to $|g\rangle$ by the two-photon resonant interaction is allowed for. This means that the denominator Ω'^{-1} that appears [see (2.77)] in the power-dependent "third-order susceptibility" (cf. Sect.2.6.4) has the significance of a level-shifted resonance. When expanded, as in (2.66), the level shift is seen to be made up of $\chi^{(5)}$, $\chi^{(7)}$... contributions, which account for the shifts that are due to the intensity-dependent refractive index, and optical Kerr effect terms such as β'^2 and $\delta_f\delta_g$. [The optical Kerr effect itself, in the $\chi^{(3)}$ approximation of Table 2.1, would give rise to shifts in the two-level atom perturbed by two fields, of the same order as the self-induced refractive index correction $\chi^{(3)}$, (2.66c).] These points concerning the meaning of Δ' and Ω' do not appear to have been made in the literature, with the result that Δ' alone is commonly referred to as *the* shifted detuning from resonance.

To conclude this subsection, brief mention should be made of adiabatic rapid passage in the two-photon resonant case. By analogy with the single-photon case, it is obvious that the states ϕ_g' and ϕ_f' can be inverted if Δ' is swept through resonance. Unlike the two-level atom, however, Δ' is actually field dependent; this led GRISCHKOWSKY and LOY [2.79] (see also [2.80]) to suggest the possibility of self-induced adiabatic rapid passage, in which suitably tailored incident pulses can cause the relative sizes of δ_f and δ_g to change in such a way that Δ' is swept through resonance by the pulses themselves. LOY [2.81] has since reported an experimental demonstration of two-photon adiabatic rapid passage (Sect.5.5.2).

2.6.8 Damping

The use of the density matrix allows a more realistic approach to the treatment of damping than that adopted in Sect.2.5. In the two-level atom model, the Bloch equations (2.71a) become, with damping (e.g., [2.26])

$$\dot{u} = \Delta v - u/T_2 \tag{2.79a}$$

$$\dot{v} = -\Delta u - \beta w - v/T_2 \tag{2.79b}$$

$$\dot{w} = \beta v - (w - w_0)/T_1 \quad , \tag{2.79c}$$

where T_1 and T_2 are respectively, the longitudinal and transverse homogeneous (dephasing) relaxation times, and w_0 is an equilibrium value of w. The steady-state solution of (2.79) is straightforward,

$$u + iv = -\beta w/(\Delta - i/T_2) \tag{2.80a}$$

$$w = w_0(1 + \Delta^2T_2^2)/(1 + \Delta^2T_2^2 + T_1T_2\beta^2) \quad . \tag{2.80b}$$

In the weak-field limit, $w \simeq w_0$, so that from (2.80a) we see that in (2.15) resonant denominators should take $\Gamma = 1/T_2$, where T_2 is the dephasing time for the resonant transitions.

The choice $w_0 = -1$ is usually made (atoms initially in their ground state); but, for (2.80) to reduce to the adiabatic solution (2.74) as $T_2 \to \infty$ (no damping) requires $w_0 = -\Delta/\Omega$ to be taken. This can be understood if it is recalled that for an adiabatically applied field, it is, strictly speaking, not possible to consider the atom without the field; therefore, w can relax only to the adiabatic ground-state population $w_0 = -\Delta/\Omega$. NAYFEH and NAYFEH [2.7], and CRISP [2.71], LEHMBERG and REINTJES [2.82], NAYFEH [2.83] investigated the adiabatic-following solutions in the presence of damping, and LEHMBERG et al. [2.68], NAYFEH and NAYFEH [2.84] have considered the two-photon resonant case. See also PUELL and VIDAL [2.28]. One requirement for adiabatic following in the presence of damping is that the incoherently absorbed energy should be small, in the sense that the transition probability between the adiabatic states remains small. The latter quantity is proportional to the pulse energy, see for example (2.56).

The solution (2.80) with $w_0 = -1$ is well-known in magnetic resonance [2.70], and was used by JAVAN and KELLY [2.85] to study self-defocussing of laser light; similarly to (2.63), (2.80) yields

$$\chi\Big(\omega;E_\omega(t)\Big) = \frac{\mathcal{N}\mu_{ab}^2T_2}{\hbar\varepsilon_0}\,\frac{\Delta T_2+i}{1+\Delta^2T_2^2+T_1T_2\beta^2} \quad . \tag{2.81}$$

The power-broadened linewidth $(T_1\beta^2/T_2)^{\frac{1}{2}}$ is due to saturation of the homogenously broadened resonant transition, and may be written in terms of a saturation intensity (Sect.4.6.1).

An example for the two-photon resonant case is furnished by WARD and SMITH [2.67]. These authors were concerned with two-photon resonant third-harmonic generation, for which, ignoring level shifts, the denominator analogous to that in (2.81) is $1 + (\Omega_{fg} - 2\omega)^2 T_2^2 + T_1 T_2 |\alpha_{fg}(\omega;\omega) E_\omega^2 / 4\hbar|^2$. A similar result is used in Sect.5.4, for an account of saturation in stimulated Raman scattering.

When the pulse length is long compared to T_2 but comparable in length to T_1, the rate-equation approximation can be used, in which u and v are assumed to come to equilibrium with w very rapidly, i.e., they follow w adiabatically. Then u and v are given in terms of w by (2.80a), but w is found from the rate equation (2.79c). We refer to ALLEN and EBERLY [2.26]; and to ACKERHALT and SHORE [2.86] for comparison of rate-equation and transient solutions to the Bloch equations. BUTYLKIN et al. [2.5] have given a detailed account of the two-photon resonant susceptibility, for a variety of cases, (see also the above references to adiabatic following).

2.7 Material Excitation

2.7.1 Spontaneous Raman Scattering

We now turn to the description of resonant nonlinear processes from the point of view of an excitation set up in the medium by the applied fields. This approach is very useful, in that it formulates the problem in classical terms, and, therefore, allows full use to be made of intuitive concepts. Thus, in this section, the main ideas are reviewed by reference to purely classical treatments of spontaneous [2.87] and stimulated [2.14,88,119] Raman scattering. (The discussion of resonant processes in Sect.2.5 made implicit use of the results of this section.)

In the classical theory of spontaneous Raman scattering ([2.87,89,90], and many texts on infrared Raman spectroscopy), the polarisability α of a molecule is defined as the electronic dipole moment induced by a unit field. Thus $\alpha = \alpha^E(Q)$, where Q denotes the nuclear conformation and the superscript E denotes an electronic polarisability. For simplicity, we take Q to represent the displacement of a single vibrational coordinate from equilibrium, in which case, for small displacements,

$$\alpha^E(Q) = \alpha_o^E + (\partial\alpha^E/\partial Q)_o Q \quad , \tag{2.82}$$

where the subscript o denotes evaluation at the equilibrium conformation Q = 0. Thus, the dipole moment induced by an incident field E_p leads to Rayleigh scattering, through $\alpha_o^E E_p$. Spontaneous Raman scattering occurs through $(\partial\alpha^E/\partial Q)_0 QE_p$, because Q vibrates (or rotates) at a frequency Ω, say, and thus QE_p oscillates at the Stokes (sideband) frequency $\omega_s = \omega_p - \Omega$ (and $\omega_{as} = \omega_p + \Omega$, which we will ignore here). A correspondence-principle extension of this theory replaces Q by the matrix element $<v_f|Q|v_g>$, where v_f, v_g are vibrational-state labels, with $\Omega = \Omega_{fg}$.

$(\partial\alpha_E/\partial Q)_0$ is easily related to quantum-mechanical quantities if the idea of a transition moment is introduced. Recall that in (2.13) the induced dipole moment $\underline{\mu}(t)$ was defined in the usual way as an expectation value of the dipole-moment operator $\underline{Q}$ (not to be confused with the vibrational coordinate Q), taken over the perturbed ground state. Now we assert [2.87,91] that the (real) induced transition moment defined by

$$\underline{\mu}_{fg}(t) = \frac{1}{2} e^{i\Omega_{fg}t} {}'<f|\underline{Q}|g>' + \text{c.c.} \tag{2.83}$$

has the physical significance of a classical dipole moment. The diagonal moment $\underline{\mu}_{gg}(t)$ is just the $\underline{\mu}(t)$ given in (2.13)[14]. The moment $\underline{\mu}_{fg}(t)$ may be inserted into the classical expression for the intensity I_ω at a distance R of a wave of polarisation $\underline{\varepsilon}_\omega$ radiated by a dipole $\underline{\mu}$ oscillating at frequency ω,

$$I_\omega = \omega^4 |\underline{\varepsilon}_\omega^* \cdot \underline{\mu}_\omega|^2 / (32\pi^2 \varepsilon_0 c^3 R^2) \tag{2.84}$$

(see, e.g., [2.19]). This expression is to be considered as having been integrated over the emission linewidth, centred on ω. As an example, in the absence of an external field, (2.83,84) give $\underline{\mu}_\omega = \underline{Q}_{fg}$, $\omega = \Omega_{fg}$, and this is the basis of the correspondence-principle theory of spontaneous emission, provided that $\Omega_{fg} > 0$ [2.91].

---

[14]Because the state $\exp(-i\Omega_g t)|g>' = \exp(-i\Omega_g' t)\phi_g$, in the notation of Sect.2.6, and similarly for $|f>'$, then in (2.83) we could write $\exp(i\Omega_{fg}' t)$ $<\phi_f|\underline{Q}|\phi_g>$. In this way, the transition frequency is strictly Ω_{fg}'. However, the compact general expression for the transition moment given in Section 2.7.2 uses the states $|g>'$, $|f>'$. Just as the discussion of secularity in Sect.2.6.5 showed for nonlinear susceptibilities, the price paid for this compact formula is that the Stark shift for the state ϕ_g is expanded in $|g>'$, etc. Unlike the rigorous expectation value $\underline{\mu}(t) = \underline{\mu}_{gg}(t)$ used in (2.13), however, the expanded Stark-shift factors do not cancel identically in the off-diagonal matrix elements in (2.83), but yield the factor $\exp[i(\Omega_{fg}' - \Omega_{fg})t]$; therefore, in general they must be considered carefully. For two- and three-photon resonant processes it is easy to show that, to lowest order, there are no secular terms, provided that the diagonal matrix elements of $\underline{Q}$ vanish, as they do for free atoms and rotating molecules. Hence, to these orders $\phi_g = |g>'$ and $\Omega_g = \Omega_g'$, etc., and (2.83) is perfectly straightforward.

The general expression for $\mu_{fg}(t)$ will be given in Sect.2.7.2; but we first complete the classical description of Raman scattering by anticipating the result that to first order, a pump wave induces a moment that oscillates at the Stokes frequency $\omega_s = \omega_p - \Omega_{fg}$;

$$\underline{\mu}_{fg}(t) = \frac{1}{2}\,\underline{\mu}_s\, e^{-i\omega_s t} + \text{c.c.}$$

where

$$\mu_s \equiv \underline{\varepsilon}_s^* \cdot \underline{\mu}_s = \alpha_{fg}(-\omega_s;\omega_p)E_p \equiv \alpha_R E_p \quad . \tag{2.85}$$

The quantity μ_s is the induced Stokes moment, and $\alpha_R \equiv \alpha_{fg}(-\omega_s;\omega_p)$ is the Raman polarisability, a special case of the first-order transition hyperpolarisability, defined in (2.19). Comparison of (2.85,82) shows that α_R and $(\partial\alpha^E/\partial Q)_0$ must be taken to be equal. Without loss of generality, we choose Q to be dimensionless and normalised to unity; then the relation

$$\alpha_R = (\partial\alpha^E/\partial Q)_0 \tag{2.86}$$

ensues. This choice of definition of Q, leading to (2.86), is implicit in much of the literature.

Before leaving (2.86) it is advisable to comment further on the definition of the classical electronic polarisability $\alpha^E(Q)$. Under the Born-Oppenheimer approximation, for vibrational Raman scattering we can set $|g\rangle = |e(Q)\rangle|v_g\rangle$ and $|f\rangle = |e(Q)\rangle|v_f\rangle$, where $|e(Q)\rangle$ is the electronic ground state. Then α_R may be expressed as

$$\alpha_R = \langle f|\hat{\alpha}_R(-\omega_s;\omega_p)|g\rangle = \langle v_f|\alpha_R^E(-\omega_s;\omega_p;Q)|v_g\rangle \quad , \tag{2.87a}$$

where $\hat{\alpha}_R(-\omega_s;\omega_p)$ is a scattering operator that may be written by comparison with (2.19); $\alpha_R^E(-\omega_s;\omega_p;Q)$ is here called the electronic Raman polarisability. Expanding $\alpha_R^E(-\omega_s;\omega_p;Q)$ as in (2.82), noting that $\langle v_f|v_g\rangle = 0$, and choosing $\langle v_f|Q|v_g\rangle = 1$, we have

$$\alpha_R = \left[\partial\alpha_R^E(-\omega_s;\omega_p;Q)/\partial Q\right]_0 \quad . \tag{2.88a}$$

On the other hand, the quantity that is strictly analogous to the $(\partial\alpha^E/\partial Q)_0$ in (2.82) is

$$[\partial\alpha^E(-\omega_p;\omega_p;Q)/\partial Q]_0 \quad , \tag{2.88b}$$

where the electronic Rayleigh polarisability $\alpha^E(-\omega_p;\omega_p;Q)$ is, similarly to (2.87a), given by

$$\alpha = \langle g|\hat{\alpha}(-\omega_p;\omega_p)|g\rangle = \langle v_g|\alpha^E(-\omega_p;\omega_p;Q)|v_g\rangle \quad . \tag{2.87b}$$

Hence, to reconcile the classical- and transition-moment approaches completely requires $\alpha^E(-\omega_p;\omega_p;Q) = \alpha_R^E(-\omega_s;\omega_p;Q)$. A condition for this is seen from (2.19) to be that ω_p should lie well below any electronic resonances, so that both types of electronic polarisability are essentially independent of frequency. This is usually the case in conventional Raman spectroscopy.

Naturally these considerations do not affect the central idea of a vibrational coordinate and the existence of a relation (2.86) that connects classical and quantum-mechanical quantities.

2.7.2 Transition Hyperpolarisabilities

If we assume for simplicity that the electric field is a superposition of monochromatic waves, then in nth order the component of $\underline{\mu}_{fg}(t)$ that oscillates at frequency ω_σ' with polarisation vector $\underline{\varepsilon}_\sigma'$ is found to be [2.21], $(\frac{1}{2}\mu_{fg;\sigma'}^{(n)}\exp-i\omega_\sigma' t + \text{c.c.})$, where

$$\mu_{fg;\sigma'}^{(n)} \equiv \underline{\varepsilon}_\sigma'^{*}\cdot\underline{\mu}_{fg;\sigma'}^{(n)}$$

$$= n!^{-1}\sum{}' K(-\omega_\sigma';\omega_1\ldots\omega_n)\gamma_{fg}^{(n)}(-\omega_\sigma';\omega_1\ldots\omega_n)E_1\ldots E_n \tag{2.89}$$

The prime on the summation sign has the same meaning as in (2.7). It is important to realise that the frequency ω_σ' is, in general, arbitrary; however, in this subsection, the correspondence-principle interpretation of $\underline{\mu}_{fg}(t)$ will be used, and requires the resonance condition

$$\omega_\sigma' = \omega_1 + \ldots + \omega_n - \Omega_{fg} \quad . \tag{2.90}$$

Equation (2.90) is analogous to the δ-function constraint in (2.2). The nth-order transition hyperpolarisability $\gamma_{fg}^{(n)}$ is

$$\gamma_{fg}^{(n)}(-\omega_\sigma';\omega_1\ldots\omega_n) = \frac{1}{\hbar^n}\mathcal{S}_T\sum_{b_1\ldots b_n}\frac{\underline{\varepsilon}_\sigma'^{*}\cdot\underline{Q}_{fb_1}\underline{\varepsilon}_1\cdot\underline{Q}_{b_1b_2}\cdots\underline{\varepsilon}_n\cdot\underline{Q}_{b_ng}}{(\Omega_{b_1g}-\omega_1-\ldots-\omega_n)(\Omega_{b_2g}-\omega_2\ldots-\omega_n)\ldots(\Omega_{b_ng}-\omega_n)} \quad . \tag{2.91}$$

The notation in (2.89) is the same as that in (2.9) and similarly for (2.91,15). It is evident from (2.91) that $\gamma_{fg}^{(n)}$ possesses overall permutation symmetry, which is as expected in the absence of damping. In order that (2.89,91) reduce in special cases to the conventional forms found in the literature, the factor n! is placed in (2.89), and ε_0 is omitted from both, cf. the analogous (2.9b,15). The nonlinear susceptibility (2.15) is clearly related to the (diagonal) hyperpolarisabilities through

$$\chi^{(n)}(-\omega_\sigma;\omega_1\ldots\omega_n) = \frac{\mathcal{N}}{\varepsilon_0 n!}\sum_g \rho(g)\,\overline{\gamma_{gg}^{(n)}(-\omega_\sigma;\omega_1\ldots\omega_n)} \quad , \tag{2.92}$$

where $\omega_\sigma' = \omega_\sigma$, because $\Omega_{gg} \equiv 0$. There are two averages in (2.92) —an orientation average, denoted by the bar, and a weighting over the initial states.

The moment induced in a molecule by a static field is given by

$$\mu_{DC} = \mu_{DC}^{o} + \alpha E + \frac{1}{2}\beta E^2 + \frac{1}{6}\gamma E^3 + \ldots \quad , \tag{2.93}$$

where $\mu_{DC}^{o} = \underline{\varepsilon}_{DC} \cdot \underline{Q}_{gg}$ is the permanent moment; $\alpha = \gamma_{gg}^{(1)}(0;0)$ is the polarisability; $\beta = \gamma_{gg}^{(2)}(0;00)$, like μ_{DC}^{o}, it vanishes for atoms and many molecules; $\gamma = \gamma_{gg}^{(3)}(0,000)$ is "the" hyperpolarisability. Equation (2.93) agrees in notation and numerical conventions with, for example, BUCKINGHAM [2.92] and BOGAARD and ORR [2.93]. For the refractive index, the familiar relation $\chi^{(1)}(-\omega;\omega) = \mathcal{N}\alpha/\varepsilon_0$ follows, where the atom is in its ground state, i.e., $\alpha = \overline{\gamma_{gg}^{(1)}(-\omega;\omega)}$. Finally, the explicit forms of (2.91) for n = 1 and n = 2 have been given in (2.19,22), where the simpler notations $\alpha_{fg} = \gamma_{fg}^{(1)}$, $\beta_{fg} = \gamma_{fg}^{(2)}$ were adopted. The theory of spontaneous hyper-Raman scattering can parallel that of spontaneous Raman scattering, Sect.2.7.1, using the hyper-Raman polarisability $\beta_{HR} = \beta_{fg}(-\omega_s;\omega_p\omega_p)$; see e.g., LONG and STANTON [2.94], KONINGSTEIN [2.95].

Equations (2.89,91) are essentially elements of the scattering matrix; as such they are closely related to transition-rate theory. For example, the two-photon absorption rate of a single frequency is proportional to $|E_\omega \mu_{fg;-\omega}|^2 = |\alpha_{fg}(\omega;\omega)E_\omega^2|^2$. The diagrammatic representation of (2.91) is identical to that described in Sect.2.5.

A further application of the idea of a transition moment is to note that (2.83) or (2.89), with (2.60), defines an effective interaction hamiltonian for a system described by the two states |g> and |f>; hence it is possible to define a generalised Rabi frequency appropriate to n-photon resonant problems. We will return to this point in the next subsection.

2.7.3 Bloch Equations

We first consider the classical derivation of the equations that govern stimulated Raman scattering (SRS). They were due originally to GARMIRE et al. [2.88]; since then they have been used extensively to describe SRS, for example SHEN and BLOEMBERGEN [2.14], WANG [2.90], MAIER et al. [2.96], AKHMANOV et al. [2.97,98], CARMAN et al. [2.99]; see also the review by WANG [2.100]. These equations are then related to the Bloch equations for two-photon resonance discussed in Sect.2.6; finally some generalisations, and the relation to the susceptibility formalism, are briefly discussed.

Again employing the classical electronic polarisability, we have from (2.60,82) that the classical Raman-interaction hamiltonian is

$$H = -\frac{1}{4}\left(\frac{\partial\alpha^E}{\partial Q}\right)_0 Q_\Omega^* E_s^* E_p + \text{c.c.} \quad , \tag{2.94}$$

where E_s and E_p can be slowly varying amplitudes, and we have written $Q = Q_\Omega(t)\exp[-i(\Omega - \Delta)t] + \text{c.c.}$, with $\Delta = \Omega + \omega_s - \omega_p$. Therefore, although in Sect.2.7.1 the excitation Q was assumed to oscillate freely with frequency Ω, causing a Stokes wave of frequency $\omega_p - \Omega$ to be radiated, here we anticipate that an external Stokes wave ω_s of frequency $\neq \omega_p - \Omega$ may be applied. In conjunction with the pump field this can drive the excitation at a frequency $\Omega - \Delta$ slightly removed from Ω.

The external fields act via (2.94) to apply a force $-\partial H/\partial Q$ to the coordinate Q. Because the motion of Q is approximately harmonic, we have[15]

$$\frac{\partial^2 Q}{\partial t^2} + 2\Gamma \frac{\partial Q}{\partial t} + \Omega^2 Q = -\frac{2\Omega}{\hbar}\frac{\partial H}{\partial Q} \quad . \tag{2.95}$$

Some damping is allowed for by the parameter Γ. Equation (2.95) can be simplified by rewriting it in terms of the slowly varying quantities Q_Ω, etc. from which we find

$$\frac{\partial Q_\Omega^*}{\partial t} - i(\Delta + i\Gamma)Q_\Omega^* = \frac{i}{\hbar}\frac{\partial H}{\partial Q_\Omega} = \frac{-i}{4\hbar}\left(\frac{\partial\alpha^E}{\partial Q}\right)_0^* E_s E_p^* \quad , \tag{2.96}$$

---

[15] A real oscillator of mass M would have a coordinate $(\hbar/2M\Omega)^{1/2}Q$, and would be driven by a force $-(2M\Omega/\hbar)^{1/2}\partial H/\partial Q$; this leads to the factor $2\Omega/\hbar$ in (2.95).

making use of the fact that $\Omega >> |\Delta|$. Using (2.86) gives

$$\frac{\partial Q_\Omega^*}{\partial t} - i(\Delta + i\Gamma)Q_\Omega^* = -\frac{i\alpha_R^*}{4\hbar} E_s E_p^* \quad . \tag{2.97}$$

The macroscopic polarisation is

$$P_s = \mathcal{N}\left(\frac{\partial\alpha^E}{\partial Q}\right)_0 Q_\Omega^* E_p = \mathcal{N}\alpha_R Q_\Omega^* E_p \quad . \tag{2.98}$$

Equation (2.98) differs from the spontaneous case in that Q_Ω is given by (2.97) rather than unity. In conjunction with Maxwell's equations [e.g., (3.8b)], (2.97,98) describe SRS. Many authors, e.g., CARMAN et al. [2.99], consider Raman amplifiers, in which the Stokes wave is incident; generation from noise is then accounted for by a fictitious Stokes-wave noise source; see e.g., Sect.3.3. AKHMANOV et al. [2.97,98] treated the initiation of the excitation Q by introducing a stochastic force into the RHS of (2.97).

A first connection with the susceptibility formalism may be made if the steady-state limit of (2.97) is taken, for example when E_s and E_p are constant. Then the time derivative may be neglected, and from (2.98) we obtain

$$P_s = \frac{1}{4\hbar}\mathcal{N}\alpha_R E_p\left(\frac{\alpha_R^* E_s E_p^*}{\Delta+i\Gamma}\right) = \frac{\mathcal{N}|\alpha_R|^2/4\hbar}{\Delta+i\Gamma}|E_p|^2 E_s \quad ; \tag{2.99a}$$

i.e., the Raman susceptibility is

$$\chi^{(3)}(-\omega_s;\omega_p,-\omega_p,\omega_s) = \mathcal{N}|\alpha_R|^2/6\hbar\varepsilon_0(\Delta + i\Gamma) \quad , \tag{2.99b}$$

which is just that given by (2.15), see Table 2.2.

Equation (2.98) shows quite clearly how the Stokes polarisation arises from the modulation of the polarisability of the medium due to an internal motion Q. Recalling the discussion in Sect.2.5, the factorised structure of resonant susceptibilities is seen to be a consequence of an expression like (2.98), a polarisability derivative (transition hyperpolarisability) times the excitation, which itself is proportional to the derivative.

The description of four-wave processes adopted in Sect.2.5 can now be justified, because it is apparent from (2.98) that the excitation Q_Ω^* alone modulates the pump wave, and hence creates the Stokes polarisation independently of the way in which Q_Ω^* itself was generated. Therefore, we can set up Q_Ω^* by one pair of waves; then Q_Ω^* modulates the third wave to create a polarisation at the fourth. The second pair of waves can contribute to

Q_Ω^*, and thus act back on the first pair, therefore, the right-hand side of (2.97) consists of two parts. As an example, consider biharmonic pumping (coherent Raman mixing), Figs.2.2b and 3.1. We take two pump waves ω_1 and ω_2, and their respective Stokes waves ω_{1s} and ω_{2s}, and assume exact two-photon resonance $\omega_1 - \omega_{1s} = \omega_2 - \omega_{2s} = \Omega_{fg}$. From (2.97,98) the steady-state polarisation at ω_{1s} is

$$P_{1s} = \frac{\mathcal{N}\alpha_{fg}(-\omega_{1s};\omega_1)E_1}{4\hbar i\Gamma}\left\{\alpha_{fg}(-\omega_{1s};\omega_1)^*E_{1s}E_1^* + \alpha_{fg}(-\omega_{2s};\omega_2)^*E_{2s}E_2^*\right\} \quad . \tag{2.100}$$

The first part of (2.100) is ordinary SRS, and the second part is the four-wave term. Therefore, we see that the two processes are inseparable (the consequences of this are discussed in Sect.3.4). Such mixing effects occur in all resonant processes.

It is evident that an excitation Q_Ω^* set up by one pair of waves will last for a time on the order of $T_2 = 1/\Gamma$ after the waves are removed. This excitation may be probed at a later time by another pair of waves; therefore, information on the relaxation time T_2 can be found. Examples of this are given in the reviews by WANG [2.100], KAISER and LAUBEREAU [2.101], MAIER [2.102], SACCHI [2.103]. NAKATSUKA et al. [2.104], MATSUOKA [2.105], and MATSUOKA et al. [2.106] investigated the decay of the material excitation (see below), in atomic rubidium and calcium vapours, in which two-photon absorption rather than Raman resonances were used.

Although (2.97) has clear advantages for the description of time-dependent processes, it is entirely equivalent to the susceptibility approach. That is, if (2.97) is solved by taking the Fourier transform, then

$$Q_\Omega^*(\omega) = \alpha_R^*\{E_sE_p^*\}_\omega/2\hbar(\omega + \Delta + i\Gamma) \quad ,$$

where $\{E_sE_p^*\}_\omega$ denotes the transform of the product $E_sE_p^*$. This result may be inserted into (2.98); then, making use of the fact that we have assumed α_R to be nonresonant, by a few manipulations we reduce the polarisation to precisely the form (2.2), with $n = 3$.

We now examine the connection between the foregoing and the two-photon Bloch equations discussed in Sect.2.6.7. The latter may be written

$$\dot{\sigma}'_{fg} - i(\Delta' + i\Gamma)\sigma'_{fg} = i\beta'^*w'/2 \tag{2.101a}$$

$$\dot{w}' = 2\mathrm{Im}(\beta'\sigma'_{fg}) - (w - w_o)/T_1 \quad , \tag{2.101b}$$

where the primes denote the two-photon equations, and $\sigma'_{fg} = u'tiv'$ is the off-diagonal density-matrix element, in the model illustrated in Fig.2.8b. Ignoring the Stark shifts in Δ', and taking $w' = w'_0 = -1$, i.e., the atoms in their ground-state, we find that (2.101) is identical to (2.97)[16]. Therefore, σ'_{fg} is the well-defined generalised excitation that is appropriate to problems in any medium. For example, it represents the polarisability of electrons in an atom, which arises from their being in a coherent superposition of the states $|g>'$, $|f>'$ (or ϕ_g, ϕ_f, in the notation of Fig.2.8), or of course the vibrational coordinate Q^*_Ω for a molecule.

It is interesting to consider the calculation of the polarisation when (2.101) is used. The polarisation is, quite generally, $\underline{P}(t) = \mathcal{N}\,\mathrm{Tr}(\rho\underline{Q})$, and may be evaluated in any basis. However, the choice of the basis $|g>'$, $|f>'$ has the advantage that it is sufficiently complete for the resonant process; therefore,

$$\underline{P}(t) = \mathcal{N}(\sigma'_{fg}\underline{\mu}'_{gf}\, e^{i\Omega'_{fg}t} + cc) + \mathcal{N}(\rho'_{ff}\underline{\mu}'_{ff} + \rho'_{gg}\underline{\mu}'_{gg}) \quad . \tag{2.102}$$

The first bracketed term has already been discussed at length; for instance, it leads to (2.98). The second term is simply the refractive index contribution, where $\underline{\mu}'_{gg}$ is the induced-dipole moment of an atom in state $|g>'$; ρ'_{gg} is the population of that state; similarly, for $\rho'_{ff}\underline{\mu}'_{ff}$. In GRISCHKOWSKY et al. [2.3], the polarisation due to two fields ω_1, ω_2 was split into three parts p_I, p_{II}, and p_{III}. The first, p_I, is just the above refractive-index contribution; p_{II} is the pure SRS or TPA polarisation [e.g., (2.98)]; and p_{III} arises from mixing terms of the form $2\omega_1 + \omega_2$, $\omega_1 + 2\omega_2$. The last contribution is analogous to (2.100); therefore, with p_{II}, it arises from the first part of (2.102).

The level shifts in (2.101a) are not in the classical (2.97); this is expected because the classical model does not allow for an internal change of the frequency Ω. However, the energy considerations of Sect.2.6.4 allow these shifts to be calculated, and, therefore, inserted into (2.97) on a plausible, if not rigorous, basis.[14] Similarly, (2.101b) is just a rate equation that can be deduced from energy-flow considerations. It is therefore

---

[16]The two-photon Rabi frequency β' was taken as real, for simplicity, in Sect.2.6.7. Clearly, we have here, in general, a complex $\beta' = \alpha_{fg}(\omega_1;\omega_2)\, E_1E_2/2\hbar$. In the literature, the fields are frequently expanded into real-amplitude and phase form; then, $|\beta'|$ is the (real) parameter of importance.

satisfying to see that an established classical model gives such complete insight into the two-photon Bloch equations, the latter a source of much attention in the current literature.

To conclude this section, we consider briefly the extension of the Bloch equations to n-photon-resonance problems. The extension of the classical model is clear: we have only to calculate the nth-order force driving (2.96), and similarly for (2.101). This leads to a generalised complex Rabi frequency $\beta^{(n)} (n \geq 2)$,

$$\beta^{(n)} = \frac{2}{\hbar}\frac{\partial H^{(n)}}{\partial \sigma'_{fg}} = \frac{1}{2\hbar}\int_0^{E_{\omega_1}} \mu^{(n-1)}_{fg;-\omega_1} dE_{\omega_1} \qquad (2.103)$$

where $H^{(n)}$ is the nth-order interaction hamiltonian, and $\mu^{(n-1)}_{fg;-\omega_1}$ is given by (2.89). The resonance is of the form

$$\Delta = \Omega_{fg} - \omega_1 - \ldots - \omega_n \quad , \qquad (2.104)$$

where, of course, any of the frequencies may be negative. In full, (2.92,93, 103) give

$$\beta^{(n)} = \frac{1}{2(n-1)!\hbar}\sum{}' K(\omega_1;\omega_2\ldots\omega_n)\gamma^{(n-1)}_{fg}(\omega_1;\omega_2\ldots\omega_n)\int_0^{E_1} dE_1\cdot E_2\ldots E_n \quad . \qquad (2.105)$$

Note that $\beta^{(n)}$ is independent of the field component used as the integration variable; this follows from overall permutation symmetry. The following table gives some examples.

Process	$\beta^{(n)}$
SRS (n = 2)	$\alpha_R E_s^* E_p/2\hbar$
n-photon absorption of a single frequency	$[2^{n-1} n!\hbar]^{-1}\gamma^{(n-1)}_{fg}(\omega;\omega\ldots\omega)E^n_\omega$
n-photon absorption of n different frequencies	$[2^{n-1}\hbar]^{-1}\gamma^{(n-1)}_{fg}(\omega_1;\omega_2\ldots\omega_n)E_1\ldots E_n$
Hyper-Raman scattering, single pump frequency (n = 3)	$\beta_{fg}(-\omega_s;\omega_p\omega_p)E_p^2/8\hbar$

These results may be checked by comparing the susceptibility obtained from the steady-state solution of (2.101) (w' = -1) such as in (2.99), with a resonant $\chi^{(2n-1)}$ obtained from (2.15). The Stark shifts are given by a diagonal form of (2.103), i.e., (2.60). Equation (2.105), with (2.101), defines the n-photon Bloch equations.

The Rabi frequency for n-photon absorption of a single frequency, given above, has been found via an ab initio derivation of the Bloch equations by MILONNI and EBERLY [2.107], which, in their notation, is $2\lambda_{ab}^{(n)}$. Multiphoton Bloch equations have been discussed in detail by BUTYLKIN et al. [2.6], who relax the restriction that there be only one strong resonance, and by FRIEDMANN and WILSON-GORDON [2.108].

2.8 Symmetry

The tensor nature of the nonlinear susceptibilities shows immediately that the symmetry of the nonlinear medium will have an important influence on the polarisation, and hence on nonlinear processes. A full survey of symmetry properties involves irreducible-tensor decompositions, with respect to the point group of the atom or molecule, and then orientation averaging with respect to the rotation group. However, for free atoms and rotating molecules, the point group is itself the rotation group. This allows a complete description in terms of angular-momentum algebra; for example, (2.16) performed the microscopic statistical averaging over the degenerate initial state *and* the orientation average. Because of this, the symmetry theory of nonlinear susceptibilities for atoms and rotating molecules reduces to a number of special cases of the general theory; we are, therefore, not justified here to go into any detail of the formalism; instead, we will quote some formulae that are useful for the calculations indicated in the later sections. YURATICH [2.21] covered, in detail, the application of irreducible-tensor methods and the generalised Wigner-Eckart theorem to general nonlinear susceptibilities, and gives results for the finite point groups and the rotation group. The permutation group S_n on n objects and its relation to the group SU(3) is used in conjunction with spherical tensors to exploit the permutation-symmetry properties of the susceptibility tensors (Sect.2.2). In this way, selection rules, line strengths, the angular dependence on polarisation vectors and the summations over intermediate-state degeneracies can all be given in compact (and practical) forms.

As discussed in Sect.2.5, two-photon resonant third-order susceptibilities factorise into a form $\chi^{(3)} \sim \tilde{\chi} \equiv \sum_{\mathrm{deq}\ f,g} \alpha_{fg}(\omega_1;\omega_2)^* \alpha_{fg}(\omega_1';\omega_2')$. By writing the various polarisation vectors in the complex spherical basis [2.65,109,110],

$$\underline{e}_0 = \underline{e}_z \quad , \quad \underline{e}_{\pm 1} = \mp (\underline{e}_x \pm i\underline{e}_y)/\sqrt{2} \quad ,$$

where $\underline{e}_x$, $\underline{e}_y$, $\underline{e}_z$ are unit cartesian vectors, the product of polarisabilities becomes

$$\tilde{\chi} = \sum_K (2K+1)^{-1} \langle \gamma_f J_f || \hat{\alpha}^{(K)} || \gamma_g J_g \rangle^* \langle \gamma_f J_f || \hat{\alpha}^{(K)\prime} || \gamma_g J_g \rangle \Theta^{(K)} \quad . \tag{2.106a}$$

The label K denotes a spherical-tensor rank, and $\Theta^{(K)}$ is an angular factor

$$\Theta^{(K)} = (-1)^K (\underline{\varepsilon}_1^* \times \underline{\varepsilon}_2^*)^{(K)} \cdot (\underline{\varepsilon}_1' \times \underline{\varepsilon}_2')^{(K)} \quad . \tag{2.106b}$$

The notation in (2.106b) denotes a tensor coupling, which involves Clebsch-Gordan coefficients, and the reduced matrix elements are

$$\langle \gamma_f J_f || \hat{\alpha}^{(K)} || \gamma_g J_f \rangle$$

$$= (2K+1)^{\frac{1}{2}} (-1)^{J_f - J_g + K} \sum_{\gamma_i J_i} \begin{Bmatrix} J_f & K & J_g \\ 1 & J_i & 1 \end{Bmatrix} \langle \gamma_f J_f || \underline{Q} || \gamma_i J_i \rangle \langle \gamma_i J_i || \underline{Q} || \gamma_g J_g \rangle$$

$$\times \frac{1}{\hbar} \left(\frac{1}{\Omega_{ig} - \omega_1} + \frac{(-1)^K}{\Omega_{ig} - \omega_2} \right) \quad . \tag{2.106c}$$

The rank K is subject to the selection rule that (J_f, K, J_g) should obey the triangle rule. The K = 0, 1, 2 terms are known as the isotropic, antisymmetric, and anisotropic parts, respectively.

Equation (2.106) enables the two-photon resonant third-order susceptibility to be calculated with considerable generality for any atom or rotating molecule. Those familiar with angular-momentum techniques will have no difficulty in applying (2.106). As an example, take Raman scattering, in which $\underline{\varepsilon}_1 = \underline{\varepsilon}_1' = \underline{\varepsilon}_s^*$, $\underline{\varepsilon}_2 = \underline{\varepsilon}_2' = \underline{\varepsilon}_p$, and $\omega_1 = \omega_1' = -\omega_s$, $\omega_2 = \omega_2' = \omega_p$. Then

$$\tilde{\chi} = \sum_K \left| \sum_{\gamma_i J_i} \begin{Bmatrix} J_f & K & J_g \\ 1 & J_i & 1 \end{Bmatrix} \langle \gamma_f J_f || \underline{Q} || \gamma_i J_i \rangle \langle \gamma_i J_i || \underline{Q} || \gamma_g J_g \rangle \right.$$

$$\left. \times \frac{1}{\hbar} \left(\frac{1}{\Omega_{ig} - \omega_p} + \frac{(-1)^K}{\Omega_{ig} + \omega_s} \right) \right|^2 |(\underline{\varepsilon}_s^* \times \underline{\varepsilon}_p)^{(K)}|^2 \quad , \tag{2.107}$$

and similarly for two-photon absorption. This result has been used to discuss, for example, spin-orbit effects in resonantly enhanced SRS, by YURATICH [2.21], YURATICH and HANNA [2.111], and COTTER and HANNA [2.112];

see Sect.5.3. Forms of (2.107) have also been treated by, for example, KONINGSTEIN and SONNICH MORTENSEN [2.113], IL'INSKII and TARANUKHIN [2.114], and GRYNBERG et al. [2.115].

It is useful to note that $(\underline{\varepsilon}_1 \times \underline{\varepsilon}_2)^{(K)}$ equals $-\underline{\varepsilon}_1 \cdot \underline{\varepsilon}_2/\sqrt{3}$ for K = 0, and $i\underline{\varepsilon}_1 \times \underline{\varepsilon}_2/\sqrt{2}$ for K = 1, where the usual dot and cross products are implied. Furthermore, for alkali atoms (states $|nls = \frac{1}{2}\, J\rangle$) the various dipole-matrix elements that arise are

$$\langle\gamma J||\underline{Q}||\gamma' J'\rangle = [(2J+1)(2J'+1)l_>]^{\frac{1}{2}}(-1)^{l_> + J' + 3/2} \begin{Bmatrix} J & J' & 1 \\ l' & l & 1/2 \end{Bmatrix}$$

$$\times \langle nl| - er|n'l'\rangle \quad , \qquad (2.108)$$

where $l_>$ is the greater of l and l'. The radial matrix element $\langle nl|r|n'l'\rangle$ is denoted $\langle nl||r||n'l'\rangle$ by EICHER [2.116], and is related to the z-component matrix element [1.15] by

$$\langle nl|r|n'l'\rangle = (4l_>^2 - 1)^{\frac{1}{2}} l_>^{-1} \langle nlm_l = 0|z|n'l'm_{l'} = 0\rangle \quad . \qquad (2.109)$$

The last two references give a considerable number of matrix elements for the alkalis; more accurate absolute values can often be obtained from alternative oscillator-strength data. The required relation, not restricted to the alkalis, is

$$\langle\gamma J||\underline{Q}||\gamma' J'\rangle^2 = \frac{3\hbar e^2}{2m}(2J+1)\, f_{\gamma J;\gamma' J'}\, \Omega_{\gamma' J';\gamma J} \quad . \qquad (2.110)$$

If there is only one important intermediate state, $\omega_p \simeq \Omega_{ig}$, then (2.107,110) give

$$(2J_g + 1)^{-1} \tilde{\chi} = \left(\frac{3e^2}{2m}\right)^2 \frac{f_{fi} f_{gi}}{\Omega_{if}\Omega_{ig}} \frac{1}{(\Omega_{ig} - \omega_p)^2} (2J_f + 1) \sum_K \begin{Bmatrix} J_f & K & J_g \\ 1 & J_i & 1 \end{Bmatrix}^2 \Theta_R^{(K)} \quad . \qquad (2.111)$$

The angular factor $\Theta_R^{(K)} = |(\underline{\varepsilon}_s^* \times \underline{\varepsilon}_p)^{(K)}|^2$ has been tabulated in YURATICH and HANNA [2.111], for linearly polarised light, with an angle β between $\underline{\varepsilon}_s$ and $\underline{\varepsilon}_p$, it is found to be $\Theta_R^{(0)} = (1/3)\cos^2\beta$, $\Theta_R^{(1)} = (1/2)\sin^2\beta$, $\Theta_R^{(2)} = (3 + \cos^2\beta)/6$. An example of the use of (2.111) is in (5.6b), where it is specialised to the case of an atom with a singlet ground state, i.e., $J_g = 0$, so that the triangle rule on K permits only $K = J_f$.

3. Propagation of Plane Waves in a Nonlinear Medium

3.1 The Wave Equation

We now turn to the problem of coupling the nonlinear polarisation to Maxwell's equations for the fields. In order to show the physical ideas most clearly, we will consider only plane waves in any detail; a more realistic treatment, involving transversely varying fields, such as gaussian beams, is considerably more complicated—some aspects of this are examined in Chaps.4, 5.

Given these restrictions, the steps that are necessary in order to obtain expressions for the fields are quite straightforward; for accounts (in varying detail) of the derivation, the reader can refer to a number of sources, such as ARMSTRONG et al. [1.69], BLOEMBERGEN [2.15], BUTCHER [1.70], DUCUING [1.71], and MAKER and TERHUNE [2.23]. Thus, in what follows, after first establishing our notation, we shall proceed quickly to the most important results.

In a dielectric nonmagnetic medium, Maxwell's equations lead to the wave equation

$$\nabla \times \nabla \times \underline{E} = -\mu_0 \frac{\partial^2 \underline{D}}{\partial t^2} \quad , \tag{3.1}$$

where the displacement vector $\underline{D}$ may be written

$$\underline{D} = \varepsilon_0 \underline{E} + \underline{P}^{L} + \underline{P}^{NL} \quad . \tag{3.2}$$

Here $\underline{P}^{L}$ and $\underline{P}^{NL}$ are, respectively, the polarisations that are linear and nonlinear in $\underline{E}$. By use of the definitions (2.1-3), (3.1) may be Fourier transformed into the frequency domain, so that

$$\nabla \times \nabla \times \underline{E}(\underline{r}\omega) - \frac{\omega^2}{c^2}\left(\underline{1} + \underline{\chi}^{(1)}(-\omega;\omega)\right) \cdot \underline{E}(\underline{r}\omega) = \omega^2\mu_0 \sum_{n=2}^{\infty} \underline{P}^{(n)}(\underline{r}\omega)$$
$$\equiv \omega^2\mu_0 \underline{P}^{NL}(\underline{r}\omega) \quad , \tag{3.3}$$

where $\underline{1}$ is the unit dyadic, $\underline{1} = \underline{\varepsilon}\underline{\varepsilon}^*$, and $\underline{E}(\underline{r}\omega) = \underline{\varepsilon}E(\underline{r}\omega)$. In (3.3), $\underline{P}^L$ has been incorporated into the LHS as the term that involves the linear susceptibility [n = 1 in (2.2)]. On the other hand, the nonlinear-polarisation contributions on the RHS of (3.3), are each in the form of integrals that contain powers of the field transform $\underline{E}(\underline{r}\omega)$, as can be seen from (2.2). Thus, in this general form, (3.3) allows arbitrary time-dependence of the field $\underline{E}$. For monochromatic waves, however, the simpler expression (2.7), can be used directly, where $\underline{E}(\underline{r}\omega) \to \underline{E}_\omega(\underline{r})$, etc., in (3.3).

To simplify the spatial analysis, we consider $\underline{E}(\underline{r}t)$ to be a plane wave that propagates in the z direction, so that $\underline{E}(\underline{r}\omega)$ is characterised by a wave vector $\underline{k}_\omega = \pm k_\omega \hat{\underline{z}}$ directed along the z axis (the + sign refers to forward propagation). Because we are concerned with isotropic media, $\underline{E}(\underline{r}\omega)\cdot\underline{k}_\omega = 0$; therefore, $\nabla \times \nabla \times \underline{E}(r\omega)$ reduces to $-\partial^2\underline{E}(z\omega)/\partial z^2$ (the field depends only on the z coordinate). Using this fact, taking the scalar product with $\underline{\varepsilon}^*$, and using the scalar form (2.9a), we reduce (3.3) to the inhomogeneous Helmholtz equation

$$\frac{\partial^2}{\partial z^2} E(z\omega) + k_\omega^2 E(z\omega) = -\mu_0\omega^2 P^{NL}(z\omega) \quad , \tag{3.4}$$

where $\underline{k}_\omega$ has magnitude

$$k_\omega = \frac{\omega}{c}\left(1 + \chi^{(1)}(-\omega;\omega)\right)^{\frac{1}{2}} \equiv \eta_\omega\omega/c \quad , \tag{3.5}$$

and η_ω is the refractive index at frequency ω. The electric field is now taken in the form of a superposition of running waves,

$$E(z\omega) = \tilde{E}(z\omega)e^{\pm ik_\omega z} \quad , \tag{3.6}$$

where $\tilde{E}(z\omega)$ is a slowly varying, possibly complex, envelope function. By slowly varying is meant that the condition

$$\left|\frac{1}{\tilde{E}(z\omega)} \frac{\partial\tilde{E}(z\omega)}{\partial z}\right| << k_\omega \tag{3.7}$$

is satisfied, so that the second derivative $\partial^2\tilde{E}(z\omega)/\partial z^2$ may be dropped when (3.6) is inserted into (3.4). (In words, the growth of the envelope in a distance on the order of a wavelength should be small.) Therefore, for forward-propagating waves [+ sign in (3.6)],(3.4) becomes

$$\frac{\partial E(z\omega)}{\partial z} = \frac{i\omega}{2\varepsilon_0 c\eta_\omega} P^{NL}(z\omega)e^{-ik_\omega z} \quad , \tag{3.8a}$$

where, to simplify the notation, we have dropped the tilde on $E(z\omega)$ in (3.8a). In the rest of this volume, the symbol $E(z\omega)$ will be taken to mean the slowly varying envelope defined in (3.6). [For backward-wave propagation, change the sign of the RHS of (3.8a), and of the exponent.]

Equation (3.8a) is the starting point for most plane-wave calculations of the growth of waves as they propagate through a nonlinear medium. In the next few sections, we shall consider examples of its use in problems of increasing complexity. Before doing so, however, a few more comments are appropriate.

Equation (3.8a) is one of an infinite set of coupled equations for the Fourier components $E(z\omega)$. Part of the art of solving these equations is to restrict consideration to the coupling of a few waves only, often those that are most important by virtue of the particular relation of their frequencies to energy levels in the medium, as for example in two-photon resonance.

Equation (3.8a) is almost invariably restricted to the treatment of monochromatic waves, through the changes $E(z\omega)$, $P(z\omega) \rightarrow E_\omega(z)$, $P_\omega(z)$ and the use of (2.9b) for the polarisation. This is not necessary because, with (2.2), it provides an integro-differential formulation of time-dependent processes. [When (2.2) or (2.9) are used in (3.8a) the transformation (3.6) must be borne in mind.] However, it is usually the case that time-dependent problems are formulated by writing the field $E(\underline{r}t)$ as (in the transverse plane-wave limit) $(1/2)\tilde{E}_\omega(zt)\exp[i(k_\omega z - \omega t)]$ + c.c. . Here $\tilde{E}_\omega(zt)$ is a field envelope function, which is slowly varying in both the space and time variables. The propagating part $\exp[i(k_\omega z - \omega t)]$ is made up of a characteristic frequency ω (for example the centre frequency of the wave as it enters the nonlinear medium) and its associated wave vector k_ω, defined in (3.5). Inserting this form for the field into (3.1) leads to an equation for $\tilde{E}_\omega(zt)$ analogous to (3.8a),

$$\frac{\partial E_\omega(zt)}{\partial z} + \frac{1}{\upsilon}\frac{\partial E}{\partial t}(zt) = \frac{i\omega}{2\varepsilon_0 c\eta_\omega} P_\omega^{NL}(zt)e^{-ik_\omega z} \quad , \tag{3.8b}$$

where the tildes have again been omitted and $\upsilon = \omega/k_\omega = \eta_\omega c$ is a characteristic velocity. The polarisation $P_\omega^{NL}(zt)$ may be constructed from (2.2) but is usually obtained more directly—for example from the vector-models (Sects.2.6,7), which, with (3.8b), are known as Maxwell-Bloch equations.

Also, because there are often terms of importance that lead to intensity-dependent refractive-index changes (Sect.2.6) the separation of the polarisation into P^L and P^{NL} is not always useful. In this case, P^{NL} may be replaced by the total polarisation P in each of (3.8), and k_ω redefined as the vacuum wave vector ω/c.

So far, we have treated the field in two ways; in (3.8a), the time variable was transformed into the frequency domain, but the space variable was only partially transformed, by use of (3.6). In (3.8b), both variables were treated similarly by use of the slowly varying envelope approximation. A third procedure is to Fourier transform the field in both the time and space variables. Thus, if $E(zt) = \int E(k\omega) \exp[i(kz - \omega t)]d\omega dk$, then (3.4) leads to

$$(k^2 - k_\omega^2)E(k\omega) = (\omega^2/\varepsilon_0 c^2)P^{NL}(k\omega) \quad . \tag{3.8c}$$

It is readily seen that $E(k\omega)$ has maxima when $k \simeq \pm k_\omega$. When $\omega > 0$, the forward-propagating part of the field is associated with the region $k \simeq + k_\omega$; whence

$$(k - k_\omega)E(k\omega)^{\text{forward}} \simeq (\omega/2\varepsilon_0 c n_\omega)P^{NL}(k\omega) \quad , \tag{3.8d}$$

and similarly for the backward-propagating part; equations (3.8a,b,d) all describe forward-propagating waves. Equations (3.8c,d) give $E(zt)$ directly, as an integral over k and ω. This method, extended to transversely varying beams, has been used by WARD and NEW [1.12] and BJORKLUND [3.1] to study the effects of focussing on parametric processes (Chaps.4, 6). However, they restricted the frequency dependence to that of monochromatic waves.

It should be noted that (3.8c) is exact in both space and time within the plane-wave restriction, whereas (3.8a) approximates the spatial dependence of the solution and (3.8b) approximates the solution with respect to both variables. In compensation for this, of the three approaches, (3.8b) may be expected to handle complicated nonlinear polarisations most easily.

3.2 Parametric Processes

We begin with third-harmonic generation, for which (3.8a) becomes

$$\frac{\partial E_3}{\partial z}(z) = \frac{i3\omega}{2\varepsilon_0 c n_3}\left[\frac{1}{4}\,\varepsilon_0 \chi^{(3)}(-3\omega;\omega\omega\omega)E_1(z)^3\right]e^{i\Delta kz} \quad , \tag{3.9a}$$

where the *phase mismatch* Δk is

$$\Delta k = 3k_1 - k_3 = \frac{3\omega}{c}(n_1 - n_3) \qquad (3.9b)^1$$

and we have used (2.9b), for monochromatic waves. Equation (3.9a) should be coupled with a similar equation for the fundamental field. For this first example, and indeed often in practice, it suffices to consider the fundamental to be undepleted over the interaction length, i.e., $E_1(z) \simeq E_1(0) = E_{1o}$. Then the pair of coupled-wave equations reduce to (3.9a) alone. The phase mismatch Δk is a measure of the difference between the phase velocities of the induced polarisation wave and the radiated third-harmonic electric field. Therefore, after a distance $|\pi/\Delta k|$, the phases of the polarisation and the radiated electric field will be out of step by 180°. This distance is known as the coherence length L_c,

$$L_c = |\pi/\Delta k| \quad . \qquad (3.10)$$

The solution of (3.9a), when an undepleted fundamental is assumed, and for the initial condition $E_3(0) = 0$, is readily found, and gives a third-harmonic intensity after a length L of medium,

$$I_3(L) = \frac{(3\omega)^2}{16\varepsilon_0^2 c^4 n_1^3 n_3}|\chi^{(3)}(-3\omega;\omega\omega\omega)|^2 I_{1o}^3 L^2 \operatorname{sinc}^2\left(\frac{\Delta kL}{2}\right) , \qquad (3.11)$$

where $\operatorname{sinc}(x) \equiv \sin(x)/x$, so that $\operatorname{sinc}(0) = 1$. The argument $\Delta kL/2$ of this function in (3.11) may be written as $\frac{1}{2}\pi L/L_c$, showing that the intensity reaches a maximum when L equals odd multiples of the coherence length L_c. Clearly, the maximum rate of conversion of fundamental to harmonic intensity is achieved when exact phase matching occurs: $\Delta k = 0$; as will be seen in Chaps.4, 6, achievement of this condition is one of the most important experimental requirements. The harmonic intensity then increases as L^2. With focussed beams, it turns out (Chap.4) that $\Delta k = 0$ is not the optimum condition; nevertheless, there is an optimum Δk to be sought, which maximises the generated harmonic power.

A convenient classification of processes can be based on whether they depend on a phase mismatch Δk. If they do (such as THG) then they are re-

---

[1]This definition is the negative of that used in Chap.4.

ferred to as "parametric"; those that are independent, of a Δk, are referred to as "nonparametric" (e.g., Raman scattering). Historically, the term parametric arose in connection with purely reactive (i.e., lossless) systems, in which case a precise definition of the term parametric can be given.

This can be illustrated with THG, where in the absence of any resonances, the susceptibility may be taken to be real. Then, from (2.57), it follows that the irreversible energy flow W from the fundamental and harmonic fields to the medium vanishes,

$$
\begin{aligned}
W &= \frac{1}{2}\,\mathrm{Im}\Big[\omega\Big(\frac{3\varepsilon_0}{4}\Big)\chi^{(3)}(-\omega;3\omega-\omega-\omega)E_1^{*3}E_3e^{-i\Delta kz} \\
&\quad + 3\omega\Big(\frac{\varepsilon_0}{4}\Big)\chi^{(3)}(-3\omega;\omega\omega\omega)E_1^3E_3^*e^{i\Delta kz}\Big] \\
&= \frac{3\varepsilon_0\omega}{8}\chi^{(3)}(-3\omega;\omega\omega\omega)\,\mathrm{Im}\Big(E_1^{*3}E_3e^{-i\Delta kz} + \mathrm{c.c.}\Big) = 0 \quad . \qquad (3.12)
\end{aligned}
$$

[In the first line, the first 3 is due to the K factor, the second is due to the harmonic field frequency 3ω; in the second line, overall permutation symmetry (Sect.2.2) and the reality condition (2.2b) have been used.] Thus, the medium acts only as a catalyst to the process, mediating the flow of energy between the fundamental and harmonic fields, a feature that is characteristic of parametric processes. A second characteristic feature is that the field of the generated radiation is in a different radiation mode from those of the driving waves (i.e., it is not a stimulated-emission process). A third feature to note is that the generated intensity depends on $|\chi^{(3)}(-3\omega;\omega\omega\omega)|^2 = (\chi')^2 + (\chi'')^2$, so that both the real (χ') and the imaginary (χ'') parts of the susceptibility contribute in the same manner. If there is an imaginary part, then, from (3.12), it is seen that the energy flow to the medium is no longer zero.

By contrast, for nonparametric processes, the real and imaginary parts of the susceptibility can be given separate physical meanings. This can be illustrated with the case of stimulated Raman scattering. As will be seen in Sect.3.3, the imaginary part of the Raman susceptibility leads to Stokes gain, and the real part leads to the optical Kerr effect. It is also shown there that the phase mismatch Δk is automatically zero.[2]

---

[2]The treatment of Raman scattering in solids requires some comment, because the Raman transition corresponds to a nonlocalised excitation (phonon). Treating the phonon as a wave, on the same footing as the electric fields, leads to a phase-matching condition that involves the phonon wave vector,

Under resonant conditions, such as the two-photon resonant processes discussed in Sect.3.4, the situation is rather more complicated. First, [see (3.26)] the polarisations (and hence the growth of the various fields) are each governed by two terms; one of those terms depends on a Δk; it may therefore, be classed as parametric; the other term is independent of Δk (and is therefore nonparametric). However, the distinction is less useful because it suggests a separability, in origin, of the two parts. Equation (3.27) shows that the two parts are in fact inseparable, which, for reasons touched on in Sect.2.7, is a feature of resonant processes. A further consequence of the resonance is that the real and imaginary parts of the susceptibilities no longer have simple meanings.

It is not practicable to present here intensity formulae for all of the parametric processes that are likely to be encountered. It is, however, useful to have an expression in terms of the polarisation. Under the conditions that none of the input fields are depleted, and that only one nonlinear process contributes to the polarisation at frequency ω_σ, we can write

$$P^{NL}(z\omega_\sigma)e^{-ik_{\omega_\sigma}z} = P^{(n)}(\omega_\sigma)e^{i\Delta kz} \tag{3.13}$$

where Δk is the phase mismatch for the process, so that from (3.8a)

$$I(L\omega_\sigma) = \frac{\omega_\sigma^2}{8\varepsilon_0 c n_{\omega_\sigma}} |P^{(n)}(\omega_\sigma)|^2 L^2 \operatorname{sinc}^2(\Delta kL/2) \quad . \tag{3.14}$$

For monochromatic waves this leads, via (2.9b), to

$$I_\sigma(L) = \frac{K^2}{4}\left(\frac{2}{\varepsilon_0 c}\right)^{n-1}\left(\frac{\omega_\sigma}{c}\right)^2 |\chi^{(n)}(-\omega_\sigma;\omega_1\ldots\omega_n)|^2 \frac{I_{1o}\ldots I_{no}}{n_\sigma n_1\ldots n_n} L^2 \operatorname{sinc}^2(\Delta kL/2) \quad , \tag{3.15a}$$

where

$$\Delta k = k_1 + \ldots + k_n - k_\sigma \quad . \tag{3.15b}$$

---

in which case the process can be thought of as parametric. Alternatively, we can consider the electric fields alone, as in the foregoing, in which case the process is thought of as nonparametric. However, the nonlinear susceptibility then contains the phase-matching condition as a selection rule for the excitation of the Raman transition.

In the general case, (3.15b) must be rewritten as a vector relationship.

To summarise, these expressions apply to processes in which all of the interacting waves (except the generated one) are provided as input and are undepleted. It is, therefore, possible that those expressions will break down as L increases, if the generated intensity becomes sufficiently large; it is then necessary to perform the analysis by use of coupled equations. However, there are also parametric processes in which more than one of the waves starts from noise, i.e., they are of comparable intensity at the input; then a coupled-wave analysis becomes essential. A well-known example of this for three-wave processes (involving a $\chi^{(2)}$ nonlinearity) is embodied in the optical parametric oscillator [3.2]. Here an input wave ω_p produces outputs at signal and idler frequencies ω_s, ω_i, respectively, where $\omega_p = \omega_i + \omega_s$ and where the signal and idler waves are generated from noise. Analogous processes occur in four-wave interactions via the $\chi^{(3)}$ nonlinearity. For example, two input pump waves ω_{p1}, ω_{p2} generate ω_s and ω_i from noise, such that $\omega_{p1} + \omega_{p2} = \omega_s + \omega_i$ (e.g., Fig.2.2c). Analysis of such processes follows the same lines as for the three-wave case, in which an effective starting-noise intensity is assumed and coupled-wave equations are used. This will be discussed in Sect.3.4. The question of how to calculate the effective starting noise will be left to Sect.3.5.

To conclude this section, we consider briefly the photon flow in third-harmonic generation. Again, we take the susceptibility to be real; then, from (3.9), following steps similar to those in (3.12), we find

$$\left(\frac{n_3}{3\omega}\right)\frac{d|E_3|^2}{dz} + \frac{1}{3}\left(\frac{n_1}{\omega}\right)\frac{d|E_1|^2}{dz} = 0$$

i.e.,

$$I_3(z) + I_1(z) = \text{constant} = I_{3o} + I_{1o} \quad . \tag{3.16}$$

This is an example of the Manley-Rowe relations for parametric amplification, and expresses the fact that in the lossless medium the total field energy is conserved. As long as $I_3(z) \ll I_{1o}$, the uncoupled solution (3.11) does not violate (3.16) significantly; this, therefore, provides a criterion for the validity of (3.11).

3.3 Nonparametric Processes

We now examine stimulated Raman scattering, an important example of a nonparametric process. Equations (3.8a) and (2.10) yield

$$\frac{\partial E_s(z)}{\partial z} = \frac{3i\omega_s}{4cn_s}\chi_R|E_p(z)|^2E_s(z) \quad , \tag{3.17a}$$

where χ_R is the Raman susceptibility, $\chi_R = \chi^{(3)}(-\omega_s;\omega_p - \omega_p\omega_s)$.

Notice that the phase mismatch $\Delta k = k_p - k_p + k_s - k_s$ vanishes identically, consistent with its classification as a nonparametric process. This implies that the relation between the phases of the Stokes polarisation and the Stokes field depends only on the argument of the (complex) Raman susceptibility.

In the small-signal region, we can put $E_p(z) = E_p(0) = E_{po}$ in (3.17), which then gives the Stokes field after a distance L,

$$E_s(L) = E_{so}\exp\left(\frac{3i\omega_s}{2\varepsilon_o c^2 n_s n_p}\chi_R I_{po}L\right) \quad . \tag{3.17b}$$

Thus the Stokes intensity becomes

$$I_s(L) = I_{so}\exp(G_R L) = I_{so}\exp(g_R I_{po}L) \quad , \tag{3.18}$$

where the Raman power-gain coefficient G_R is given by

$$G_R = \frac{-3\omega_s}{\varepsilon_o c^2 n_s n_p}\chi''_R I_{po} \quad , \tag{3.19}$$

and $g_R\,[\mathrm{mW^{-1}}]$ is the Raman power gain per unit pump intensity. Using the susceptibility expression for a two-photon resonance, from (2.19) and Table 2.3, we find that when $\Omega_{fg} + \omega_s - \omega_p = 0$, the susceptibility is purely imaginary and

$$G_R = \frac{\omega_s}{2\varepsilon_o^2 c^2 \hbar n_s n_p}\overline{|\alpha_{fg}(-\omega_s;\omega_p)|^2}\left(\frac{NI_{po}}{\Gamma}\right) \quad . \tag{3.20}$$

N is the number density of atoms associated with the transition, and the bar denotes an orientation average and summation over final-state degeneracy (see Sect.2.5); the formulae of Sect.2.8 may be substituted directly into (3.20).

Notice that (3.18) requires some input intensity I_{so} at the Stokes frequency. This could be achieved by deliberately injecting radiation at the Stokes frequency. In practice, however, the approach usually adopted is to rely on the weak noise intensity at the Stokes frequency produced by spontaneous Raman scattering at the input end of the medium. The high gain that is available in typical experimental conditions will amplify this noise to a readily detectable level, or indeed to a level at which the pump becomes depleted by the stimulated process. In the small-signal-gain region, i.e., before this saturation region is reached, the generated Stokes intensity I_s shows a very rapid dependence on incident pump intensity by virtue of the exponential gain factor in (3.18). The process is, therefore, described as showing a threshold, i.e., until the pump intensity reaches a particular value there will be little Stokes intensity. The definition of this threshold intensity is of course arbitrary, because it depends on the detector sensitivity. However, one simple definition, which is particularly useful in theoretical work, is to take the threshold pump intensity as that required to give a small-signal Stokes gain that would amplify the Stokes noise I_{so} (in the absence of saturation) to a level that implies total conversion of the pump wave, i.e., to $I_s(L) = (\omega_s/\omega_p)I_{po}$ (see opposite). From (3.18), this gives an implicit relation for I_{po}

$$(G_R L)_{th} = (g_R I_{po} L)_{th} = \ln(\omega_s I_{po}/\omega_p I_{so}) \quad . \tag{3.21}$$

Estimates of I_{so} are given in Sect.3.5. Taking $I_{so} \sim 1\,\mathrm{Wm}^{-2}$ and $I_{po} \sim 10^{12}\,\mathrm{Wm}^{-2}$ as typical values, we find $(G_R L)_{th} \sim 30$. As an alternative definition of threshold we could take the pump intensity required to produce a detectable Stokes signal. For typical experimental situations, this would give only a slightly different value of $(G_R L)_{th}$; in the sequel we shall use $(G_R L)_{th} = 30$ as our criterion. Thus, despite the widely differing Stokes intensities implied by these two definitions of threshold, the effect of the logarithm in (3.21) is to give comparable values for the gain exponent $(G_R L)_{th}$.

In passing from (3.17b) to (3.18), we made no mention of the effect of the real part of the Raman susceptibility, χ_R'. However, it is readily seen from (3.17b) that it causes a change of refractive index for the Stokes field,

$$\delta n_s = 3\chi_R' I_{po}/2\varepsilon_0 c n_s n_p$$

(the optical Kerr effect), which should be incorporated into the definition of Stokes intensity in (3.18), i.e., $I_s = \varepsilon_0 c(n_s + \delta n_s)|E_s|^2/2$. Now, as just discussed, we have in mind growth from noise, when it can be assumed that the frequency of the observed Stokes photons will be that for which the Raman gain is largest. This condition is satisfied when exact Raman resonance occurs, $\Omega_{fg} + \omega_s - \omega_p = 0$, and as noted above χ_R' is then zero, so that $\delta n_s = 0$; the intensity-dependent refractive-index contribution thus vanishes; we do not consider it again in Chap.3. (Similar changes of refractive index occur for all nonparametric processes.) This argument needs some qualification, however, because as will be pointed out in Chap.5, focussing can, in principle, result in maximum gain when the Raman resonance occurs, $\Omega_{fg} + \omega_s - \omega_p = 0$, and as noted above χ_R' is then zero, so dependent analyses, the frequency spread of the pulses means that χ_R' must be taken into account, when it contributes to pulse reshaping and chirping.

From (3.17a) and the analogous equation for $E_p(z)$, we obtain the relation that corresponds to (3.16),

$$\frac{I_p(z)}{\omega_p} + \frac{I_s(z)}{\omega_s} = \frac{I_{po}}{\omega_p} + \frac{I_{so}}{\omega_s} \simeq \frac{I_{po}}{\omega_p} \quad , \tag{3.22}$$

where the approximation will be valid for generation from noise, $I_{po} \gg I_{so}$. This equation expresses photon conservation; that is, when a pump photon is annihilated, a Stokes photon is created and vice versa. Note, however, that it does not imply that the energy of the field is conserved because, when the pump is fully depleted, we have $I_s(\infty) = (\omega_s/\omega_p)I_{po} + I_{so}$, giving a net loss of field energy that is proportional to $(1-\omega_s/\omega_p)I_{po}$. Indeed from (2.57) the power flow from the Stokes field into the medium, in the small-signal region, is

$$W \simeq \frac{3}{4}\varepsilon_0\omega_s\,|E_s|^2|E_p|^2\chi_R'' \quad . \tag{3.23a}$$

This implies a volume rate of Stokes photon generation of

$$\frac{1}{\tau} = \frac{-W}{\hbar\omega_s} = \frac{G_R I_s}{\hbar\omega_s} \quad , \tag{3.23b}$$

where (3.19) has been used. This is a well-known result [2.19] and may also be obtained by setting $\tau^{-1} = (\hbar\omega_s)^{-1}(dI_s/dz)$, where $dI_s/dz = G_R I_s$, which follows from (3.17) or (3.18). The latter method indicates the resemblances between approximations that lead to the slowly varying amplitude equation (3.8a), and the transition-rate approach in the time domain.

From (3.17,22) it is a straightforward matter to solve the coupled wave equations for SRS (see e.g., [2.19], who includes noise by means of a quantised-field treatment); we quote the result for later reference

$$I_s(z) = I_{so} e^{G_R z} \Big/ \left(1 + \frac{\omega_p}{\omega_s} \frac{I_{so}}{I_{po}} e^{G_R z}\right) , \tag{3.24}$$

where the limit $I_{po} >> I_{so}$ has been assumed, appropriate to build-up of $I_s(z)$ from noise. This expression reduces to the small-signal formula (3.18) when $z \to 0$ and gives the saturated intensity $(\omega_s/\omega_p)I_{po}$ when $z \to \infty$. There is marked transition between these two limits when $G_R z$ is varied; the transition point is accurately given by (3.21); in fact, this threshold is rigorously the point at which $I_s(z)$ decreases, because of saturation, to 50% of the value predicted by the small-signal formula (3.18).

Another important example of a nonparametric process is provided by two-photon absorption. The close formal similarity between SRS and TPA means that, with a few simple changes, the SRS formulae may be converted into the corresponding ones for TPA. Thus, if we wish to consider the absorption of wave ω_1 in the presence of a second wave ω_2 (the latter assumed to be the more powerful, so that its depletion may be neglected), then the absorption coefficient K_{TPA} for ω_1 is equal to G_R in (3.19,20), provided that we make the changes - $\omega_s \to \omega_1$, $\omega_p \to \omega_2$ and change the sign of Γ. Thus $dI_1/dz = -K_{TPA}I_1$. The different sign of the $i\Gamma$ term (see Table 2.2) leads to absorption, instead of amplification, as in SRS. By analogy with the G_R and g_R of SRS it is also useful to define k_{TPA} such that $K_{TPA} = k_{TPA}I_{2o}$. If the absorption is from a single beam ($\omega_1 = \omega_2 = \omega$) then $dI(z)/dz = -kI^2/2$, where k is given by k_{TPA} with ω_1 and ω_2 but equal to ω; the 1/2 follows from the new K factor (Table 2.1). The exponential decay is altered, because the absorption coefficient $kI(z)$ decreases with increasing z. The result is $I = I_o/(1 + kI_oz/2)$ which is much weaker than exponential decay when $kI_oz >> 1$.

Finally, as an example of a fifth-order nonparametric process, we consider stimulated hyper-Raman scattering. The gain coefficient is found by a calculation analogous to that which led to (3.18,19) and is given by

$$G_{HR} = \frac{-15\omega_s}{2\varepsilon_o^2 c^3 n_s n_p^2} \operatorname{Im}\chi^{(5)}(-\omega_s;\omega_p\omega_p - \omega_p - \omega_p\omega_s)I_{po}^2$$

$$= \frac{\omega_s}{16\varepsilon_0^3 c^3 \hbar n_s n_p^2} \overline{|\beta_{fg}(-\omega_s;\omega_p\omega_p)|^2} \left(\frac{NI_{po}^2}{\Gamma}\right) , \tag{3.25}$$

where a single pump is used, and $\Omega_{fg} + \omega_s - 2\omega_p = 0$. For two pumps, it is only necessary to change $I_{po}^2 \rightarrow I_{p_1o} I_{p_2o}$, etc., throughout (3.25), and multiply by 4. The three-photon absorption coefficient follows from G_{HR}, in the same way as K_{TPA} was obtained from G_R.

3.4 Coupled-Wave Equations

We have so far considered some simple examples of processes that could be categorised as parametric (e.g., THG) or nonparametric (e.g., SRS); by neglecting depletion of the incident pump wave, it was possible to describe the growth of the generated radiation by a single equation. However, for a number of processes, the situation is much more complicated than this. For example, the description of parametric amplification involves a coupled-wave analysis, because the signal wave produces a polarisation at the idler frequency and vice versa.

Other even more complicated problems arise when, in addition to the coupled-wave analysis made necessary by the growth of two or more waves from noise, more than one nonlinear process has to be considered. This is typified by so-called "biharmonic pumping" [3.3], or "coherent Raman mixing", as it has also been named [1.62]. Loosely speaking, this can be described as a process in which stimulated Raman scattering and four-wave mixing occur simultaneously, i.e., the coupled behaviour of a parametric and non-parametric process. In fact, whenever a two-photon resonant nonlinear process is involved, this mixture of parametric and nonparametric needs to be considered. The various possible two-photon resonant processes are shown in Fig.2.2. Rather than attempt a general treatment that deals simultaneously with all of these, we shall for the most part illustrate the ideas with the particular example of biharmonic pumping. With a few obvious changes, the analysis can be applied to other processes. In Sect.3.4.1, we give a brief description of the nature of the biharmonic-pumping scheme and why it is of interest, leaving the details of experimental results obtained in H_2 gas till Sect.7.6. The coupled-wave analysis in Sect.3.4.1 is restricted to the small-signal case. Because a potential use of biharmonic pumping is in the efficient generation of tunable infrared radiation, we also examine some aspects of the large-signal behaviour in Sect.3.4.2.

3.4.1 Biharmonic Pumping — Small Signal

The biharmonic-pumping process is illustrated in Fig.3.1. The input frequencies are the pumps ω_1 and ω_2. The output frequencies are the Stokes frequencies ω_{1s} and ω_{2s}, associated with these pumps. Of particular interest is the situation where the available intensity of the ω_2 pump is sufficient (with no other input) to exceed the threshold for SRS generation of ω_{2s}, but the intensity of the ω_1 pump is insufficient (without the ω_2 input) for SRS generation of ω_{1s}. However, if both ω_2 and ω_1 are input, then ω_{1s} is generated by four-wave mixing, $\omega_{1s} = \omega_1 - \omega_2 + \omega_{2s} = \omega_1 - \Omega_{fg}$. The wave ω_{1s} then acts as a substantial input for Raman amplification by the pump ω_1.

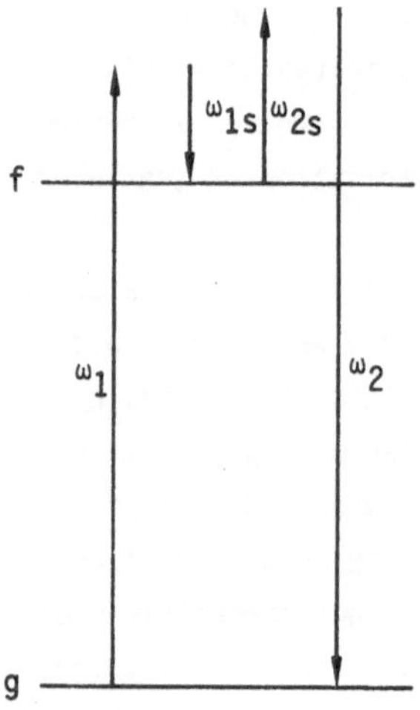

Fig.3.1. Biharmonic pumping (cf. Table 2.2a). Pump fields at frequencies ω_1 and ω_2 are input, generating ω_{1s}, ω_{2s}, respectively, by SRS, simultaneously with the parametric process $\omega_{1s} = \omega_1 - (\omega_2 - \omega_{2s})$

Another aspect of this process is revealed by considering the response of a single atom to the applied fields (Sect.2.5,7). Thus, the SRS produced by the pump ω_2 sets up a polarisability in the medium, oscillating at frequency Ω_{fg}. The interaction of the pump ω_1 with the polarisability produces a polarisation at the difference frequency $\omega_1 - \Omega_{fg}$, which then radiates. This point of view emphasises the fact that even if ω_1 enters the medium after ω_2 has left it, ω_{1s} can be still generated, provided the induced polarisability has not had time to decay. By monitoring the intensity of the generated ω_{1s} as a function of delay between ω_2 and ω_1 pump pulses, it is possible to measure the dephasing time of the polarisability, with time resolution in the picosecond region when mode-locked pulses are used. This technique has been explored in depth by KAISER et al. (see, e.g., [2.101]). The analysis of this time-dependent process is normally based on (3.8b) and the material-excitation theory sketched in Sect.2.7.

A more recent interest in biharmonic pumping has been motivated by the possibility of generating stimulated Stokes radiation ω_{1s} at infrared wavelengths. The reduction of Raman gain with increasing Stokes wavelength means that it may become difficult, if not impossible, to reach threshold, either because the available pump power at ω_1 is insufficient or because the gain for ω_{1s} is insufficient even at intensities of ω_1 that produce material breakdown. However, if SRS threshold can be reached with a different pump ω_2 (for example where ω_{2s} is a higher frequency than ω_{1s} and hence has higher gain) then the simultaneous use of ω_1 and ω_2 can generate a useful intensity at ω_{1s}. Because the primary interest in this scheme is to realise efficient generation of ω_{1s}, we now examine how this conversion efficiency can be calculated from a coupled-wave analysis.

The small-signal theory to be sketched in this subsection is straightforward; analogous solutions for various four-wave processes are well-known in the literature. We note here the example in which a pump laser ω_p generates, by SRS, a Stokes wave $\omega_s = \omega_p - \Omega_{fg}$, and the anti-Stokes wave $\omega_{as} = \Omega_{fg} + \omega_p$ is simultaneously observed, generated parametrically by the process $\omega_{as} = 2\omega_p - \omega_s$. [This is the same parametric process that is used in CARS (Table 2.1), with the experimental difference that here the pump ω_p is the only input, whereas in CARS the ω_p intensity is not sufficient to generate ω_s and an input ω_s is, therefore, provided. This means that, for CARS, use of (3.15) is adequate.] The coupled-wave equations for this process were first studied by SHEN and BLOEMBERGEN [2.14] and were reviewed in great detail by BLOEMBERGEN [2.15]; see also GIORDMAINE and KAISER [3.4].

To return to biharmonic pumping, it follows from (3.8a) that the four coupled-wave equations are

$$\frac{dE_1}{dz} = \frac{3i\omega_1}{4cn_1}\left(\chi_{4wm}E_{1s}E_{2s}^*E_2\, e^{i\Delta kz} + \chi_{SRS1}^*|E_{1s}|^2E_1\right) \tag{3.26a}$$

$$\frac{dE_{1s}}{dz} = \frac{3i\omega_{1s}}{4cn_{1s}}\left(\chi_{4wm}^*E_1E_{2s}E_2^*\, e^{-i\Delta kz} + \chi_{SRS1}|E_1|^2E_{1s}\right) \tag{3.26b}$$

$$\frac{dE_{2s}}{dz} = \frac{3i\omega_{2s}}{4cn_{2s}} \left(\chi_{4wm} E_1^* E_{1s} E_2 \; e^{i\Delta kz} + \chi_{SRS2} |E_2|^2 E_{2s}\right) \qquad (3.26c)$$

$$\frac{dE_2}{dz} = \frac{3i\omega_2}{4cn_2} \left(\chi_{4wm}^* E_1 E_{1s}^* E_{2s} \; e^{-i\Delta kz} + \chi_{SRS2}^* |E_{2s}|^2 E_2\right) \qquad (3.26d)$$

$$\Delta k = (k_2 - k_{2s}) - (k_1 - k_{1s}) \quad , \qquad (3.26e)$$

where the resonance condition $\Omega_{fg} + \omega_{1s} - \omega_1 = \Omega_{fg} + \omega_{2s} - \omega_2 = 0$ has been taken, so that

$$\chi_{4wm} \equiv \chi^{(3)}(-\omega_1 ; \omega_{1s} - \omega_{2s} \omega_s) = (iN/6\hbar\varepsilon_0\Gamma)\overline{\alpha^* \beta} \qquad (3.27a)$$

$$\chi_{SRS1} \equiv \chi^{(3)}(-\omega_{1s} ; \omega_1 - \omega_1 \omega_{1s}) = (-iN/6\hbar\varepsilon_0\Gamma)\overline{|\alpha|^2} \qquad (3.27b)$$

$$\chi_{SRS2} \equiv \chi^{(3)}(-\omega_{2s} ; \omega_2 - \omega_2 \omega_{1s}) = (-iN/6\hbar\varepsilon_0\Gamma)\overline{|\beta|^2} \quad . \qquad (3.27c)$$

Here $\alpha = \alpha_{fg}(-\omega_{1s};\omega_1)$ and $\beta = \alpha_{fg}(-\omega_{2s};\omega_2)$ are Raman polarisabilities, and N is the effective atomic-number density (Sect.3.3). The form of the susceptibilities in (3.26) follows from the reality condition (2.2b) and permutation symmetry of the polarisabilities (2.19). In each of (3.26) it can be seen that the growth (or decay) of each field is due to the combined effect of a parametric contribution (the first term on the RHS) and an SRS contribution. Such combined processes have been anticipated in (2.7,9) by the primed sum [see also Sect.2.7, e.g., (2.100)].

From (3.26), the photon-conservation relations are, in terms of intensities,

$$\frac{\omega_1}{\omega_{1s}} I_{1s} + I_1 \simeq I_{1o} \quad ; \quad \frac{\omega_2}{\omega_{2s}} I_{2s} + I_2 \simeq I_{2o} \quad , \qquad (3.28)$$

where the approximations result from $I_{1o} \gg I_{1so}$, $I_{2o} \gg I_{2so}$. It follows from these relations that the parametric process, by linking $I_{1,1s}$ to $I_{2,2s}$, can mediate the flow of energy within these pairs of fields, but the actual energy of $I_{1,1s}$ cannot be given to $I_{2,2s}$, and vice versa.

To see more clearly the relative magnitudes of the terms in (3.26), we normalise the fields so as to give them the dimensions of (Intensity)$^{1/2}$; we also set the refractive indices equal to unity, although this is not essential. Then, with the replacement

$$E_i[Vm^{-1}] \rightarrow (2/\varepsilon_0 c)^{\frac{1}{2}}\left[E_i[Wm^{-2}]^{\frac{1}{2}}\right] ,$$

(3.26) becomes

$$\frac{dE_1}{dz} = -\frac{g_1}{2}\frac{\omega_1}{\omega_{1_s}} (\sigma_1 E_{1s} E_{2s}^* E_2 \; e^{i\Delta kz} + |E_{1s}|^2 E_1) \tag{3.29a}$$

$$\frac{dE_{1s}}{dz} = \frac{g_1}{2} (\sigma_1^* E_1 E_{2s} E_2^* \; e^{-i\Delta kz} + |E_1|^2 E_{1s}) \tag{3.29b}$$

$$\frac{dE_{2s}}{dz} = \frac{g_2}{2} (\sigma_2 E_1^* E_{1s} E_2 \; e^{i\Delta kz} + |E_2|^2 E_{2s}) \tag{3.29c}$$

$$\frac{dE_2}{dz} = \frac{-g_2}{2}\frac{\omega_2}{\omega_{2s}} (\sigma_2^* E_1 E_{1s}^* E_{2s} \; e^{-i\Delta kz} + |E_{2s}|^2 E_2) \quad , \tag{3.29d}$$

where $G_2 = g_2 I_{2o}$, $G_1 = g_1 I_{1o}$ are the Raman power gains $[m^{-1}]$ for ω_{2s} and ω_{1s}, respectively, [see (3.20)], and the dimensionless parameters σ_1, σ_2 are defined by

$$\sigma_1 = \overline{\alpha^* \beta}/\overline{|\alpha|^2} \quad ; \quad \sigma_2 = \overline{\alpha^* \beta}/\overline{|\beta|^2} \quad . \tag{3.30}$$

To simplify further, we introduce the parameter

$$\sigma = (\overline{|\beta|^2}/\overline{|\alpha|^2})^{\frac{1}{2}} = (\chi_{SRS2}/\chi_{SRS1})^{\frac{1}{2}} \quad , \tag{3.31a}$$

and the angular correlation function F,

$$F = \overline{\alpha^* \beta}/(\overline{|\alpha|^2} \; \overline{|\beta|^2})^{\frac{1}{2}} \quad . \tag{3.31b}$$

In terms of these, σ_1 and σ_2 become $\sigma_1 = F\sigma$ and $\sigma_2 = F/\sigma$. It is readily shown, by use of the theory presented in Sect.2.8, that if the two-photon transition is such that only the isotropic parts of the transition polarisabilities α,β are nonzero, then $F \equiv 1$ (to within a constant phase that is unimportant). This circumstance arises in a number of important cases. For example, ns-n's transitions in the alkali atoms, singlet transitions in the alkaline earths, and on polarised Raman transitions in molecules ($F = 1$ is really the definition of a polarised line). Another situation in which $F \simeq 1$ is when $\alpha \simeq \beta$, i.e., when $\sigma \simeq 1$. This will be seen in Chap.7

to apply to biharmonic pumping in homonuclear diatomic molecules such as H_2, N_2 where the pump frequencies ω_1 and ω_2 are far removed from the lowest electronic levels. Then α and β are nondispersive, i.e., insensitive to the values of ω_1 and ω_2, and are, to a good approximation, equal (this assumes that the geometries of $\underline{\varepsilon}_1$, $\underline{\varepsilon}_{1s}$ and $\underline{\varepsilon}_2$, $\underline{\varepsilon}_{2s}$ are the same, for example all linearly polarised in the z direction). Indeed, α and β will always become approximately equal when ω_1 and ω_2 are sufficiently close together, for any medium (with the aforementioned geometrical restriction). Thus, we have seen that F often equals unity either through symmetry, in which case σ may have any value, or in particular cases, when $\sigma \simeq 1$. Because this covers most situations of interest, we shall in the sequel take $F = 1$. (It is of interest to note that this is, in fact, also equivalent to ignoring all geometrical properties.) Then $\sigma_1 = \sigma_2^{-1} = \sigma$, and also $\chi''_{4wm} = (\chi''_{SRS1}\chi''_{SRS2})^{\frac{1}{2}}$. Moreover, $g_2 = g_1\sigma^2\omega_{2s}/\omega_{1s}$ independently of the value of F.

Equation (3.29) is easily related to the uncoupled-wave analyses given earlier. For example, if waves at frequencies ω_1, ω_2 and ω_{2s} were input, then in the small-signal region we could put $E_1, E_2, E_{2s} \to E_{1o}, E_{2o}, E_{2so}$ in (3.29b), and drop the other equations. This gives an uncoupled equation for E_{1s}, involving a parametric term $(g_1\sigma/2)E_{1o}E_{2so}E^*_{2o}\exp(-i\Delta kz)$ and a Raman term $(g_1/2)|E_{1o}|^2E_{1s}$. Then if $I_{1o}I_{1so} \ll I_{2so}I_{2o}$, the raman term can be dropped, giving a single equation for E_{1s} due to a parametric process alone; this type of equation was discussed in Sect.3.2. For the inverse inequality the SRS equation (3.17) is regained.

We return now to the situation of interest in this section when only waves at frequencies ω_1 and ω_2 are input and ω_{1s} and ω_{2s} are generated from noise. In the small-signal region, (3.29) can be approximated to (3.29b,c) for E_{1s}, E_{2s}, where the pump fields are taken to be undepleted. This pair of coupled equations is easily reduced to a second-order differential equation with constant coefficients, for E_{1s} or E_{2s}. The solution is therefore characterised by two growth constants, which we call $G_\pm$; it is of the form

$$E_{1s,2s} = A_{1,2}e^{G_+z/2} + B_{1,2}e^{G_-z/2} , \tag{3.32}$$

where $G_\pm$ are

$$G_\pm = \frac{1}{2}(G_1 + G_2) - i\Delta k \pm \left[\left(\frac{1}{2}(G_1 - G_2) + i\Delta k\right)^2 + G_1G_2\right]^{\frac{1}{2}} . \tag{3.33}$$

G_+ gives the growing solution if $(G_1 - G_2)\Delta k > 0$, i.e., if the imaginary part of the expression within the square root is positive; conversely, G_-

gives the growing solution when $(G_1 - G_2)\Delta k < 0$.

Some insight into the physical meaning of (3.33) can be obtained by comparison with the solution for a pair of equations coupled solely by parametric interaction; therefore, we digress for awhile in order to discuss these; we return to biharmonic pumping on page 104. Consider then the four-wave parametric process for the generation of $\omega_4 = \omega_1 + \omega_2 - \omega_3$, where ω_1 and ω_2 are input, and both ω_3 and ω_4 are generated internally from noise. Figure 2.2c illustrates the arrangement, but for the moment no two-photon resonance enhancement is assumed, so as not to obscure the general properties of the parametric process. The two coupled equations for this process are readily found from (3.8) to be

$$\frac{dE_{3,4}}{dz} = \frac{3i\omega_{3,4}}{2\varepsilon_o c^2} \chi_{4wm3,4} E_{1o} E_{2o} E^*_{4,3} e^{-i\Delta kz} \quad , \tag{3.34a}$$

where $\Delta k = -k_1 - k_2 + k_3 + k_4$ is the phase mismatch; the fields have been normalised as in (3.29), and the input beams are taken to be undepleted. The susceptibilities are $\chi_{4wm,3} = \chi^{(3)}(-\omega_3;-\omega_4\omega_1\omega_2)$ and $\chi_{4wm,4} = \chi^{(3)}(-\omega_4;-\omega_3\omega_1\omega_2)$. In the absence of a strong single-photon resonance for ω_3 or ω_4 these two susceptibilities are equal because of overall permutation symmetry; this is the case that we consider at present. Then $\chi_{4wm,3} = \chi_{4wm,4} = \chi_{4wm}$. The solutions of these equations are again of the form (3.32); but now the growth constants are

$$G_\pm(\text{parametric}) = -i\Delta k \pm (G_p^2 - \Delta k^2)^{\frac{1}{2}} \quad , \tag{3.34b}$$

where the parametric gain G_p is given by

$$G_p^2 = 9\mu_o^2\omega_3\omega_4|\chi_{4wm}|^2 I_{1o} I_{2o} \quad . \tag{3.34c}$$

The exponential factors in the solution may be combined, to give the form

$$I_3 \sim |\sin[(\Delta k^2 - G_p^2)^{\frac{1}{2}} z/2 + \phi]|^2 \quad , \tag{3.34d}$$

where ϕ is a constant, dependent on the values of E_{3o}, E_{4o}. The parametric gain G_p is a measure of the strength of the coupling of the two waves ω_3, ω_4. [For $G_p \simeq 0$ and $E_{4o} >> E_{3o}$, (3.34d) reduces to (3.15).]

Thus, (3.34d) shows that if $G_p < |\Delta k|$ then the coupling increases the effective coherence length, to $L_c' = \pi/(\Delta k^2 - G_p^2)^{\frac{1}{2}}$. When $G_p = |\Delta k|$, the parametric gain over a coherence length $L_c = \pi/\Delta k$ exactly balances the loss of intensity due to the phase interference (for $\Delta k \neq 0$); so I_3 does not increase. When $G_p > |\Delta k|$, the oscillatory character of the solution disappears and I_3 (and I_4) increase exponentially as e^{Gz}, where the gain $G = (G_p^2 - \Delta k^2)^{1/2}$. The idea of a coherence length therefore, loses its meaning as the maximum effective interaction length. The increase of I_3 and I_4 now takes place throughout the medium, although the phase mismatch Δk still has the effect of reducing the effective gain. Maximum gain is obtained with $\Delta k = 0$. (This behaviour is closely connected with the concept of phase-locking, to be discussed in Sect.3.4.2.)[3]

In the situation of greatest practical interest, however, use is made of two-photon resonance, $\Omega_{fg} - \omega_1 - \omega_2 \simeq 0$, Fig.2.2c. Therefore, the nonlinear susceptibility may be factorised; similarly to (3.27) and the discussion following (3.31), χ_{para} can be related to the susceptibilities for two-photon absorption of ω_1, ω_2 and ω_3, ω_4, and thus to the two-photon absorption coefficients. It is, therefore, impossible to consider this parametric process independently of two-photon absorption, because a large G_p may imply strong TPA.[4]

Following this digression, we now return to the analogous situation in biharmonic pumping. Here, the parametric process $\omega_{2s} = \omega_2 - (\omega_1 - \omega_{1s})$ and the SRS processes are considered together, because of the two-photon resonance. The product G_1G_2 in $G_\pm$ (3.33) is the parametric contribution, i.e., it corresponds to the parametric gain G_p^2 discussed above, with suitable changes of notation. This shows clearly that the parametric process is inseparable from the nonparametric processes (now SRS) because of the two-photon resonance, and, moreover, is of comparable importance. It is, however, now necessary that only G_1^2 or G_2^2 rather than G_1G_2 be large compared to Δk^2 for there to be a strong, exponentially growing solution, reflecting the

---

[3]These equations have their analogues in second-order parametric down conversion (e.g., [1.10]), in which a pump ω_p generates from noise the signal (ω_s) and idler (ω_i) waves, where $\omega_s = \omega_p - \omega_i$ and $\chi_{para} = \chi^{(2)}(-\omega_s;-\omega_i\omega_p)$. In the present example, the two pump beams I_{1o}, I_{2o} together act as a pump at frequency $\omega_p = \omega_1 + \omega_2$.

[4]The situation can become further complicated where a single-photon resonance of, say, ω_4 is involved. This is equivalent to a three-photon resonance of $\omega_1 + \omega_2 - \omega_3$, which is also that appropriate to generation of ω_3 by stimulated hyper-Raman scattering of the pumps ω_1 and ω_2. The competition between SHRS and the parametric process discussed in the foregoing will be discussed in Sect.5.5.1.

SRS contributions to the gain. Because the radical in (3.33) is always complex when $\Delta k \neq 0$, the two parts of (3.32) together act to give an oscillatory behaviour for I_{1s} and I_{2s}. However, this is important only in the relatively uninteresting early stages ($z \to 0$) of the growth of the waves; as z increases, the growing part of the solution dominates and the effect of the oscillations reduces to a negligible modulation of the intensity. This is in contrast to the purely parametric process considered above, for which the solution is either oscillatory *or* exponentially growing.

The case of most interest is when $\Delta k = 0$, i.e., the parametric process is phase matched. Then, $G_- = 0$ and $G_+ = G_1 + G_2$, so that

$$I_{1s,2s} \sim \exp(G_1 + G_2)L \quad . \tag{3.35}$$

Thus the parametric coupling transfers the high Raman gain of the strong pump at ω_2 to the Stokes wave ω_{1s}; this is what makes biharmonic pumping interesting, in view of the envisaged applications discussed in the introduction.

From (3.33), it is possible to derive the correction to (3.35) that is needed when Δk is nonzero, but small compared to $(G_1 + G_2)/2$. A little algebra to find the coefficients in (3.32) gives the conversion efficiency for the growing solution,

$$\frac{I_{1s}(L)}{I_{1o}} = \left(\frac{\omega_{1s}}{\omega_{2s}}\right)^2 \left|\frac{\chi''_{SRS1}}{\chi''_{SRS2}}\right| \left[1 + \left(\frac{2\Delta k}{G_1+G_2}\right)^2\right]^{-1} \frac{I_{2s}(L)}{I_{2o}} \quad . \tag{3.36}$$

Because $G_2 >> G_1$, the reduction in conversion due to Δk is roughly $[1 + (2\Delta k/G_2)^2]^{-1}$. In the special case that $\chi_{SRS1} = \chi_{SRS2}$, (equivalent to $\sigma = 1$, see earlier) then (3.36) becomes equivalent to the expression given by BYER and HERBST [3.2]. The ratio $I_{2s}(L)/I_{2o}$ can be measured experimentally (it is the stimulated-Raman conversion efficiency) and may easily reach tens of percent, in favourable circumstances. The conversion efficiency $I_{1s}(L)/I_{1o}$ can then be estimated by use of (3.36). It is to be noted that this conversion efficiency is rather small if $\omega_{1s} << \omega_{2s}$. We are ignoring here the effects of higher-order processes (such as second, third Stokes generation, etc.), which can play very significant roles when the Raman conversion efficiency is high.

Equation (3.36) may also be derived more directly, using some physically obvious approximations, as shown in (3.37); this approach will be extended to the large-signal region in Sect.3.4.2, where it will be shown that (3.36) is still approximately correct [(3.40)]. This therefore provides some justification for the use of (3.36) in typical experimental situations in which saturation is significant.

All of the two-photon resonant processes (Fig.2.2) have analogues to (3.36); as will be seen later (Chap.6), the use of resonance enhancement can cause the ratio of susceptibilities to differ considerably from unity, thereby having a strong influence on the conversion efficiencies. Similarly, as the generated frequency goes further into the infrared, the conversion efficiencies decrease rapidly.

In what will be called the Raman-driven-parametric approximation, use is made of the fact that SRS of the strong pump (ω_2) proceeds almost independently of the parametric coupling; conversely, generation of ω_{1s} by SRS of the weak pump (ω_1) may be ignored compared to its generation by parametric coupling. Therefore, (3.29) is replaced by the parametric parts of (3.29a,b) and the SRS parts of (3.29c,d). Restricting attention to the small-signal region reduces the equations still further, to

$$\frac{dE_{1s}}{dz} \simeq \frac{g_1\sigma}{2} E_{1o}E_{2o}^{*}E_{2s}e^{-i\Delta kz} \tag{3.37a}$$

$$\frac{dE_{2s}}{dz} \simeq \frac{G_2}{2} E_{2s} \quad . \tag{3.37b}$$

The SRS equation has solution $E_{2s} = E_{2so}\exp(G_2z/2)$, which is to be inserted into (3.37a), giving

$$\frac{I_{1s}(L)}{I_{1o}} = \frac{1}{\sigma^2}\left(\frac{\omega_{1s}}{\omega_{2s}}\right)^2\left[1+\left(\frac{2\Delta k}{G_2}\right)^2\right]^{-1}\frac{I_{2so}}{I_{2o}}\,e^{G_2L} \quad . \tag{3.37c}$$

This derivation shows clearly how the gain of the strong pump is transferred to I_{1s}. Using (3.31) for σ and the solution for $I_{2s} = |E_{2s}|^2$ then gives (3.36) in the limit $G_2 \gg G_1$.

3.4.2 Biharmonic Pumping — Large Signal

The theory of the large-signal behaviour of the coupled-wave equations (3.29), and the corresponding equations for other two-photon resonant four-wave mixing processes, has been treated in detail by BUTYLKIN et al. [3.5], KROCHIK and KHRONOPULO [3.6], VENKIN et al. [3.7] and YURATICH [3.8]. In the general case, no solutions are available, but by means of an extension of the Raman-driven parametric approximation (3.37) to the large-signal region, closed-form expressions for the intensities in biharmonic pumping have been obtained; for $\Delta k = 0$ by VENKIN et al. [3.7], and for arbitrary Δk by YURATICH [3.8]. Moreover, when $\Delta k = 0$ it is possible to treat all four coupled-wave equations in great detail, to the extent of finding for particular initial conditions a few exact solutions, and an essentially complete description in the remaining cases. This is due to the phenomenon of phase locking [3.5], in which the two flows of energy between I_1 and I_{1s}, and between I_2 and I_{2s} are locked in phase by the parametric interaction (and similarly for other two-photon resonant four-wave processes). The phase locking is maintained even for $\Delta k \neq 0$, provided that $(2\Delta k/G)^2 < 1$, where G is a typical growth constant, for example $G_1 + G_2$ in (3.36). Hence, the coherence length becomes, in effect, infinite and the process is insensitive to the value of Δk, in marked contrast to the length in (3.10). When $(2\Delta k/G)^2$ becomes very large, the parametric process becomes ineffective; that is, the phase locking breaks down, so that the process is characterised by the coherence length $|\pi/\Delta k|$, which is then very short. The ordinary SRS equations for the separate pairs I_1, I_{1s} and I_2, I_{2s} then apply. Such behaviour has already been noted, in the small-signal discussion of Sect.3.4.1. In the remainder of this section, we will pursue these points for the particular process of biharmonic pumping, drawing freely on the results contained in the foregoing references.

Equations (3.29) may be cast into a different form if the fields E_i are written in terms of their real amplitudes A_i and phases ϕ_i; that is, $E_i = A_i e^{i\phi_i}$ and $I_i = |E_i|^2 = A_i^2$. The result is

$$\frac{-1}{\omega_1 A_{1s}} \frac{dA_1}{dz} = \frac{1}{\omega_{1s} A_1} \frac{dA_{1s}}{dz} = \frac{g_1}{2\omega_{1s}} (\sigma A_2 A_{2s} \cos\theta + A_1 A_{1s}) \tag{3.38a}$$

$$\frac{1}{\sigma\omega_2 A_{2s}} \frac{dA_2}{dz} = \frac{-1}{\sigma\omega_{2s} A_2} \frac{dA_{2s}}{dz} = -\frac{g_1}{2\omega_1} (\sigma A_2 A_{2s} + A_1 A_{1s} \cos\theta) \quad , \tag{3.38b}$$

where the generalised phase $\theta(z)$ is defined by

$$\theta(z) = \Delta kz + \phi_{1s} - \phi_1 + \phi_2 - \phi_{2s} \tag{3.38c}$$

$$\frac{d\theta}{dz} = \Delta k + W \sin\theta \tag{3.38d}$$

$$W = \frac{g_1\sigma}{2} A_1 A_{1s} A_2 A_{2s} \left(\frac{\omega_1}{I_1} - \frac{\omega_{1s}}{I_{1s}} + \frac{\omega_2}{I_2} - \frac{\omega_{2s}}{I_{2s}}\right) \quad . \tag{3.38e}$$

These equations show that the entire process is characterised by the field amplitudes and the generalised phase θ, i.e., it is independent of the individual phases ϕ_i. Consider (3.38d) when $\Delta k = 0$. Then it is straightforward to show that $\theta = 0$ is a stable solution if $W < 0$ and conversely $\theta = \pi$ if $W > 0$; for example, in regions where $W < 0$, the phase tends, as e^{Wz}, to π. When $(2\Delta k/G_2)^2$ is small, then for $W > 0$, $\theta \simeq \pi + \Delta k/W$ is the stable solution, and for $W < 0$, $\theta = \Delta k/W$. It is certainly the case that $W(z = 0) < 0$ when I_{1so}, I_{2so} are noise intensities; therefore, it is to be expected that in the small-signal region the phase θ will rapidly fall to zero if $\Delta k = 0$, or to $\Delta k/W$ when Δk is small; this may be verified by direct calculation from the small-signal two-coupled-wave solutions of Sect.3.4.1. Confining our attention to $\Delta k = 0$, we see that, initially, the growth of the waves is optimised when $\theta(0) = 0$; hence, it is only necessary to consider those noise fields with the phases to ensure this, i.e., we can use $\theta(z = 0) = 0$ as an initial condition in (3.38a). When $\theta = 0$ in (3.38d), it remains 0 while z increases, until $W = \infty$, i.e., until there is a singularity of W. This is a singular point of the transformation from (3.29-38), and corresponds to one or both of $A_{1,2}$ decreasing to zero. It is then readily shown [3.8] that a jump of θ from 0 to π occurs; this corresponds simply to a change of sign of a field $E_{1,2}$ as it passes through zero. It can also be shown that now $W < 0$, so that $\theta = \pi$ is stable for z beyond this critical point. For simplicity, we assume here that there is only one jump of θ. Then, because $\cos\theta = -1$ in this region, a steady state for the four equations (3.38a,b) is reached, when $A_2A_{2s} = A_1A_{1s}$. By use of this equality, with conservation relations (3.28) and with implicit relations between A_2 and A_{1s} [obtainable from (3.38a,b) by integration] it is possible to deduce the steady-state intensities and thus the conversion efficiencies. Because the temporally steady-state material excitation is proportional to the RHS of (3.38a,b) [see Sect.2.7, Eq. (2.100)], this spatial steady-state condition corresponds to balance of the driving and generated contributions, to give zero net excitation.

As is hinted by the phase jump and consequent discontinuity of (3.38) (which, in general, occurs when $W = \pm\infty$, i.e., when any of the fields pass through zero), the analysis is facilitated by working with the real fields E_i, where reality follows from the initial condition $\theta = 0$, and $\Delta k = 0$. Equations (3.38) are then needed only to establish the initial phase condition. In the work of BUTYLKIN et al. [3.5], and KROCHIK and KHRONOPULO [3.6], great emphasis was laid on the connection between the phase jump and the sign of W; as we have seen this is misleading, because jumps occur at the singularities of W, not its zeros. Indeed, the singularities of W can occur at either sign of W. Thus, besides greatly complicating the analysis, attention focussed on the sign changes of W leads to incorrect conclusions. This matter has been discussed at length by YURATICH [3.8].

We now digress a little and discuss the Raman-driven-parametric approximation, because later this will provide a useful point of reference. Recall from Sect.3.4.1 that this approximation consists of retaining only the SRS terms for the strong waves I_2, I_{2s}, and only the parametric coupling of the weak waves I_1, I_{1s}. Then for a large-signal analysis, (3.37a,b) are replaced by two equations for E_1, E_{1s} [the simplified (3.29a,b)] and the solution for E_2, E_{2s}. The latter follows from (3.24,28), with suitable changes of notation. Then, because at exact Raman resonances the field phases are unchanged by SRS, we have $E_2 = I_2^{1/2}\, e^{i\phi_{2o}}$, and similarly for E_{2s}.

For the phase-matched case, $\Delta k = 0$, VENKIN et al. [3.7] found the exact solution of these equations. When a negligible term proportional to I_{1so} is dropped, this solution is

$$\alpha \equiv \frac{\omega_1}{\omega_{1s}} \frac{I_{1s}(L)}{I_{1o}} = \sin^2 R\beta \quad , \tag{3.39a}$$

where α is the fraction of input pump photons (ω_1) converted to Stokes photons (ω_{1s}) after a distance L, R is a frequency factor defined by

$$R = \frac{1}{\sigma}\left(\frac{\omega_1\omega_{1s}}{\omega_2\omega_{2s}}\right)^{1/2} \quad , \tag{3.39b}$$

and the angle β is given by

$$\beta = \tan^{-1}\left[\left(\frac{\omega_2 I_{2so}}{\omega_{2s} I_{2o}}\right)^{1/2} e^{G_2 L/2}\right] \quad . \tag{3.39c}$$

It is seen from (3.39a,c) and Fig.3.2a that as β increases from very nearly zero at $z = 0$ to its maximum of $\pi/2$ at $z = L \to \infty$, the photon-conversion factor α oscillates between 0 and 100%. With the help of (3.28), the weak pump intensity is found to be given by $I_1(L)/I_{1o} = \cos^2 R\beta$; we also note that in the present notation, $\omega_2 I_{2s}/\omega_{2s} I_{2o} = \sin^2\beta$ and $I_2/I_{2o} = \cos^2\beta$. It is interesting to observe that $\beta = \pi/4$ corresponds to the SRS threshold condition (3.21) applied to the strong pump; this is a convenient measure of the scale of β related to $G_2 z$. When $\beta \to 0$, i.e., the small-signal region, then (3.39) reduces to the $\Delta k = 0$ version of (3.37c), as should happen.

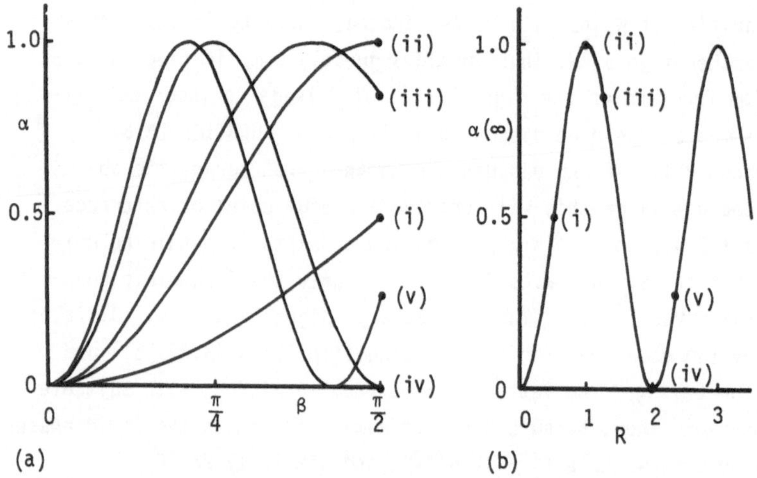

Fig.3.2a,b. Raman-driven parametric approximation to biharmonic pumping. [Curves (i)-(v) on (a) correspond to R values marked on (b).] (a) shows the solution (3.29) for the photon conversion α from ω_1 to ω_{1s} as a function of the angle β. The values $\beta = 0$, $\pi/4$, $\pi/2$ correspond to gains G_2L of 0, threshold for SRS of ω_2, ∞ respectively. The solution for $\omega_s I_{2s}/\omega_{2s} I_{2o}$ coincides with the curve marked (ii). (b) shows the saturated values of α at $\beta = \pi/2$ ($G_2L = \infty$) as a function of R

The argument $R\beta$ of the $\sin^2$ function in (3.39a) sweeps out an angle 0 to $\pi R/2$; therefore, the steady-state photon conversion depends strongly on R: $\alpha(L \to \infty) = \sin^2(\pi R/2)$. This is illustrated in Fig.3.2. If R is small, as could be the case when ω_{1s} is in the medium to far infrared, the angle $\pi R/2$ is small, i.e., α does not reach 100%. Indeed (3.39a) may now be approximated, to give

$$\frac{I_{1s}(\infty)}{I_{1o}} \simeq \frac{\omega_{1s}}{\omega_1}\frac{R^2\pi^2}{4} = \frac{\pi^2}{4}\frac{1}{\sigma^2}\left(\frac{\omega_{1s}}{\omega_{2s}}\right)^2\left(\frac{\omega_{2s}}{\omega_2}\right) .$$

The last factor on the RHS of this expression may be replaced by $I_{2s}(\infty)/I_{2o}$, because 100% photon conversion of I_{2o} to I_{2s} is achieved by SRS as $L \to \infty$. We then have

$$\frac{I_{1s}(\infty)}{I_{1o}} = \frac{\pi^2}{4}\frac{1}{\sigma^2}\left(\frac{\omega_{1s}}{\omega_{2s}}\right)^2\frac{I_{2s}(\infty)}{I_{2o}} . \qquad (3.40)$$

This differs only by $\pi^2/4$ from the small-signal relation (3.37c) ($\Delta k = 0$).

Recently YURATICH [3.8] has obtained the general solution to the Raman-driven parametric equations (arbitrary Δk,R) in terms of hypergeometric functions. One particularly simple conclusion is that as $L \to \infty$,

$$\alpha(L \to \infty) = \sin^2(\pi R/2)\operatorname{sech}^2(\pi\Delta k/G_2) \quad , \qquad (3.41)$$

i.e., the correction of (3.39a) in this limit is merely the sech^2 factor. This result shows that the phase mismatch Δk reduces the effective gain much more markedly than was suggested by the earlier small-signal analysis. Thus, with increasing Δk, the term $\operatorname{sech}^2(\pi\Delta k/G_2) \sim 4\exp(-2\pi\Delta k/G_2)$ decreases much more rapidly than the term $[1 + (2\Delta k/G_2)^2]^{-1}$ found in (3.37c).

VENKIN et al. [3.7] considered a special case ($\Delta k \neq 0$, $R = 2$) of the Raman-driven-parametric process and gave a good illustration of the phase-locking phenomenon, i.e., in passing from the small- to large-signal regions, θ jumps from 0 to π. That this is a very special case follows from the $\Delta k = 0$ solution (3.39a), $\alpha = \sin^2 2\beta$, so that as z increases from 0 to ∞, α goes from 0 to 100% at $\beta = \pi/4$ to 0 again at $\beta = \pi/2$. The solution for arbitrary Δk is, in the present notation, $\alpha = [\sin^2 2\beta/1 + (2\Delta k/G_2)^2]$, which can be contrasted with (3.37c).[5] The phase θ may be put into the form $\theta = \tan^{-1}[(2\Delta k/G_2)\sec 2\beta]$; a graph of this is given in Fig.3.3. The jump in θ is clearly seen when $2\Delta k/G_2$ is small; as Δk increases, the jump passes into a smooth variation of the phase, indicating breakdown of phase locking, thus illustrating the points made in the introduction to this section.

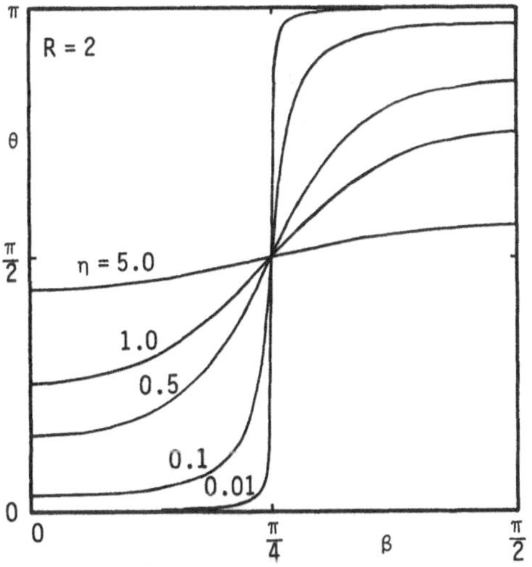

Fig.3.3. Illustration of phase-locking. For small $\eta = 2\Delta k/G_2$, the phase is initially locked to $\theta = \eta$, but as the waves propagate, a jump in θ eventually occurs, to a value $\theta = \pi - \eta$, in which region a steady state is obtained for the intensity. The expression for θ as a function of β is given in the text, for arbitrary Δk, but such that $R = 2$ and the conditions of the Raman-driven-parametric approximation are valid

[5]There is an important typographical error in VENKIN et al. [3.7], Eq.(18), noted by YURATICH [3.8]; the graphs illustrating it are, however, correct.

We are now in a position to discuss the solution of the full set of coupled equations (3.29) without making any approximations, but with the restriction that $\Delta k = 0$. The solutions $\alpha = \sin^2 R\beta$, $I_1(L)/I_{1o} = \cos^2 R\beta$, and correspondingly, for I_{2s}, I_s, the solutions $\sin^2\beta$, $\cos^2\beta$, still apply, but now β is a somewhat more complicated function of distance than (3.39c). In fact [3.8] it is given by

$$\frac{d\beta}{dz} = \frac{1}{4} G_2 \, [\sin^2\beta + (\lambda/R)\sin 2R\beta] \ , \tag{3.42}$$

where $\lambda = G_1/G_2$. [The first and second parts of the RHS of (3.42) arise from the products E_2E_{2s} and E_1E_{1s}, respectively.] The Raman-driven-parametric approximation consists of setting $\lambda = 0$ in this equation, whereupon (3.39c) is regained. The RHS of (3.42) is initially positive, so β increases with z, until the RHS equals zero (the zero at $\beta = 0$ is not important because $\beta > \beta_o > 0$, where β_o is a very small angle that corresponds to the noise input). This is the steady-state condition, where $z = \infty$, and corresponds to the relation $A_2A_{2s} = A_1A_{1s}$ that was discussed at the beginning of this section. Hence, the steady-state value of α is $\alpha(z \to \infty) = \sin^2(R\beta_1)$, where β_1 is the first positive zero of the RHS of (3.42). As $\lambda \to 0$ we find $\beta_1 = \pi/2$, as given by (3.39c). A general solution, in a useful form, of (3.42) does not appear possible, but we note that for some choices of λ and R, simple results have been found. Thus if R = 1, we need only replace G_2 in (3.39c) by $(G_1 + G_2)$, with some other minor changes that introduce I_{1so}.

Other two-photon resonant four-wave processes (see Table 2.2) are analysed similarly, but in those cases the solutions and the RHS of (3.42) are replaced by combinations of trigonometric and hyperbolic functions of β; therefore, there is a considerable variety in behaviour of the various processes.

3.5 Noise Power

A number of important nonlinear processes are initiated by noise that is subsequently amplified. Among such processes, we have already mentioned stimulated Raman and hyper-Raman scattering. Parametric amplification, most familiar as a three-wave process, but which has a four-wave analogue, is an example of a parametric process that produces its own noise. In order to calculate the intensity generated in any of these processes, it is necessary to know the starting noise intensity. In practice, however, a precise knowledge of the noise intensity is not needed, because it undergoes exponential amplification and the final intensity is usually determined by saturation. We will, therefore, content ourselves with a simple heuristic derivation of the noise input; we find that whatever the process considered and whatever the intensity of the driving fields, the noise from which the process starts is equivalent to one photon per mode of the generated field. Because we are ultimately interested in a noise intensity, we first reexpress this result in such terms.

We consider the nonlinear medium to be irradiated by the pump beam (or beams) over a cylindrical region of radius a and length L. Although noise power is radiated in all directions, we need only consider the noise emitted in such a direction that it travels along the length of the cylinder, because only that noise will be effective in producing the detected output, i.e., will undergo the full amplification provided by the medium. Also, despite the fact that the noise is emitted over a wide range of frequencies, we need only consider that emitted into the bandwidth $d\omega$ of the finally detected output. The calculation of effective starting-noise intensity is therefore reduced to a calculation of the number of modes contained in this bandwidth and contained within the solid angle subtended by the irradiated cylindrical region. It is a well-known result of radiation theory that the density of modes in the frequency range between ω and $\omega + d\omega$ is given by $\omega^2 d\omega/\pi^2 c^3$ (see, e.g., [2.19]); when expressed as a density of modes per steradian and of given polarisation, this becomes

$$\rho(\omega) \equiv \frac{\omega^2 d\omega}{8\pi^3 c^3} \quad .$$

The cylindrical region subtends a solid angle $d\Omega \sim \pi a^2/L^2$ and has a volume $V = \pi a^2 L$, thus the total number of modes that contribute to the effective noise is $V\rho(\omega)d\Omega = \omega^2 d\omega a^4/8\pi c^3 L$. Because the noise energy per mode is $\hbar\omega$, and radiation travels the length L in a time L/c, the effective starting noise power P_n is

$$P_n = \hbar\omega \cdot \frac{c}{L} \cdot \frac{\omega^2 d\omega a^4}{8\pi c^3 L} = \frac{\hbar\omega^3 d\omega a^4}{8\pi c^2 L^2} \quad . \tag{3.43}$$

From this expression, it can be seen that the noise depends on the geometry via the factor a^4/L^2, but is independent of the intensity of the driving fields. To obtain a rough estimate of a typical noise intensity, we shall assume that the cylindrical region is produced by a gaussian beam focussed so as to be confocal over the length L, i.e., $2\pi a^2/\lambda = L$, where we identify a as the gaussian spot size. Substitution of this expression for L into (3.43), gives the noise power as $P_n \sim h\nu d\nu$. This particularly simple form results from our choice of confocal focussing, which has, in fact, led to a cylindrical region that contains just one gaussian spatial mode. If we now assume $h\nu = 10^{-19}$J and $d\nu = 10^{10}$Hz (i.e., $\sim 0.3\ cm^{-1}$) then the effective starting noise power is $P_n \simeq 10^{-9}$W. Expressed as an intensity, I_n, this becomes

$$I_n = P_n/\pi a^2 \sim h\nu d\nu/\lambda L = h\nu^2 d\nu/cL \quad .$$

Assuming L = 10 cm gives a noise intensity $\sim 10^{-6}\ Wcm^{-2}$. Although these numbers have been obtained by use of particular assumed value of ν, d, a and L, under conditions of high exponential gain, the final intensity is not sensitively dependent on the assumed noise level, so the figures given in the foregoing provide useful and easily remembered values, suitable for rough calculation.

We now return to the question of proving that the noise is equivalent to the presence of one photon per mode. This requires a quantised-field treatment (see [2.19]). It is then found that, for stimulated processes, the rate of emission into a given radiation mode is proportional to $(n_s + 1)$ where n_s is the number of photons already in that mode. The n_s term refers to stimulated emission, and the 1 refers to spontaneous emission, i.e., the rate of spontaneous emission into a given mode appears to be that due to stimulation by a fictitious single photon already in that mode. The relation between stimulated-emission rate and spontaneous-emission rate is the same whatever the process. Thus the ordinary single-photon emission rate is proportional to $(n_s + 1)$; the Raman-emission rate is proportional to $n_p(n_s + 1)$ where n_p is the number of photons in the mode of the pump field and n_s is the number of photons in the Stokes mode; the hyper-Raman emission rate is proportional to $n_{p_1}n_{p_2}(n_s + 1)$, and so on.

At first sight, it appears somewhat surprising that the effective starting noise power should be the same regardless of the process and intensity of the driving fields, because it might be expected that higher pump intensities would provide more spontaneous emission and hence stronger noise; also, a higher-order process, such as hyper-Raman scattering, might be expected to produce less noise. To make this point more clear we now give a simple derivation of the noise.

We again consider a cylindrical volume of medium of length L and radius a. We suppose that this region of the medium produces a spontaneous emission of photons emitted in all directions, with a rate of emission ϕ per atom. To be specific, this emission could be due to fluorescence from excited atoms, in the case of ordinary single-photon emission, or due to spontaneous Raman or hyper-Raman emission caused by a cylindrical beam of pump radiation interacting with the atoms. Let the medium have a gain G for the spontaneously emitted photons, i.e., the power gain is exp Gz over a length z along the cylinder. This gain could be due to stimulated Raman or hyper-Raman emission. Over a length 1/G of the medium the spontaneous emission is amplified by an amount e; therefore, the noise emitted from the first 1/G length of medium will, after amplification, dominate the spontaneous emission from subsequent lengths of the medium. The effective starting noise is, therefore, simply the noise emitted from the first 1/G length of the medium that subsequently propagates along the whole length L of the amplifier, i.e., is contained within the solid angle subtended by the medium. If the density of atoms is N (excited atoms in the case of single-photon emission, or ground-state atoms in the case of Raman or hyper-Raman emission), the effective starting noise power is just

$$\hbar\omega N\phi \cdot \frac{\pi a^2}{G} \cdot \frac{\pi a^2}{L^2} = \frac{N\phi}{G} \cdot \frac{\pi^2 a^4 \hbar\omega}{L^2} \quad .$$

Finally, we will show that, whatever the process, the ratio $(N\phi/G)$ takes the same value, namely $\omega^2 d\omega/\pi^2 c^2$, where $d\omega$ is the linewidth of the spontaneous emission; thus our earlier expression (3.43) for effective starting noise is recovered apart from a numerical factor 8, which is a casualty of our rather loose specification of linewidth, solid angle etc. The reason why the effective noise intensity is independent of the process or the intensity of the driving field can now be clearly seen. An increased pump intensity will increase the gain G and the spontaneous-emission rate $N\phi$ in the same ratio. The length 1/G from which the starting noise originates is thus reduced in the same proportion as the spontaneous-emission rate per atom is increased. The ratio $N\phi/G$ can be evaluated by considering the rate of increase of n_s, the number of photons in a particular mode of the generated field. As pointed out in the foregoing, this takes the form

$$dn_s/dt = K(n_s + 1) \quad , \tag{3.44}$$

whatever the process. Considering just the stimulated-emission term in (3.44), the equation can be converted to one that describes the change of intensity with distance, due to amplification. This equation is

$$dI_s/dz = (K/c)I_s \quad . \tag{3.45}$$

Therefore, we can identify G as being equal to K/c. However, from (3.44) the spontaneous-emission rate per mode is K = Gc, and because there are $\omega^2 d\omega/\pi^2 c^3$ modes per unit volume, the total spontaneous-emission rate per unit volume in the spontaneous-emission bandwidth $d\omega$ is $N\phi = Gc\omega^2 d\omega/(\pi^2 c^3)$, i.e., we have $N\phi/G = \omega^2 d\omega/(\pi^2 c^2)$, as required.

We have, so far, confined our discussion to nonparametric processes. The question of parametrically generated noise needs a more careful treatment. We simply quote the result here. For a detailed discussion of parametric noise, the reader is referred to GIALLORENZI and TANG [3.9] and KLEINMAN [3.10]. They considered the noise generation in a three-wave process, where pump photons ω_p produce noise at signal and idler frequencies ω_s, ω_i, with $\omega_p = \omega_s + \omega_i$. The same treatment can be readily extended to four-wave processes, $\omega_p + \omega_p = \omega_s + \omega_i$ and so on. The result is that the effective starting noise can be considered to arise from an energy $\hbar\omega_s/2$ per signal mode and $\hbar\omega_i/2$ per idler mode. Thus, apart from niceties that involve the factor 1/2, which are carefully considered by Kleinman, but which do not materially alter the final intensity when large gains are involved, we can summarise by saying that whatever the process, parametric or nonparametric, the effective starting noise can be considered to arise from the presence of one photon per mode of the generated field.[6]

[6]When considering the generation of very long wavelengths, the number of photons per mode due to blackbody radiation may exceed unity, thus giving a higher "noise" level than quoted above.

4. Sum Frequency and Harmonic Generation

4.1 Background Material

Optical third-harmonic generation was first observed when a Q-switched ruby laser beam (6943 A°) was focussed into various transparent solids [4.1,2,2.23]. Typical coherence lengths were only ~ 1μm; tight focussing was, therefore, necessary in order to produce detectable third-harmonic (2313A°) signals, with the consequence that the fundamental power densities involved were close to the threshold for dielectric breakdown. The best results were obtained in calcite, for which phase-matching could be achieved by an appropriate choice of crystal orientation. With a 4mm-thick calcite crystal and 1MW ruby laser pulse, the best conversion efficiency was 3×10^{-6} [2.23]. From measurements of third harmonic and laser intensities it was deduced that the bulk susceptibility for THG was $\sim 10^{-15}$esu (units cm^3/erg) ($\sim 10^{-23} m^2 V^{-2}$), thus implying a microscopic susceptibility (per molecule of $CaCO_3$) of $\sim 5 \times 10^{-36}$esu (units cm^6/erg, or in SI units $\sim 10^{-49} m^5 V^{-2}$). (See Appendix for a discussion of units.)

The first observation of THG in gases was reported in a letter by NEW and WARD [1.11] and further results and a detailed elaboration of this work appeared later [1.12]. They too used a Q-switched ruby laser, and third-harmonic signals were obtained from several gases. Third-harmonic susceptibilities were measured for He, Ne, Ar, Kr, Xe, H_2, CO_2, and N_2, ranging from values of 4×10^{-39}esu per atom (units cm^6/erg) to about 10^{-36}esu per atom for Xe.

The motive for studying these gases was that (at least for the inert gases) their nonlinear susceptibilities are relatively amenable to quantum-mechanical calculation. Theoretical estimates were derived in a number of ways. SITZ and YARIS [4.3] performed a time-dependent-perturbation variational calculation for He, and the result, $\chi_{THG}(He) = 4.0 \times 10^{-39}$esu per atom, is thought to be good to about 1%, although it is unlikely that this type of calculation can be extended to more complex atoms. The experimentally measured absolute value of $\chi_{THG}(He)$ was $7.0 \times 3^{\pm 1} \times 10^{-39}$esu per atom. For the other inert gases, the experimental values were normalised relative to the value of $\chi_{THG}(He) = 4.0 \times 10^{-39}$ esu per atom, thus removing the large absolute uncertainty. Agreement between these experimental values and theoretical estimates was good, e.g., for Xe (in units of 10^{-39} esu per atom) the experimental value was 979 ± 190, whereas the theoretical values were 987 and 962 by two different methods of calculation.

These calculations can best be explained by reference to the expression (2.18) for the third-harmonic susceptibility, which we repeat here, but simplified to the case of linearly polarised light:

$$\chi_{THG} \equiv \chi^{(3)}(-3\omega;\omega\omega\omega) = \frac{\mathcal{N}}{\hbar^3\varepsilon_0}\sum_{gabc}\rho(g)\left(\frac{\mu_{ga}\mu_{ab}\mu_{bc}\mu_{cg}}{(\Omega_{ag}-3\omega)(\Omega_{bg}-2\omega)(\Omega_{cg}-\omega)}+\dots\right) \quad . \tag{4.1}$$

It can be seen that if the frequency ω of the applied optical field is small compared to the frequencies Ω characteristic of the atom (as for a ruby laser and inert gases), then the third-harmonic susceptibility χ_{THG} is related in a simple way to the static susceptibility $\chi^o \equiv \chi^{(3)}(0;000)$ and to the Kerr susceptibility $\chi^K \equiv \chi^{(3)}(-\omega;00\omega)$. WARD and NEW give

$$\chi^o \approx \frac{\chi^K}{\left[1+\frac{5\omega^2}{(2\Omega)^2}\right]} \approx \frac{\chi_{THG}}{\left[1+\frac{15\omega^2}{\Omega^2}\right]} \quad . \tag{4.2}$$

(The convenience of the factor K in the definition of susceptibility can be seen to advantage here; it ensures that the susceptibilities χ^o, χ^K and χ_{THG} are all equal when ω goes to zero.). Thus, by use of (4.2), a value for χ_{THG} can be derived from theoretically calculated values of χ^o and experimentally measured values of χ^K, both of which are available for the inert gases. Taking Xe, for example, using the measured value of χ^K [4.4] and applying the frequency correction from (4.2), WARD and NEW obtained the value 962×10^{-39} esu per atom, referred to above.

Alternatively, (4.1) can be used directly to calculate χ_{THG}, provided that enough information is available concerning the matrix elements involved in the sum (this must include knowledge of their signs, because cancellations in the sum can have significant effects). In this way, WARD and NEW obtained the value 987×10^{-39} esu per atom for Xe. DAWES [4.5] showed that a particularly simple calculation gives good agreement with the experimental results of NEW and WARD. In this calculation, the sum over intermediate states is reduced to a contribution from a single effective excited state. The same procedure can, of course, be applied to the linear susceptibility; in fact, the data that relate to this effective state are obtained by making a best fit of the linear susceptibility expression to experimental refractive index data.

The conclusion to be drawn from Ward and New's work is that, for the conditions of their experiment (modest laser power, laser frequency well away from atomic resonances), the third-harmonic susceptibilities are in good agreement with values calculated from perturbation theory. However, the third-harmonic conversion efficiencies achieved were very low. The potential of the inert gases for generation of useful harmonic power had to await the arrival of high-power coherent uv sources. This low conversion efficiency was in part due to a small coherence length (large Δk). By use of a plane-wave analysis [as in Sect.3.2, leading to (3.11)] it is found that the third-harmonic power generated from a length L of medium varies as $L^2 \mathrm{sinc}^2(\frac{\Delta kL}{2})$, thus implying that, for plane waves, ideally $\Delta k = 0$. WARD and NEW [1.12] showed, however, that when a focussed beam is used, the dependence of harmonic power on Δk can be very different; this has important consequences in a device where the optimum focussing configuration is sought in order to achieve maximum efficiency of harmonic generation. Their analysis revealed the surprising result that, if $\Delta k = 0$, no third-harmonic is generated by focussing in an infinite medium. We examine this question of optimum focussing in the next section.

4.2 Theory of Optimum Focussing and Phase Matching

First, we give a qualitative explanation for this absence of THG and then we give the results of analytical treatments of the problem. Consider the phase shift of a light wave as it travels between two points. If the wave travels through a focus somewhere between these two points, then it undergoes a smaller phase shift than does a plane wave. This phase lag becomes 180° in the limit where the two points are each many confocal parameters away from the focus. This result is discussed by BORN and WOLF [4.6] for the case of a uniform intensity beam; we shall see that it applies also to gaussian beams. Because the third-harmonic polarisation takes the form of a wave whose amplitude is proportional to $[E \exp(i\phi)]^3 = E^3 \exp(3i\phi)$, where $E \exp(i\phi)$ is the field of the focussed fundamental wave, it follows that this polarisation wave shows a phase lag of 3π compared with a plane polarisation wave. On the other hand, the polarisation wave radiates a third-harmonic wave that is also focussed and shows a phase lag of π relative to a plane harmonic wave. Thus, even if $\Delta k = 0$, i.e., the phase velocities of fundamental and harmonic waves are equal, as the harmonic wave propagates through the focus it accumulates a phase *lead* of 2π relative to the driving polarisation wave. The harmonic radiation generated before the focus then cancels that generated after the focus and no net harmonic is produced. To avoid this cancellation, WARD and NEW focussed their laser beam in front of the gas cell (Fig.4.1). The qualitative argument given above also suggests that by choosing $\Delta k \equiv k_3 - 3k_1 < 0$, then the phase lead may be compensated and optimum harmonic generation under strong focussing conditions should, therefore, occur for some negative value of Δk (i.e., the nonlinear medium should have a negative dispersion).

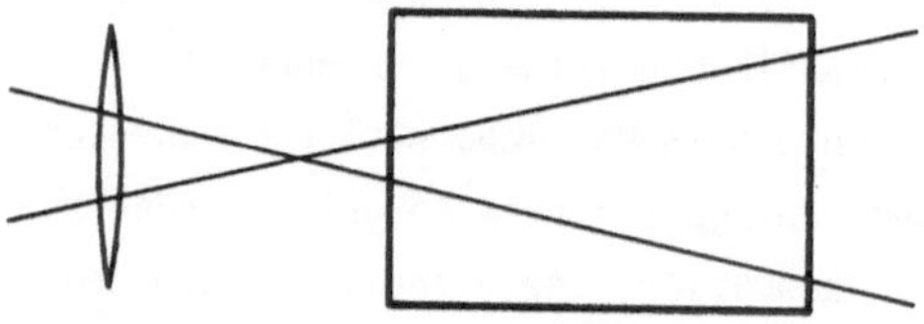

Fig.4.1. Focussing arrangement used by WARD and NEW [1.12] to avoid cancellation of third-harmonic fields

In their analysis, WARD and NEW [1.12] examined the general problem of qth-harmonic generation by a focussed laser beam. Here we quote their results for third-harmonic generation only (and convert their formulae to SI units). The laser beam is assumed to be a TEM_{00} gaussian mode propagating in the positive-z direction, having a waist located at $z = f$, and a confocal parameter b. The field of this mode is given by

$$\underline{E}_1(\underline{r}t) = \frac{1}{2}\left[\underline{E}_1(\underline{r})e^{-i\omega_1 t} + \underline{E}_1^*(\underline{r})e^{i\omega_1 t}\right] , \tag{4.3}$$

where

$$\underline{E}_1(\underline{r}) = \underline{E}_{1o}\, e^{ik_1 z}(1 + i\xi)^{-1} \exp\left[-\frac{k_1(x^2 + y^2)}{b(1 + i\xi)}\right] , \tag{4.4}$$

k_1 is the fundamental wave vector in the nonlinear medium and $\xi = 2(z - f)/b$ is a normalised z coordinate referred to $z = f$ as origin. The significance of these symbols is clarified in Fig.4.2, and some standard gaussian-beam relations are summarised there. If we consider the phase variation along the

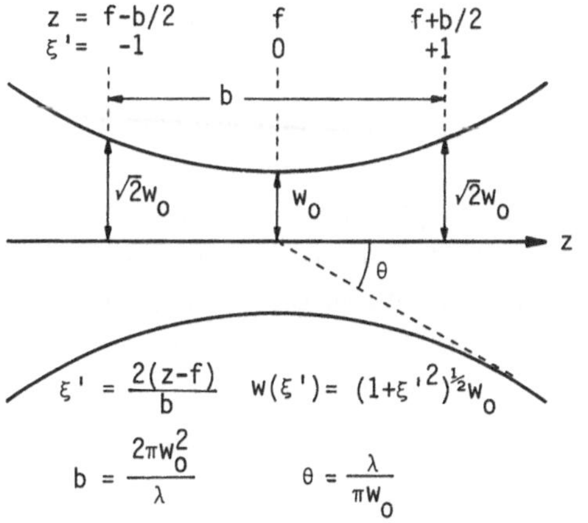

Fig.4.2. Notation used in the text to describe the incident gaussian beam, and some useful gaussian-beam relations. The nonlinear medium extends from $\xi' = -\zeta$ to $\xi' = \xi$

beam axis ($x = y = 0$), we see from (4.4) that in addition to the $\exp(ik_1z)$ term there is a contribution from the $(1 + i\xi)^{-1}$ factor.

Rewriting $(1 + i\xi)^{-1}$ as

$$(1 + i\xi)^{-1} = (1 + \xi^2)^{-\frac{1}{2}} \exp(-i\tan\xi) \quad , \tag{4.5}$$

we can see that this additional contribution to the phase goes from $+\pi/2$ at $\xi = -\infty$ to 0 at $\xi = 0$ to $-\pi/2$ at $\xi = +\infty$, thus demonstrating the phase lag of π in going through the focus.

The third-harmonic polarisation is

$$\underline{P}^{(3)}_{3\omega}(\underline{r}) = \frac{1}{4}\,\varepsilon_0\chi_{THG}[\underline{E}_1(\underline{r})]^3 \quad , \tag{4.6}$$

where χ_{THG} is the macroscopic third-harmonic susceptibility, (4.1).

From WARD and NEW [1.12], the generated third-harmonic field at a point $\underline{r}$ beyond the nonlinear medium (which extends from $\xi' = -\zeta$ to $\xi' = \xi$) is

$$\underline{E}_3(\underline{r}) = \left[\frac{i\omega_3 b\chi_{THG}E^3_{1o}}{16\eta_3 c}\right]\left[e^{i3k_1z}(1 + i\xi)^{-1}\exp\left(\frac{-3k_1(x^2 + y^2)}{b(1 + i\xi)}\right)\right]$$

$$\times\; I(\Delta k,\xi,\zeta) \quad , \tag{4.7}$$

where $\omega_3 = 3\omega_1$, η_3, and η_1 are the refractive indices of the nonlinear medium at frequencies ω_3 and ω_1, respectively, c is the velocity of light in vacuo, $\Delta k = k_3 - 3k_1 = 3\omega_1(\eta_3 - \eta_1)/c$ and

$$I(\Delta k,\xi,\zeta) = \int_{-\zeta}^{\xi} \frac{d\xi'\exp[ib\Delta k(\xi-\xi')/2]}{(1+i\xi')^2} \quad . \tag{4.8}$$

Examination of (4.7) shows that, apart from the integral I, the harmonic field is also a TEM_{00} mode, with the same confocal parameter as the fundamental and hence having a spot size and beam divergence that are both $1/\sqrt{3}$ smaller than for the fundamental. The integral $I(\Delta k,\xi,\zeta)$ describes the combined effect on the harmonic field of the phase mismatch Δk and the phase lag due to focussing; it will be discussed in more detail in the sequel.

It now remains to recast (4.7) as a relation between the total harmonic power generated and the total incident fundamental power. The cycle-averaged intensity of the fundamental beam at a point where the fundamental field is E_1, is given by $\frac{1}{2}\varepsilon_o c n_1 |E_1|^2$, (2.8), and the total power P_1 in the gaussian beam is $\frac{1}{2}(\pi w_{10}^2)$ × (on-axis intensity at the beam waist), where w_{10} is the beam radius at the waist. Hence,

$$P_1 = \frac{1}{2}\pi w_{1o}^2 \cdot \frac{1}{2}\varepsilon_o c n_1 |E_{1o}|^2 \quad . \tag{4.9}$$

Similarly, the total third-harmonic power leaving the nonlinear medium is

$$P_3 = \frac{1}{2}\pi w_3^2(\xi) \cdot \frac{1}{2}\varepsilon_o c n_3 |E_3(\xi)|^2 \quad , \tag{4.10}$$

where $E_3(\xi)$ is the on-axis third-harmonic field that leaves the nonlinear medium, and $w_3(\xi)$ is the beam-spot size (radius) at that point. From (4.7), with the help of (4.5), we see that

$$|E_3(\xi)|^2 = \left|\frac{\omega_3 b \chi_{THG} E_{1o}^3}{16 n_3 c}\right|^2 |I|^2 (1+\xi^2)^{-1} \quad . \tag{4.11}$$

But, as a standard result of gaussian-beam theory [4.7], we have

$$w_3^2(\xi)(1+\xi^2)^{-1} = w_{3o}^2 \quad , \tag{4.12}$$

where w_{3o} is the beam radius of the third-harmonic beam at its waist. As pointed out earlier,

$$w_{3o} = w_{1o}/\sqrt{3} \quad . \tag{4.13}$$

Combining (4.9-13) provides the required result (in which we have replaced all refractive indices by unity),

$$P_3 = \frac{3\pi^2}{\varepsilon_o^2 c^2 \lambda_1^4} |\chi_{THG}|^2 P_1^3 |I|^2 \tag{4.14}$$

$$= 4.2 \times 10^6 \frac{\mathcal{N}^2}{\lambda_1^4} |\chi_{THG}^{mic}|^2 P_1^3 |I|^2 \quad , \tag{4.15}$$

where P_1, P_3 are in watts, λ_1 in metres, $\mathcal{N}$ in atoms m^{-3} and χ_{THG}^{mic} is the per atom susceptibility in m^5V^{-2}. We now consider the behaviour of the dimensionless integral I. In general, I must be evaluated numerically. However, recognising that the integral has a form similar to those encountered in standard optical-diffraction problems, WARD and NEW showed that useful insight could be obtained by constructing vibration curves to represent the integral. For two limiting cases, which are of experimental interest, the integral can be evaluated analytically. In the plane-wave approximation (where the confocal parameter b is much greater than the length L of the nonlinear medium), $|I|^2$ reduces to

$$\frac{4L^2}{b^2} \operatorname{sinc}^2\left(\frac{\Delta kL}{2}\right) .$$

The expression for P_3 then reduces to the familiar form obtained by a plane-wave analysis, except for a difference of numerical factor, which accounts for the fact that the power density is not uniform across the gaussian beam. For convenience the two formulae are given together. Thus for a gaussian beam in the limit where b >> L,

$$\frac{P_3}{P_1} = \frac{3\pi^2 |\chi_{THG}|^2}{\varepsilon_o^2 c^2 \lambda_1^4} \cdot P_1^2 \cdot \left(\frac{4L^2}{b^2}\right) \operatorname{sinc}^2\left(\frac{\Delta kL}{2}\right) , \tag{4.16}$$

whereas a plane-wave analysis applied to a beam of uniform intensity (area A) yields, see (3.11)

$$\frac{P_3}{P_1} = \frac{9\pi^2 |\chi_{THG}|^2}{4\lambda_1^2 \varepsilon_o^2 c^2} \left(\frac{P_1}{A}\right)^2 L^2 \operatorname{sinc}^2\left(\frac{\Delta kL}{2}\right) . \tag{4.17}$$

Equation (4.16) can be put in a form resembling (4.17) by substituting

$$A' = \frac{1}{2} \pi w_{1o}^2 = \frac{1}{4} b\lambda_1 \quad . \tag{4.18}$$

Then (4.16) becomes

$$\frac{P_3}{P_1} = \frac{3\pi^2|\chi_{THG}|^2}{4\lambda_1^2\varepsilon_0^2c^2}\left(\frac{P_1}{A'}\right)^2 L^2 \operatorname{sinc}^2\left(\frac{\Delta kL}{2}\right) . \tag{4.19}$$

Thus A', having the dimensions of area, can be interpreted as an effective area in the sense that if, for a gaussian beam, $(P_1/A')^2$ is used in the plane wave formula (4.17), and provided that the factor 3 is used instead of 9 (corresponding to the fact that the harmonic-beam area is reduced by 3 compared to the fundamental), then the plane-wave formula correctly predicts the harmonic power in the case where $b \gg L$.

The other limiting case, of tight focussing, is where $\xi,\zeta \to \infty$; I then takes the values

$$\left.\begin{aligned} I(\Delta k,\infty,\infty) &= 0 \quad \text{if} \quad \Delta k \geq 0 \\ &= -\pi b\Delta k \exp\left(\frac{b\Delta k}{2}\right) \quad \text{if} \quad \Delta k < 0 \end{aligned}\right\} . \tag{4.20}$$

(Like MILES and HARRIS [1.15]), in this section we define $\Delta k = k_3 - 3k_1$, which is the negative of the Δk used by WARD and NEW and by us in Sect.3.2.) Equation (4.20) confirms that for a normally dispersive medium (positive dispersion, $\Delta k > 0$) and also for a nominally phase-matched medium ($\Delta k = 0$) the harmonic intensity generated by focussing in an infinite medïum (or in practical terms, focussing near the centre of a cell, with $b \ll L$) is zero.

The question now arises of what is the optimum focussing arrangement, i.e., to produce a maximum harmonic power. This proves to be a complex question to answer in general, much more so, for example, than the question of optimum focussing for second-harmonic generation in crystals. In the case of gases, there are more variables involved - e.g., the length of the medium, the number density of nonlinear atoms and the number density of any buffer atoms, in addition to the beam-focussing characteristics. We, therefore, confine ourselves to a few simple considerations on the matter and point out some of the complicating features that need to be considered in any attempt to arrive at an answer. A detailed examination of this question has been given by BJORKLUND [3.1].

From (4.20) it is found that, for a given Δk, the maximum value of I for tight focussing is achieved with $b\Delta k = -2$, in which case $|I|^2 = 5.3$. The optimum condition will, however, be different if, for example, Δk is proportional to the number density of the nonlinear species and can be varied. It is then necessary to optimise $|\chi_{THG}|^2|I|^2$ rather than simply $|I|^2$. Roughly speaking, the condition $b\Delta k = -2$ means focussing in a negatively dispersive medium so that the confocal parameter equals the coherence length L_c. ($L_c = \pi/|\Delta k|$). Under these conditions, the generated harmonic intensity is roughly that which a plane-wave analysis would indicate for one coherence length of medium. This approximate, but easily remembered, result is obtained from (4.14) by putting $|I|^2$ equal to 5.3 and noting that the beam area is $A' = \pi w_{10}^2/2 = b\lambda_1/4 = \lambda_1(2/|\Delta k|)/4 = \lambda_1 L_c/2\pi$. For confocal focussing ($b = L$) at the centre of a negatively dispersive medium ($\xi = \zeta$), $|I|^2$ takes the maximum value 2.46 when $\Delta kL = -3.5$.

These apparently simple criteria must be treated with caution, however. A parallel may be drawn here with some earlier work (in particular that of [4.8]) on optimum focussing for second-harmonic generation and parametric oscillation in crystals. The optimum focussing considered there referred to the focussing that gave maximum harmonic power or parametric gain for a given fundamental laser power. This criterion led to a tightly focussed beam and, hence, high intensity. However, there is a practical limitation to the intensity that a crystal can sustain without suffering damage. A more appropriate definition of optimum focussing under conditions in which copious laser power is available would be the focussing arrangement that gives maximum parametric gain for the maximum laser intensity that can be tolerated. Vapours and gases do not suffer irreversible damage when subjected to high intensity; nevertheless, a number of limiting and saturation processes can occur at very high intensities (these will be discussed later). Thus, although the tight-focussing scheme appears attractive (it gives a large value of $|I|^2$), it may also lead to these undesirable competing processes.

In practice, it may be better to modify Δk (by adding a dispersive buffer gas) to a smaller value, so that $L_c \simeq L$ and then use confocal focussing. (We will return to this point in more detail when we discuss the results of KILDAL and BRUECK [4.9] for THG in liquid CO/O_2 mixtures, Sect.7.3). For confocal focussing, the conversion efficiency is independent of cell length (provided that $\Delta kL = -3.5$ is satisfied) and in principle (if Δk can be adjusted at will) a very long cell and correspondingly low intensities can be used. Here again a cautionary proviso must be added. Alkali vapours, which are in many respects attractive nonlinear media, have a significant population of dimers at typical operating vapour pressures (see Sect.5.4.1). These display broad absorption bands, which over a long cell length may lead to considerable absorption of the fundamental laser beam. Finally, if we consider the situation where $b \gg L$, then Δk must be made zero for optimum harmonic efficiency. If Δk is not zero, then the maximum value that $L^2 \operatorname{sinc}^2(\frac{\Delta kL}{2})$ in (4.16) can take is $L^2/(\Delta kL/2)^2 = 4L_c^2/\pi^2$, whereas if $\Delta k = 0$, the value of $L^2 \operatorname{sinc}^2(\Delta kL/2)$ is just L^2. So, if Δk is adjusted to zero, the harmonic power increases by $(\pi L/2L_c)^2$ for the plane-wave limit. When $\Delta k \neq 0$, the efficiency is the same whether Δk is positive or negative, unlike the tight-focussing case.

For a more detailed examination of the question of optimum focussing, the reader is referred to a paper by BJORKLUND [3.1] and to a more recent paper in which TOMOV and RICHARDSON [4.10] extend the treatment to the case of fifth-harmonic generation in isotropic media. Particular examples that show good agreement between experimental results and the theoretical predictions of BJORKLUND will be given later. In passing, we mention briefly some other results of a general nature that come out of Bjorklund's analysis. He considered three types of four-wave nonlinear process; $\omega_1 + \omega_2 + \omega_3 \to \omega_4$ (of which THG is a particular example); $\omega_1 + \omega_2 - \omega_3 \to \omega_4$; $\omega_1 - \omega_2 - \omega_3 \to \omega_4$. (These are illustrated for two-photon resonance in Table 2.2.) For all three of these processes, in the tight-focussing limit, efficient generation can be achieved only for restricted ranges of Δk. For the process $\omega_1 + \omega_2 + \omega_3 \to \omega_4$, Δk must be negative; for the process $\omega_1 - \omega_2 - \omega_3 \to \omega_4$, Δk must be positive. Given the correct sign of Δk for the particular process, optimisation can be achieved by adjusting the tightness of focus. The efficiency of each of these processes depends on an integral analogous to the $|I|^2$ considered in the foregoing; this integral is maximised under tight-focussing conditions when

$$b\Delta k = \begin{cases} -2 & \text{for} \quad \omega_1 + \omega_2 + \omega_3 \to \omega_4 \\ 0 & \text{for} \quad \omega_1 + \omega_2 - \omega_3 \to \omega_4 \\ +2 & \text{for} \quad \omega_1 - \omega_2 - \omega_3 \to \omega_4 \end{cases} \quad . \tag{4.21}$$

The middle process is, therefore, unique in that $|I|^2$ can be optimised whatever the value of Δk (+ ve or - ve) simply by focussing very tightly ($b \to 0$). Again, however, a maximum value of $|I|^2$ does not necessarily imply maximum generation efficiency.

This completes our discussion of the principles of optimum focussing and phase matching. We now turn our attention to the choice of nonlinear medium and in Sect.4.4 we will indicate how phase matching is achieved in practice.

4.3 The Nonlinear Medium

HARRIS and MILES [1.14] first proposed the use of alkali vapours as media for THG from visible and near-infrared sources. Because alkali vapours have transitions with resonant frequencies in the visible region, with large oscillator strengths, the nonlinear susceptibilities are large. Also, with the fundamental laser frequency below the first principal s → p transition, and the third harmonic above this frequency, the vapour becomes negatively dispersive. Phase matching can then be achieved by adding a positively dispersive gas, such as an inert gas. These ideas were confirmed by YOUNG et al. [1.16] in generating the third harmonic of a 1.06 μm Nd laser beam in rubidium vapour. To generate shorter wavelengths, vapours such as Cd [1.19], Hg [4.11], or inert gases (e.g., [1.20, 4.12]), have been used. Although the primary consideration in choosing a nonlinear medium for a particular process is to find an atom or molecule with energy levels placed suitably to permit resonance enhancement, other factors that need to be considered are: 1) whether phase matching can be achieved either by adding another gas or choosing the frequencies appropriately, 2) whether the physical properties of the species will allow a sufficiently high pressure to be produced conveniently, 3) how to minimise the effects of various competing or saturation processes. We shall concentrate most of our discussion here on the alkali-metal vapours because they have been investigated in most detail and because the necessary data are readily available (e.g., data on the signs as well as the magnitudes of the matrix elements). Many of the considerations we shall cover are also applicable to other vapours but, in some cases, vital data needed for making calculations are not available. We shall start by examining the question of how to calculate the nonlinear susceptibility. In Sect.4.4, we consider how phase matching can be achieved, including the question of group-velocity matching, which is necessary when short (picosecond) pulses are involved. In Sect.4.5, we survey the experimental results. Finally, in Sect.4.6, we look at some of the limiting processes such as multiphoton ionisation, multiphoton absorption, breaking of the phase-matching condition, etc.

4.3.1 Nonlinear-Susceptibility Calculations

The starting point for a calculation of nonlinear susceptibility is the general expression given in (2.15). For third-harmonic generation, this expression becomes that given in (2.18) or (4.1). If the fundamental laser frequency is not in close resonance with particular levels, then it is necessary to carry out a summation over many intermediate levels and the signs, as well as the magnitudes of the matrix elements must be known, because cancellations can become important in the sum. Detailed calculations of this nature were first presented by MILES and HARRIS [1.15] for all of the alkali vapours, for third-harmonic generation of fundamental frequencies over the range 3 μm - 0.5 μm. In their paper, they provide tables of dipole matrix elements $\langle z \rangle \equiv \langle n\ell m_\ell = 0|z|n'\ell + 1\ m_\ell = 0\rangle$ (including their signs) and transition frequencies for the transitions that make the major contributions to the susceptibility. The magnitudes of the matrix elements were obtained from tabulated oscillator strengths (the calculations of ANDERSON and ZILITIS [4.13,14] were used) and the signs were obtained from calculations by BATES and DAMGAARD [4.15], with corrections of signs as pointed out by BEBB [4.16]. EICHER [2.116] has since given more extensive tables of dipole radial matrix elements. The quantity tabulated by EICHER is denoted by the symbol $\langle n\ell||r||n'\ell + 1\rangle$, and is the same as the radial matrix element $\langle n\ell|r|n'\ell + 1\rangle$. The relation between this quantity and $\langle z \rangle$ is given in (2.109).

In Eicher's work, the effect of spin-orbit splitting was included, whereas MILES and HARRIS neglected this and replaced each of the $s\text{-}p_{1/2}$, $s\text{-}p_{3/2}$ resonance doublets with weighted-average single resonance lines. Another useful and extensive tabulation of data, in which spin-orbit splittings are included, has been produced by WARNER [4.17]. Fine structure becomes significant when detunings from resonance are small, i.e., comparable to the spin-orbit splitting. EICHER also presented calculated values of the third-harmonic susceptibility, but for a more restricted range of fundamental frequencies than MILES and HARRIS; EICHER concentrated on the region around the Nd laser (1.06 μm) and I laser (1.315 μm) wavelengths and around the second, third and fourth harmonics of those lasers. A further refinement of Eicher's calculations is to consider the behaviour of the susceptibility in the region of an exact resonance (one or two photon). Resonance linewidths of 0.1 cm^{-1} and 1 cm^{-1} were assumed in these calculations.

Figures 4.3-8 show the calculated values of third-harmonic susceptibility per atom (in esu) obtained by MILES and HARRIS. To convert to SI units $[m^5V^{-2}]$, their values must be multiplied by $4\pi/9 \times 10^{-14}$ (see appendix). Figure 4.5 shows the energy levels of sodium and also indicates the photon energies for the first, second, and third harmonic of ruby (0.6943 μm) and Nd (1.06 μm) lasers. The ruby wavelength is close to two-photon resonance with the 3s-3d transition; this is indicated by the resonance of susceptibility near 0.7 μm in Fig.4.4 (labelled at the top of the figure by 3d).

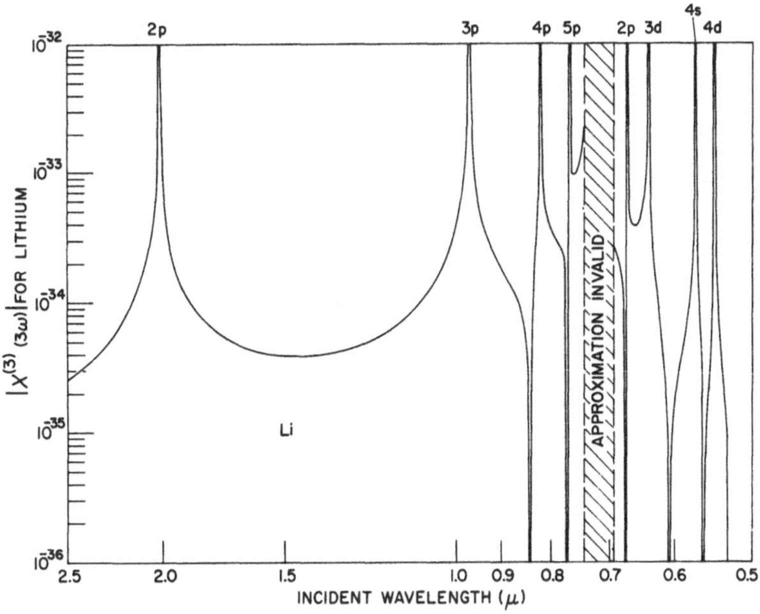

Fig.4.3. Nonlinear susceptibility of lithium versus wavelength. (After [1.15])

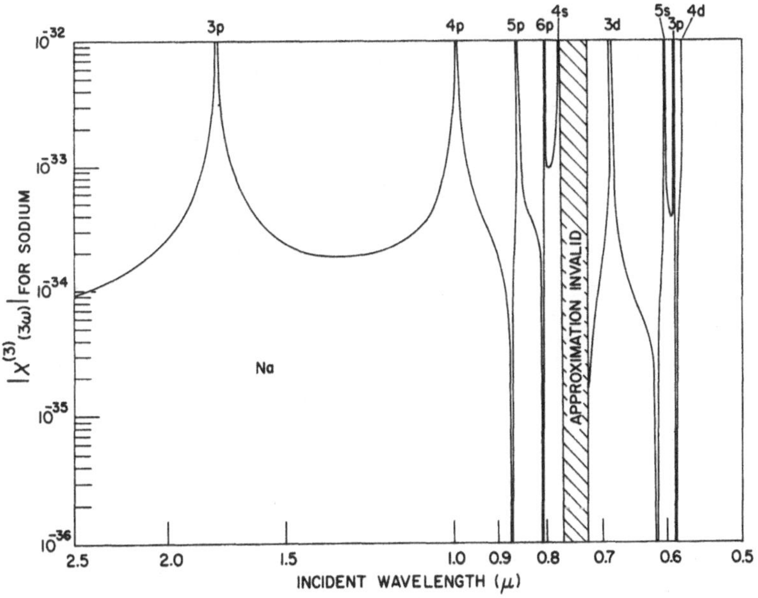

Fig.4.4. Nonlinear susceptibility of sodium versus incident wavelength. (After [1.15])

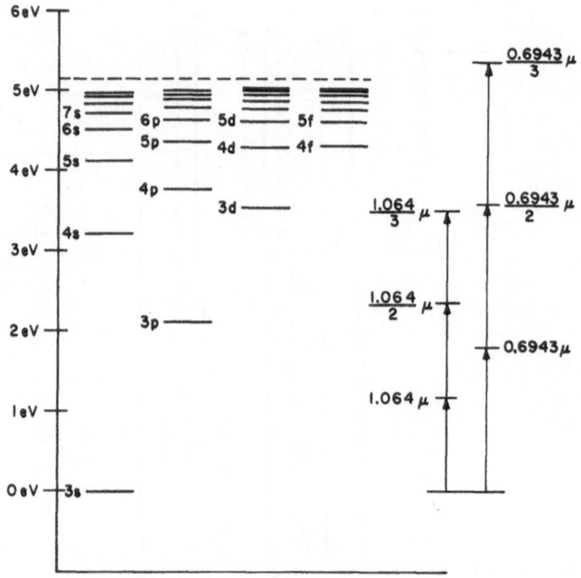

Fig.4.5. Energy levels of sodium. (After [1.15])

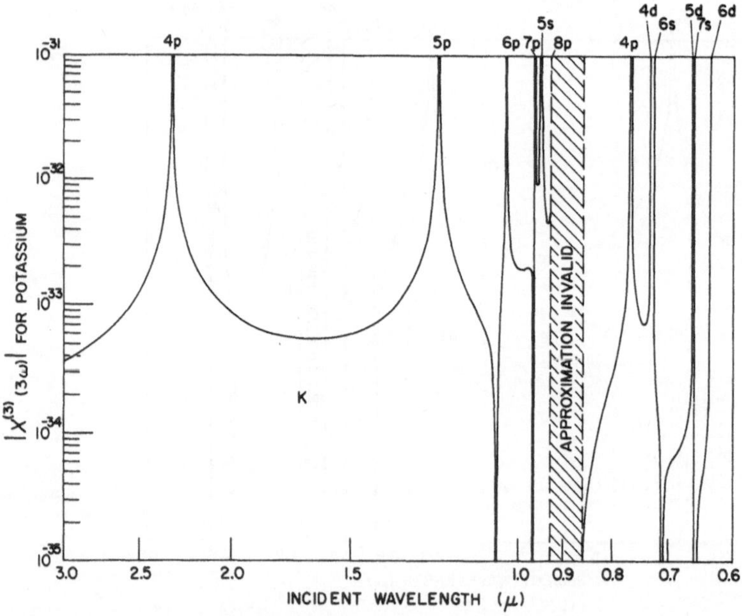

Fig.4.6. Nonlinear susceptibility of potassium. (After [1.15])

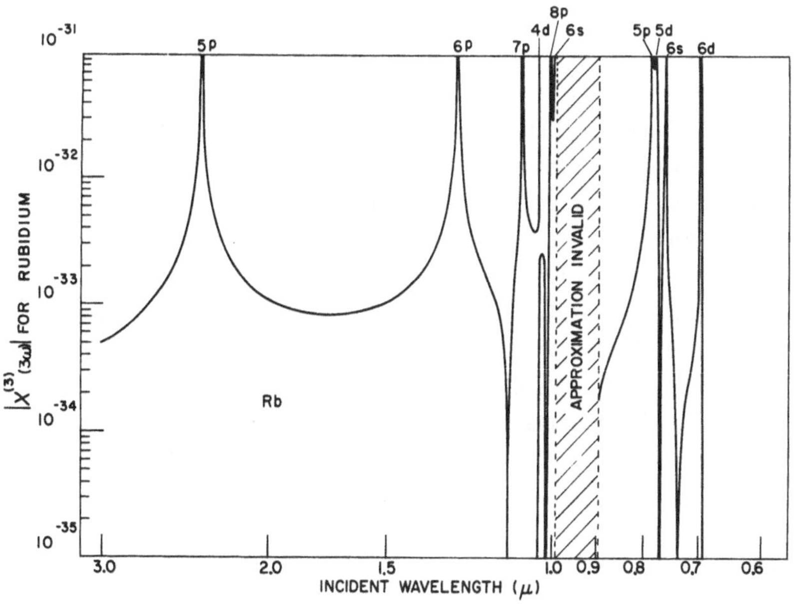

Fig.4.7. Nonlinear susceptibility of rubidium. (After [1.15])

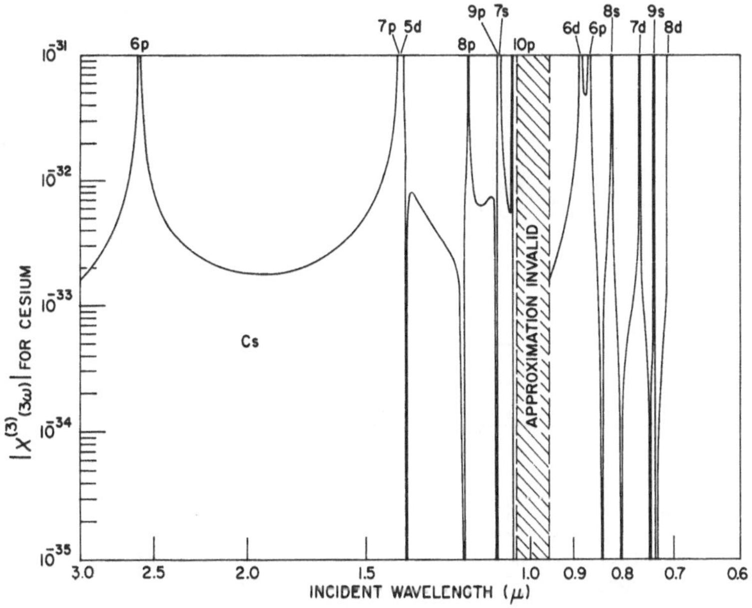

Fig.4.8. Nonlinear susceptibility of cesium. (After [1.15])

Other resonance structures are seen in Fig.4.4, such as single-photon resonance (of fundamental or third harmonic) with p levels and two-photon resonance with s or d levels.

In their calculations, MILES and HARRIS used the approximation of including summation of the contribution from only 12 levels, the lowest four of the s, p, d levels. For regions in which the first, second, or third harmonic lies above these levels and yet below the ionisation potential, that approximation is invalid; such regions are shown in the figures.

Figures 4.9-10 show results calculated by EICHER for the case of potassium (also in esu). The strong cancellation of susceptibility near 1.06 μm seen in Fig.4.6 can also be seen in Fig.4.9 (curve a), and the strong resonance around 0.66 μm (close to the second harmonic of the iodine laser) is seen in curve b_2 of Fig.4.9; it is shown in more detail, on an expanded scale, in Fig.4.10.

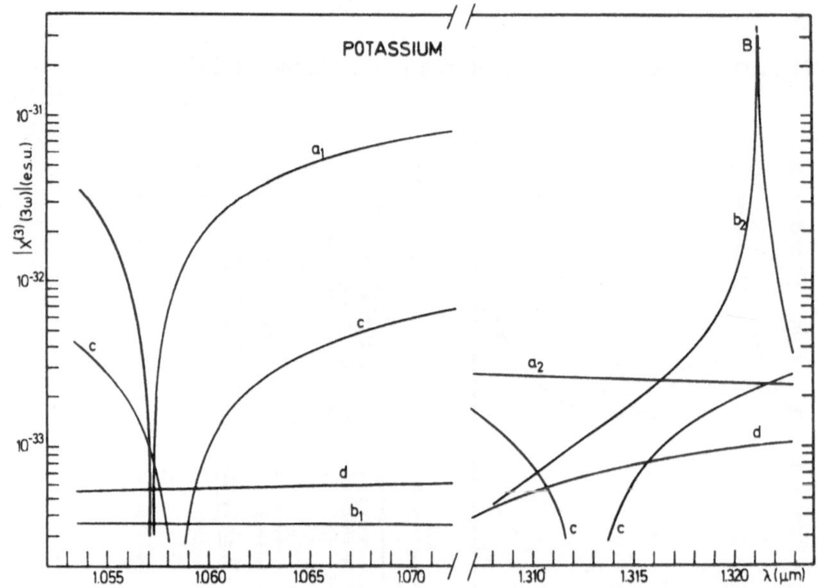

Fig.4.9. Third-order susceptibility of potassium versus wavelength in the vicinity of neodymium- and iodine-laser lines. Curves a, b, c, and d are calculated for the first, second, third, and fourth harmonics of the incident laser radiation, respectively. For each curve the logarithmic scale must be multiplied by a scaling factor: $a_1(10^{-2})$, $a_2(1)$, $b_1(10^{-2})$, $b_2(10^{-1})$, $c(10^{-4})$, and $d(10^{-3})$. The transition linewidth Γ is assumed to be 0.1 cm^{-1}. The resonance B is enlarged in Fig.4.10. (After [2.116])

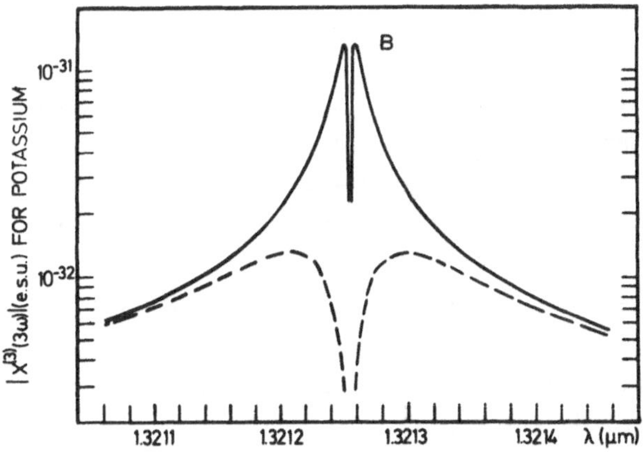

Fig.4.10. Resonant third-order susceptibility of potassium for the second harmonic of the iodine-laser line, enlarged from Fig.4.9. Solid and dashed lines correspond to the transition linewidhts $\Gamma = 0.1\ cm^{-1}$ and $\Gamma = 1\ cm^{-1}$, respectively. (After [2.116])

In comparing the calculated susceptibilites from EICHER with those of MILES and HARRIS and with experimentally measured values, it is important to note that the susceptibility defined by EICHER is a factor of two greater than that of MILES and HARRIS. Thus, for THG from 1.06 μm in rubidium vapour, MILES and HARRIS predict a χ_{THG} of 6.15×10^{-33}esu, whereas EICHER predicts 1.072×10^{-32}esu. The difference of value is not an exact factor of two because EICHER includes many more levels in these calculations. A measured value of 1.4×10^{-32} esu was obtained by HARRIS and co-workers [1.16] and in view of the difficulties generally experienced in obtaining accurate susceptibility measurements, the agreement can be considered quite good. In fact, MILES and HARRIS report that, with a later (unspecified) correction of the theory, the experimental result was found to agree within 15%. OHASHI et al. [4.18] also measured the susceptibility for tripling 1.06 μm radiation in Rb vapour and obtained a value of $(6.5 \pm 0.7) \times 10^{-33}$esu. Again, the agreement is very good and is probably helped by the absence of complicating features, such as a very close resonance, and by the fact that the third harmonic lies below the ionisation potential.

When a close resonance is involved, the calculation of harmonic-generation efficiency becomes more complicated because it must incorporate the effect of a finite linewidth of the transition and the finite linewidths of the incident and generated radiation. As pointed out in Sect.2.5, by use of a two-photon resonance, one of the frequency denominators in the susceptibility expression is made very small. The susceptibility and harmonic-generation efficiency can then become large, without excessive absorption via the two-photon transition. (The question of limitations imposed by this two-photon absorption is discussed in Sect.4.4.3). The two-photon resonance situation is, therefore, of considerable practical interest and LEUNG et al. [1.23] and STAPPAERTS et al. [4.19] have examined the effects of finite laser and transition linewidths on the harmonic-generation process.

In the experiment of LEUNG et al., ruby laser radiation was tripled in Cs, where an exact two-photon resonance ($6s^2S_{1/2} \rightarrow 9d^2D_{3/2}$) can be achieved. The energy of the third-harmonic photon is greater than the ionisation potential. They first calculated the susceptibility for a single atom and then synthesised an effective susceptibility for their experimental situation, in which they took account of the effect of Doppler broadening and the fundamental laser spectrum. The single-atom susceptibility calculation contained two refinements that were not included by MILES and HARRIS or EICHER, viz. to include the j dependence of radial matrix elements, which are significant in caesium, and to include the contribution of bound-free transitions to the susceptibility. In fact, these continuum terms were found to make only a small contribution (less than 20%) to the over-all susceptibility. The results obtained were: a synthesised theoretical value of 3×10^{-30}esu/atom (50% uncertainty) and an experimental value of 10^{-30}esu/atom (with a factor-of-five uncertainty).

STAPPAERTS et al. [4.19] considered the case of a two-photon transition that displays both pressure broadening and Doppler broadening. The finite linewidth of the incident laser radiation was modelled by a superposition of randomly phased discrete laser modes. As expected, their analysis predicts that when the laser is detuned several linewidths away from the two-photon transition frequency, the harmonic efficiency decreases as $(\text{detuning})^{-2}$. This follows because the efficiency is proportional to the square of the susceptibility, and the susceptibility is inversely proportional to the detuning [see (2.18)]. In view of this result, it is perhaps not so obvious that when the laser is tuned to exact two-photon resonance, the harmonic efficiency decreases only linearly with laser linewidth, i.e., as $(\text{linewidth})^{-1}$. However, the analysis of STAPPAERTS et al. shows this to be the case, and the explanation lies in the fact that exact two-photon resonance is achieved, not only by the pump mode whose frequency is exactly half the transition frequency, but also by any combination of two-mode frequencies whose sum lies within the transition linewidth. In practical terms, this means that efforts made to reduce the laser linewidth, often with a consequent loss of power, may not significantly improve the efficiency. As an interesting experimental demonstration of the predictions of their analysis, STAPPAERTS et al. showed that observations of harmonic-generation efficiency versus buffer-gas pressure could be used to measure the pressure broadening of the 3s → 3d two-photon transition in Na. This technique permits use of a laser linewidth (~ 2 cm^{-1}) well in excess of the transition linewidth being measured; furthermore, pressure broadenings much less than the (0.04 cm^{-1}) Doppler width could be measured.

4.3.2 Contribution of Autoionising Transitions to the Susceptibility

Although it is found that for the alkalis there is no dramatic contribution from the continuum (either to third-harmonic susceptibility or third-harmonic absorption) the same is not true of alkaline earths. This is because the alkaline earths show a strong resonance structure within the continuum due to the presence of autoionising levels.

The phenomenon of autoionisation may be explained by first considering, in the usual fashion, the atomic-orbital picture that arises from use of an approximate atomic hamiltonian. It is then possible for there to be discrete states embedded in the photoionisation continuum, corresponding to the simultaneous excitation of, say, two electrons to a bound state of higher energy than the ionisation limit that arises from the excitation of one electron alone. If higher-order terms in the atomic hamiltonian connect this discrete state to the continuum, rapid radiationless tunnelling from the bound state to the continuum can occur, during which one electron returns to its ground state and the other is detached, taking away the excess energy. The discrete line will, therefore, be broadened as a consequence of its shortened lifetime, and the atomic spectrum becomes continuous. More generally, the spectrum of eigenstates of the exact hamiltonian will display a continuum, in which there is considerable structure, with associated changes in the photoionisation rate. In the case that the orbital picture described above is realistic (for example, little configuration mixing) the discrete states lie in the vicinity of peaks in the density of the true continuum states. (An analogous situation that is important in molecules is predissociation, where the picture of a bound state embedded in a continuum may be based on Born-Oppenheimer states, and the true molecular eigenstates lead to highly structured predissociation continua.)

An example of the autoionisation resonances seen in Sr is shown in Fig.4.11. It is seen that they have, besides broadening, a characteristic

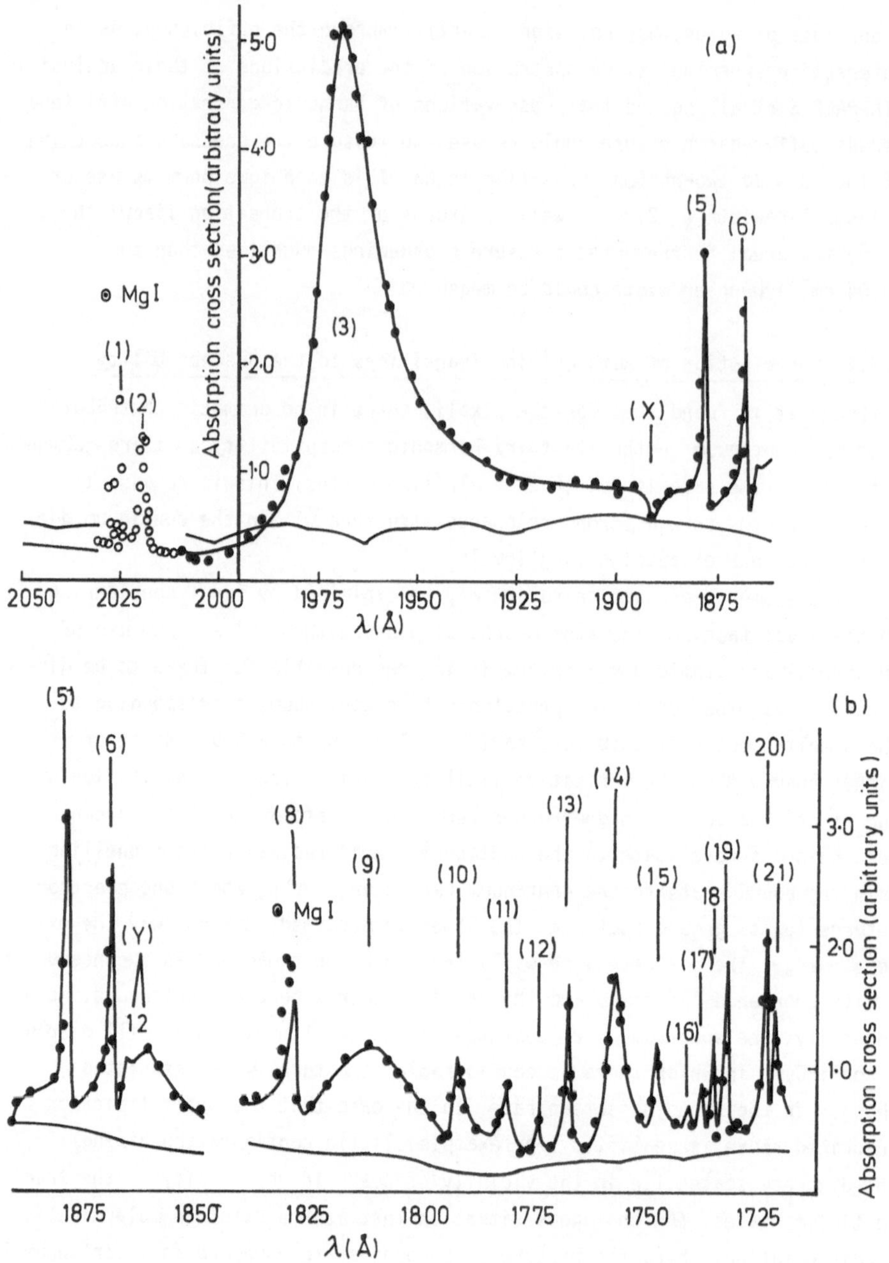

Fig.4.11. Autoionisation resonances in SrI. Plotted points are measured values of relative absorption cross section. The solid curve is a summation of Fano profiles determined by parameters from Table 4.1. (After [4.20])

asymmetric line shape. This has been explained by FANO [4.21], who represents the true atomic states near an autoionising resonance as a superposition of a discrete and continuum state, in accordance with the above description; the line shape arises as a result of interference between the two components, which is a rapid function of the energy of the true autoionising state, near the discrete level. Fano's calculations show that the ratio of absorption cross section in the region of an autoionising line to that of the uperturbed photoionisation continuum is given by the square of the ratio of dipole matrix elements

$$\left|\frac{\mu_{\psi_\Omega g}}{\mu_{\psi^o_\Omega g}}\right|^2 = \frac{(q+\varepsilon)^2}{1+\varepsilon^2} \quad . \tag{4.22}$$

Here ψ_Ω, ψ^o_Ω are, respectively, the perturbed and unperturbed continuum states, q is a dimensionless parameter that determines the line shape and ε is the frequency detuning from the Fano resonance at Ω, normalised with respect to its half-width $\Gamma/2$, ie $\varepsilon = (\omega - \Omega)/(\Gamma/2)$. Figure 4.12 shows some examples of Fano profiles, with plots of $(q + \varepsilon)^2/(1 + \varepsilon^2)$, for various values of q. The autoionising resonances shown in Fig.4.11 have been fitted with Fano profiles, i.e., the linewidth and q have been chosen to give a best fit. The results of this fitting procedure are given in Table 4.1. It can be seen that the q values cover a considerable range, both positive and negative; for the larger q values, the absorption coefficient at the peak of the resonance may be enhanced by three or four orders of magnitude. Similar enhancements have been observed by HODGSON et al. [1.22] when generating sum-frequency radiation in Sr vapour. In their experiment, one laser had its frequency (ν_1) fixed on a two-photon resonance while the other laser had its frequency (ν_2) tuned, thus producing tunable sum-frequency radiation $2\nu_1 + \nu_2 = \nu_{vuv}$, (see Fig.4.13). The vuv signal showed very large variations while was ν_2 tuned; the positions of the signal peaks corresponded to known autoionising levels. An example of this behaviour is shown in Fig.4.14, which shows the vuv-generation spectrum at wavelengths that correspond to the extreme right of the range shown in Fig.4.11. The good quality of this spectrum, with well-resolved lines, indicates the potential of this technique for vuv spectroscopy. Further examples of these vuv-generation spectra are given by WYNNE and SOROKIN [4.22], and SOROKIN et al. [4.23].

ARMSTRONG and WYNNE [4.24,25] have shown how the Fano theory of autoionising levels can be incorporated into the calculation of nonlinear susceptibility and thus lead to an expression for the frequency dependence of

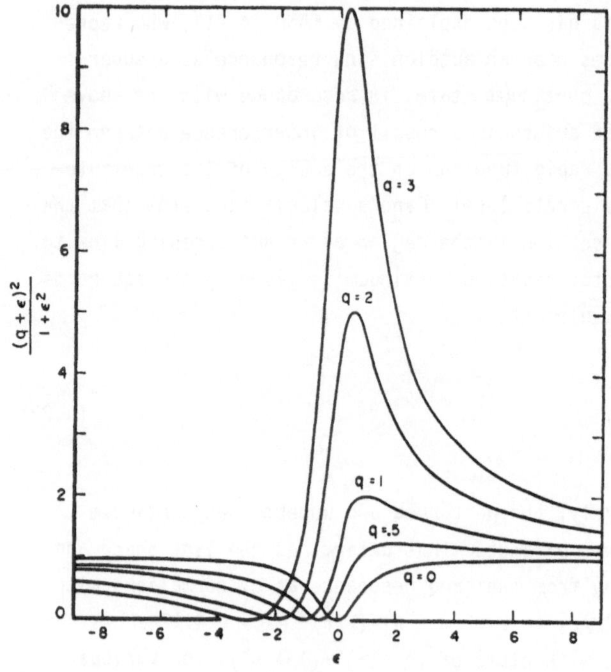

Fig.4.12. Plots of the function $(q + \varepsilon)^2/(1 + \varepsilon^2)$ for several values of the parameter q. (After [4.21])

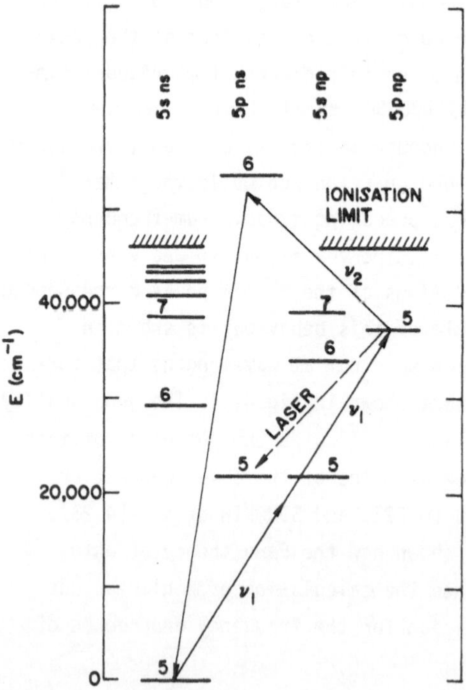

Fig.4.13. Partial energy-level diagram of Sr atom, showing resonantly enhanced tunable $2\nu_1+\nu_2$ generation using a $5p^2$ intermediate state. (After [1.22])

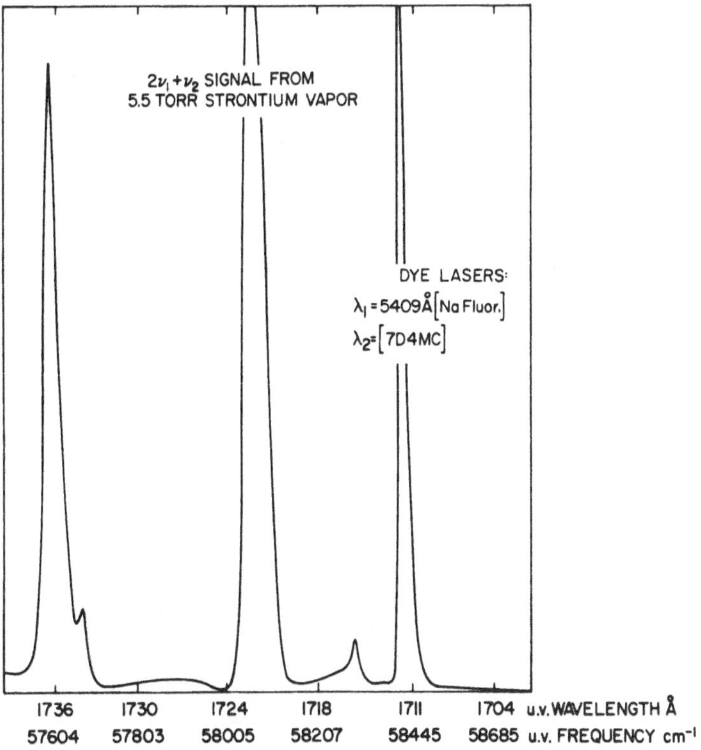

Fig.4.14. Intensity of vuv generated in Sr in a range that includes the extreme right-hand portion of Fig.4.11. (After [4.22])

the generated vuv signal in terms of the Fano q parameters. The starting point of Armstrong and Wynne's analysis is the susceptibility expression (2.15) particularised to the case $\omega_1 + \omega_1 + \omega_2 \rightarrow \omega_{vuv}$, for which the susceptibility is $\chi_{vuv} \equiv \chi^{(3)}(-\omega_{vuv};\omega_1,\omega_1,\omega_2)$. The summation in that expression is simplified by assuming a two-photon resonance $2\omega_1 \approx \Omega_{j'g}$, with one dominant intermediate level j', and a near resonance of the generated ω_{vuv} with a Fano state, (Fig.4.15). The susceptibility expression then becomes

$$\chi_{vuv} \simeq \frac{\mathscr{N}}{3\hbar^3\varepsilon_0} \frac{\mu_{j'j}\mu_{jg}}{(\Omega_{j'g} - 2\omega_1 - i\Gamma_{j'g})(\Omega_{jg} - \omega_1 - i\Gamma_{jg})}$$

$$\times \int_{-\infty}^{+\infty} \frac{\mu_{g\psi_\Omega}\mu_{\psi_\Omega j'}d\omega}{(\omega-\omega_{vuv}-i\Gamma(\omega))} \quad . \tag{4.23}$$

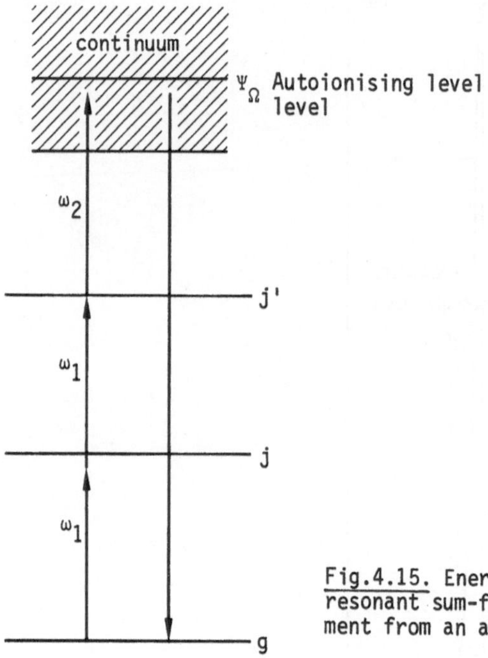

Fig.4.15. Energy-level scheme for two-photon resonant sum-frequency generation with enhancement from an autoionising resonance

The integral in (4.23) includes both the contribution to χ_{vuv} of the autoionising state and that of the continuum with which it interacts, where as described by FANO [4.21]

$$\mu_{g\psi_\Omega} = \mu_{g\psi_\Omega^o}(q_g \sin\Delta - \cos\Delta) \tag{4.24}$$

$$\mu_{\psi_\Omega j'} = \mu_{\psi_\Omega^o j'}(q_{j'} \sin\Delta - \cos\Delta) \quad . \tag{4.25}$$

In (4.24) and (4.25), $\mu_{g\psi_\Omega^o}$ is the dipole matrix element between the ground state g and the unperturbed continuum state ψ_Ω^o; q_g is the Fano q parameter appropriate to the transition between ground state and autoionising state [i.e., the same as the q in (4.22)] and $q_{j'}$ is that for the transition between the two-photon resonant state j' and the autoionising state. The quantity Δ, a function only of the autoionising state, is given by

$$\Delta = -\arctan(1/\varepsilon) \quad . \tag{4.26}$$

The main contribution to the integral in (4.23) comes from the region around $\omega = \omega_{vuv}$, so, after substitution of (4.24-26) into the integral the slowly varying quantities $\mu_{g\psi_\Omega^o}$ and $\mu_{\psi_\Omega^o j'}$ can be removed from the integral, thus giving

$$\chi_{vuv} = \frac{\mathcal{N}}{3\hbar^3\varepsilon_0} \frac{\mu_{g\psi^o_\Omega}\mu_{\psi^o_\Omega j'}\mu_{j'j}\mu_{jg}F(\omega_{vuv})}{(\Omega_{j'g} - 2\omega_1 - i\Gamma_{j'g})(\Omega_{jg} - \omega_1 - i\Gamma_{jg})} , \tag{4.27}$$

where

$$F(\omega_{vuv}) = \int_{-\infty}^{+\infty} \frac{[q_g q_{j'} + \varepsilon^2 + \varepsilon(q_g + q_{j'})]d\varepsilon}{[\varepsilon - x - 2i\Gamma(\varepsilon)/\Gamma](1+\varepsilon^2)} , \tag{4.28}$$

and

$$x = (\omega_{vuv} - \Omega)/(\Gamma/2) \quad .$$

In evaluating this integral, ARMSTRONG and WYNNE took, in effect, only the principal part of the integral; this meant that they had derived only the real part of χ_{vuv}. In fact, the generated vuv signal depends on both the real and imaginary parts, because it is proportional to $|\chi_{vuv}|^2$. ARMSTRONG and BEERS [4.26] showed that, when the imaginary part of χ_{vuv} was also included, the theoretical expression gave a much better fit with the experimental data. The final expression of Armstrong and Beers is

$$|\chi_{vuv}|^2 \sim |\mu_{g\psi^o_\Omega}|^2 |\mu_{\psi^o_\Omega j'}|^2 \left\{ \frac{q_g^2 q_{j'}^2 + [x + q_g + q_{j'}]^2}{(1+x^2)} \right\} \quad . \tag{4.29}$$

Figures 4.16 and 4.17 show the results of fitting this expression to experimental data of ARMSTRONG and WYNNE [4.24]. Both sets of data refer to uv generation around the autoionising resonance at 1867Å, belonging to the 4d4f configuration (resonance 6 in Fig.4.11). The data of Fig.4.16 were obtained by use of $5s5d\,^1D_2$ as the two-photon resonant state, whereas in Fig.4.17 the $5p^2\,^1D_2$ state was involved. From (4.29), it can be seen that as x becomes large, i.e., well away from the autoionising resonance, $|\chi_{vuv}|^2$ reduces to $\sim|\mu_{g\psi^o_\Omega}\mu_{\psi^o_\Omega j'}|^2$. For large values of q_g, $q_{j'}$, the signal at resonance ($x = 0$) is enhanced by a factor of $\sim q_g^1 q_{j'}^2$, which may be as great as $\sim 10^4$. Thus, the presence of an autoionising resonance can become a vital factor for efficient vuv generation by sum mixing (see Sect.4.4.3).[1]

---

[1] It should, however, be noted, as pointed out by SCHEINGRABER et al. [4.27], that the autoionising resonance may not enhance the third-harmonic output if the resonance also causes strong absorption of the third-harmonic.

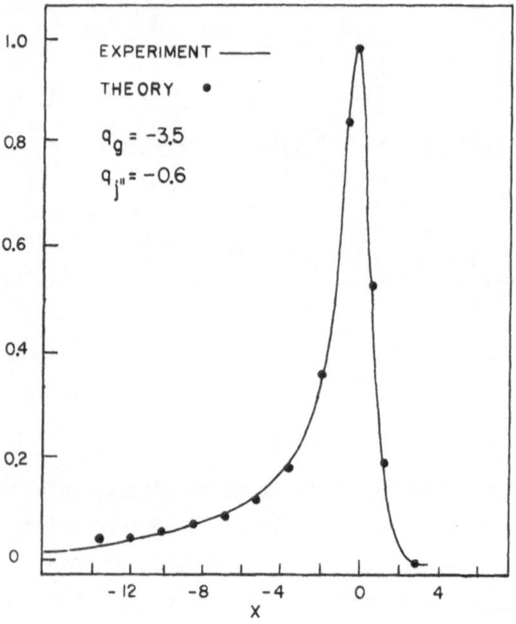

Fig.4.16. Comparison of experimental data of ARMSTRONG and WYNNE [4.24] with the theory of ARMSTRONG and BEERS [4.26], (4.29), for the intermediate state (j'), $5s5d\,^1D_2$ of Sr I. The vertical scale is arbitrarily set to unity, and $x = (\omega_{vuv} - \bar{\Omega})/(\Gamma/2)$. (After [4.26])

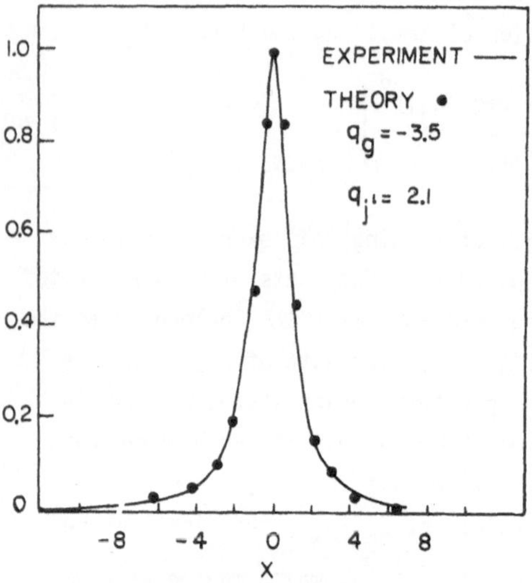

Fig.4.17. Comparison of experimental data of ARMSTRONG and WYNNE [4.24] with the theory of ARMSTRONG and BEERS [4.26], (4.29), for the intermediate state (j'), $5p^2\,^1D_2$ of Sr I. The vertical scale is arbitrarily set to unity and $x = (\omega_{vuv} - \bar{\Omega})/(\Gamma/2)$. (After [4.26])

Unfortunately, the necessary data on autoionising levels are as yet far from complete, because, although q_g values may be available from ground-state vuv-absorption data, such as in Table 4.1, the data on $q_{j'}$, are limited.

Table 4.1. Results of analysis of Sr absorption spectrum (Fig.4.11). (After [4.20])

Resonance no.	Transition	λ_r (Å)	Γ (cm^{-1})	-q
	5^1S_0 to			
1	$4d6p\,^3D_1^o$	2024	too narrow for analysis	
2	$^3P_1^o$	2018	53.1	
3	$^1P_1^o$	1970	415	- 5.2
X	$5p6s\,^3P_1^o$	1891	83.8	0.10
5	$4d(^2D_{3/2})4f[3/2]_1^o$	1878.0	17.4	4.6
6	$4d(^2D_{5/2})4f[3/2]_1^o$	1867.9	20.3	3.5
Y_1	$5p6s\,^1P_1^o$ +	1865.5	49.5	- 1.9
Y_2	$4d(^2D_{5/2})4f[3/2]_1^o$?	1859.6	291	-13
8	$4d7p\,^3P_1^o$	1827.2	51.1	2.5
9	$^1P_1^o$	1810.1	700	
10	$4d(^2D_{3/2})5f[3/2]_1^o$	1791.4	68.4	- 2.6
11	$4d(^2D_{5/2})5f[3/2]_1^o$	1780.1	75.3	4.8
12	$4d8p\,^3D_1^o$	1772.8	19.0	7.1
13	$^3P_1^o$	1766.7	36.7	27
14	$^1P_1^o$	1756.8	132	- 6.2
15	$4d(^2D_{3/2})6f[3/2]_1^o$	1747.1	47.6	- 3.1
16	$4d(^2D_{5/2})6f[1/2]_1^o$	1740.3	53.1	
17	$4d(^2D_{5/2})6f[3/2]_1^o$	1737.4	26.5	
18	$4d9p\,^3D_1^o$	1735.1	14.3	
19	$^3P_1^o$	1732.2	24.4	
20	$4d(^2D_{3/2})7f[3/2]_1^o$	1723.5	31.0	3.7
21	$4d9p\,^1P_1^o$	1722.2	59.9	- 2.5

4.3.3 Other Nonlinear Media

Nonlinear susceptibility data, experimental or theoretical, are rather sparse for atoms other than the alkalis or alkaline earths. Even for the alkaline earths, vital information is often missing, such as the value of oscillator strengths for transitions between excited states. Also, the Bates-Damgaard approach works less well than for the alkalis and data on signs of matrix elements are generally lacking. Given the necessary oscillator-strength data, even in the absence of data on signs, a rough estimate of susceptibility can be made when a single path through the atomic levels dominates the nonlinear optical susceptibility. The summation over levels [as in (2.15)] can then be dropped and the susceptibility has the form, (for THG)

$$\chi_{THG} = \frac{\mathcal{N}}{\hbar^3\varepsilon_0}\frac{\mu_{gj''}\mu_{j''j'}\mu_{j'j}\mu_{jg}}{(\Omega_{j''g}-3\omega)(\Omega_{j'g}-2\omega)(\Omega_{jg}-\omega)} \propto \frac{f^{1/2}_{gj''}f^{1/2}_{j''j'}f^{1/2}_{j'j}f^{1/2}_{jg}}{(\Omega_{j''g}-3\omega)(\Omega_{j'g}-2\omega)(\Omega_{jg}-\omega)} \quad , \quad (4.30)$$

where the f are oscillator strengths.

A number of other nonlinear media, besides alkalis and alkaline earths have been considered (and some actually investigated experimentally) for harmonic and sum-frequency generation. Table 4.2 lists these media, with references to publications that give experimental results or calculated data.

Table 4.2. Elements (other than the alkalis and alkaline earths) for which experimental or theoretical work has been published on harmonic generation and sum-frequency generation

Element	Reference
H	[4.28]
Hg	[4.11,29-31]
Cd	[1.19, 4.10,31]
Zn	[4.10,3]
Tl	[4.32]
Eu, Yb	[4.22]
He	[4.11]
He, Ne	[4.12,33,34]
Kr	[4.30,35,36]
Xe	[1.20,25, 4.37]
Ar	[4.38]

4.3.4 Higher-Order Processes

In addition to third-harmonic generation and sum mixing a number of authors have discussed fifth-order processes [1.25,4.10,12,31,34,39] and even 7th- and 15th-order processes [1.25]. So far, experimental results on these higher-order processes in vapours include generation of the fifth and seventh harmonics of 266.1nm in He and Ne [4.12,33,34,39], fifth-harmonic of 1.06 μm in Na [4.31,40] and ninth-harmonic of 1.06 μm in Na [4.41]. Some of the shortest wavelengths generated so far are; 57.0nm by third-harmonic generation in argon gas of the 171 nm output from a Xe_2 laser [4.38]; 53.2nm by fifth-harmonic generation of 266.1nm radiation in He and Ne [4.12]; 38nm by seventh-harmonic generation of 266.1nm in He [4.33]. To obtain resonance enhancement for generation of such short-wavelength radiation it is necessary to go to media with very high ionisation potentials, such as the inert gases. HSU et al. [4.11] have generated 120.28nm radiation in Hg vapour; this is particularly interesting because it is half the frequency of the $1s\,^1S_0$ - $2s\,^1S_0$ two-photon transition in He. By adding a third photon to two photons of 120.28nm in He, it should be possible to achieve resonantly enhanced sum-frequency generation tunable around 40nm.

Extension of these techniques to yet shorter wavelengths will require using ions as the nonlinear medium, because the higher ionisation potential of a singly ionised atom can provide discrete levels for resonance enhancement at much higher energies than are available from even the inert gases. One scheme suggested by HARRIS [1.25] consists of fifteenth-harmonic generation of 266nm radiation in Li^+ to produce 17.73nm. Ionisation of Li would first be achieved through photoionisation by an incident laser pulse. There is ample evidence that high concentrations of ions can be produced in this way (as reported for example by McILRATH and LUCATORTO [4.42] for the case of Li); because the ions would be produced in a tightly confined cylindrical region, the nonlinear ionic medium would have a geometry well suited to subsequent harmonic generation. WYNNE and SOROKIN [4.22] have, in fact, recently observed THG from Ca ions that were produced in the focal region of a dye-laser beam by two-photon ionisation. The same laser provided the fundamental frequency for THG.

4.4 Phase Matching Techniques

4.4.1 General Experimental Considerations

In Sect.4.2, it was shown that under tight-focussing conditions ($b \ll L$) the third-harmonic efficiency is maximised with a negative dispersion that satisfies $b\Delta k = -2$. It was also pointed out that, despite the attractive simplicity of the tight-focussing scheme, it may be desirable to adjust Δk to a smaller value so as to allow an increase of b and thus reduce the fundamental intensity below the level at which serious saturation effects occur. The value of Δk may be adjusted in three ways, by adding a dispersive buffer gas (or vapour), by choosing the interacting frequencies to exploit the dispersive characteristics of the nonlinear medium itself, or by varying the pressure of the medium (in the latter case, see Sect.6.3). First we examine the use of a buffer gas.

The principle of phase matching by means of a buffer gas is illustrated in Fig.4.18 for the particular case of THG of a 1.06 μm fundamental in rubidium vapour, with Xe as the buffer gas. Because the 1.06 μm input has a lower frequency than the 5s-5p transition (oscillator strength 1.09) and the third harmonic at 0.357 μm lies above this transition, the dispersion $\Delta k = k_3 - 3k_1$ is negative. The refractive index of rubidium as a function of wavelength is shown as the solid curve in Fig.4.18. The refractive index of Xe gas shows normal dispersion over this range of wavelengths (because both wavelengths are below any Xe transitions); its refractive index is indicated by the dotted line in Fig.4.18. By adding a sufficient pressure of Xe vapour the over-all refractive index of the mixture (the sum of the solid and dotted curves in Fig.4.18) can be made equal at 1.06 μm and 0.357 μm, hence $\Delta k = 6\pi(\eta_3 - \eta_1)/\lambda_{1.06} = 0$. To calculate the required ratio of pressures of buffer gas to metal vapour it is necessary to know the refractive index versus wavelength for both constituents. The refractive index is given by the Sellmeier equation

$$\eta(\lambda) - 1 = \frac{\mathcal{N}e^2}{8\pi^2 mc^2\varepsilon_0} \sum_{i,j} \frac{\overline{\rho(i)}f_{ij}}{\left(\frac{1}{\lambda_{ij}^2} - \frac{1}{\lambda^2}\right)} \tag{4.31}$$

$$= \frac{\mathcal{N}r_e}{2\pi} \sum_{i,j} \frac{\overline{\rho(i)}f_{ij}}{\left(\frac{1}{\lambda_{ij}^2} - \frac{1}{\lambda^2}\right)} , \tag{4.32}$$

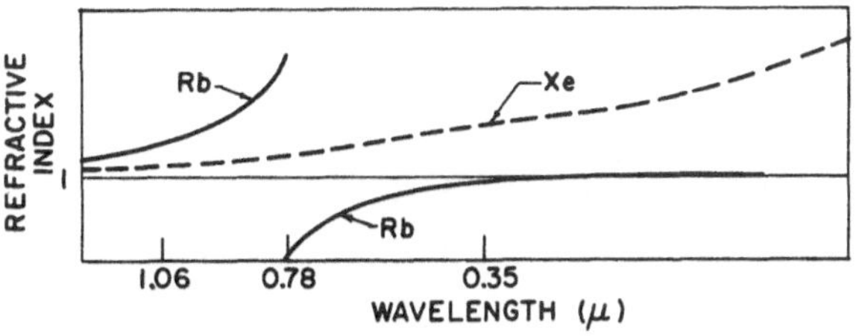

Fig.4.18. Refractive indices of rubidium and xenon versus wavelength. (After [1.15])

where $\mathcal{N}$ is the number density of the atoms, r_e is the classical electron radius (2.8×10^{-15}m), f_{ij} is the oscillator strength of the transition of wavelength λ_{ij}, from level i to level j, and $\overline{\rho(i)}$ is the fractional population of level i.

The density $\mathcal{N}$ is related to the vapour pressure p (torr) by

$$\mathcal{N} = 9.66084 \times 10^{24} \frac{p(\text{torr})}{T} \text{ atoms/m}^3 \quad . \tag{4.33}$$

A useful collection of vapour-pressure data for a wide range of materials (81 elements) has been given by HONIG and KRAMER [4.43] and some more recent data on the alkalis and alkaline earths are given by SCHINS et al. [4.44].

For the alkalis, at pressures around 1 torr, the vapour pressure p (in torr) is well approximated by the expression

$$p(\text{torr}) \simeq \exp(-\,a/T + d) \quad , \tag{4.34}$$

where the constants a and d (taken from MILES and HARRIS) are given in Table 4.3, along with some typical results. From (4.33,34), we deduce $\mathcal{N} = (9.66084 \times 10^{24}/a)(d - \ln p)p$. For the inert gases, a useful compilation and critical survey of refractive-index data has been given by LEONARD [4.45]. However, the data become sparse and less accurate as wavelengths approach the vuv. In practice, the question whether or not the medium is negatively dispersive may then need to be answered by observation of the phase matching behaviour rather than from the existing refractive-index data.

Figure 4.19 shows the calculated ratio $\mathcal{N}_{Xe}:\mathcal{N}$ of Xe number density to alkali number density, as a function of incident wavelength for phase-matched third-harmonic generation. It can be seen that typically the ratio $\mathcal{N}_{Xe}:\mathcal{N}$

Table 4.3. Vapour pressure data for the alkali metals

	Li	Na	K	Rb	Cs
a	19571.4	12423.3	10210.4	9140.07	8827.38
d	19.130	17.3914	16.539	16.0628	16.0007
Temp(^{o}C) for 10 torr SVP	890	546	444	390	373
and corresponding number density (10^{23} atoms m^{-3})	0.83	1.17	1.35	1.45	1.50

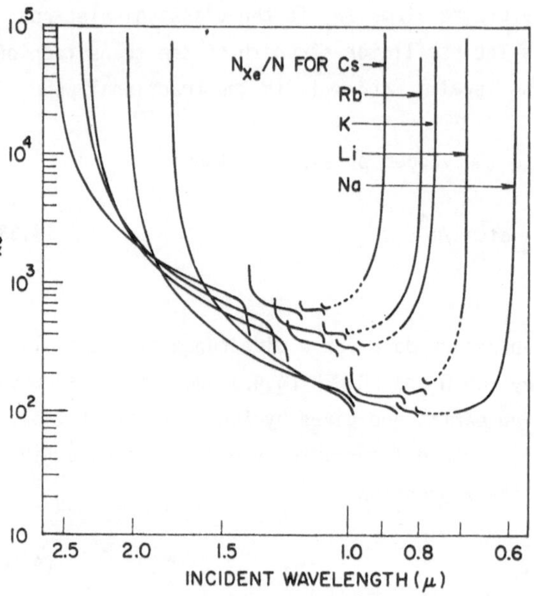

Fig.4.19. Required ratio of Xe to alkali atoms versus incident wavelength for phase-matched third-harmonic generation. (After [1.15])

is in excess of 100. Some experimental confirmation of these calculations has been obtained by YOUNG et al. [1.16], OHASHI et al. [4.18] and PUELL et al. [4.46]. PUELL and VIDAL [2.28] have turned the method around and shown that observation of the third harmonic intensity versus buffer-gas pressure can be used to determine the refractive index and hence oscillator strengths of the alkali itself. YOUNG et al. generated the third harmonic of 1.06 μm in a rubidium cell that was sealed off with 81 torr of xenon at 20°C. They calculated that phase matching should occur with a rubidium vapour pressure that corresponds to a cell temperature of 262°C. Figure 4.20 shows the results obtained for third-harmonic power versus cell temperature. The solid curve is the result of theoretical calculation using (4.15); the best fit was obtained by uniform translation of all of the experimental points 5°C toward lower temperature. Figure 4.21 shows, for comparison, the experimental data and theoretical curve for the same conditions of incident power and focussing (L = 19 cms, b = 40 cms) but with pure rubidium vapour. The shapes of the curves can be readily understood in terms of the theory given in Sect.4.2. Thus, if plane-wave focussing were used, then (4.15) would reduce to (4.17); therefore, by use of $L_c^{-1} \propto |\Delta k| \propto \mathcal{N}$,

$$P_3 \propto \mathcal{N}^2 L^2 \mathrm{sinc}^2\left(\frac{\pi}{2}\frac{L}{L_c}\right) \propto \sin^2\left(\frac{\pi}{2}\frac{L}{L_c}\right) .$$

Under plane-wave conditions, third-harmonic peaks would be expected to occur whenever the temperature is such that L is an odd integral number of L_c; successive peaks would be expected to have the same height. The first peak is higher because the plane-wave approximation is not valid in this experiment ($b/L \simeq 2$) and the slightly converging fundamental beam, with its associated phase slip, allows the phase mismatch to be compensated by the negative dispersion of the medium. In fact, the observation of this behaviour can be useful in providing confirmation that the medium is negatively dispersive when the refractive-index data are not sufficiently accurate to be relied upon, as for example, in the generation of 177.3nm by tripling 532nm radiation in Cd vapour ([1.19], see Fig.4.22). Another method of checking the sign of the dispersion has been used by HSU et al. [4.11] in a sum-frequency-generation experiment in Hg vapour, using tight focussing. Starting with pure Hg vapour, they admitted the Ar buffer gas (having a known dispersion) into the cell, and observed that the sum-frequency signal decreased monotonically with increase of Ar pressure. By comparing this with the behaviour of the focussing integral I^2 [3.1], it was deduced that the Hg vapour was negatively dispersive. Had the dispersion been positive, the sum-frequency signal would have exhibited an oscillating amplitude with comparable peak heights as the Ar pressure increased.

So far, we have discussed cases where the negative dispersion of a metal vapour is compensated by the positive dispersion of an inert gas. However, for shorter wavelengths, the dispersion of the inert gas may itself become negative, as for example when the third harmonic (or sum frequency) lies above and close to a transition of high oscillator strength. Thus, in gen-

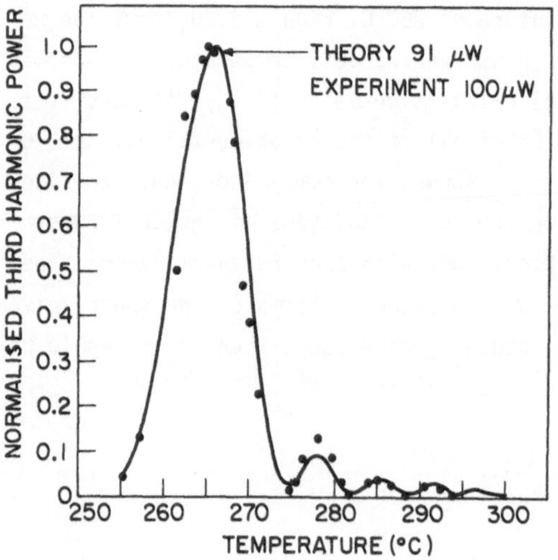

Fig.4.20. Normalised third-harmonic power versus temperature for Rb with 81 torr Xe at 20°C. Experimental points were uniformly translated 5°C toward cooler temperatures. The 50kW, 1.064μm beam was focussed on the output cell window with a confocal parameter of 40cm. (After [1.16])

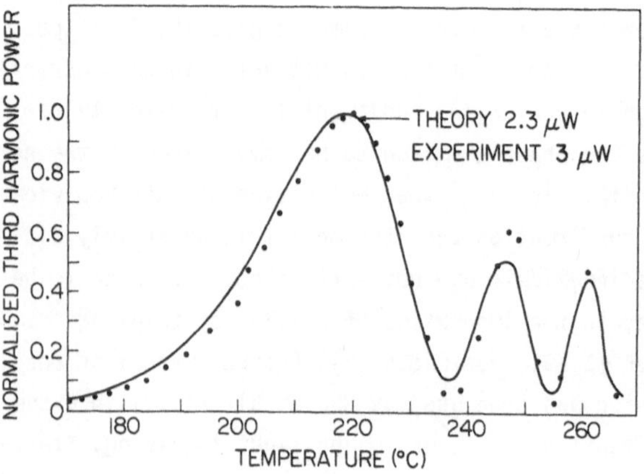

Fig.4.21. Normalised third-harmonic power versus temperature for pure Rb vapour. Incident power, focussing and confocal parameter are as in Fig.4.20. (After [1.16])

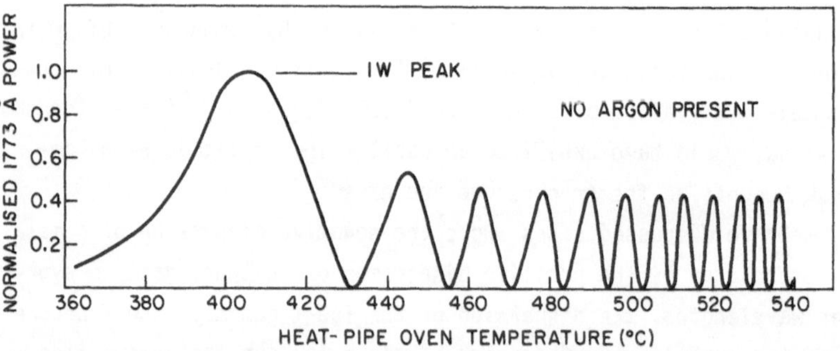

Fig.4.22. Normalised 1773Å output power versus oven temperature without phase matching. 5320Å input power 20MW. (After [1.19])

erating 118.2nm radiation in Xe (third harmonic of 354.7nm), KUNG et al. [1.20] showed that the negative dispersion of Xe (probably resulting from the 119.2nm 5p-5d transition) could be compensated by the positive dispersion of Ar gas. Because the longest-wavelength resonance line of any inert gas is the 146.95nm line of Xe, it should be possible to use such phase-matched mixtures of inert gases over parts of the region 146.9nm to 50nm, where the continuum of He begins. The experiments of KUNG [4.37] give further confirmation of the phase-matching theory developed by BJORKLUND [3.1]. In these experiments, a mode-locked Nd:YAG laser pumped an ADP parametric generator and amplifier to produce tunable picosecond pulses with signal and idler wavelengths tunable over the range 420 to 720nm. Xe gas was then used as the nonlinear medium to generate harmonics and sum and difference frequencies between the signal, idler, and 1.06 μm pump beams. By use of tight focussing, substantial portions of the region 118 to 194.6nm were spanned in this way. In Fig.4.23, the hatched regions indicate the wavelength ranges over which generation could, in principle, be achieved by various combinations of the input wavelengths. The cross-hatched regions indicated the ranges over which generation was actually observed. Because of the strong contribution to the refractive index by the continuum, Xe is probably positively dispersive unless the generated radiation is within several thousand wave numbers of one of the strongest Xe resonance lines. This means that, for harmonic generation and sum-frequency generation, a signal could be observed only over limited ranges where the dispersion is negative. These were from 140 to 147nm, corresponding to the region above the $5p^56s^3P_1$ level of Xe, from 126.5 to 129.6nm (above $5p^56s^1P_1$) and from 118 to 119.2 (above $5p^55d^3P_1$). On the other hand, for difference-frequency generation, $\omega_1 + \omega_2 - \omega_3 = \omega_4$, BJORKLUND has shown that Δk need not be negative to avoid cancellation (see Sect.4.2); this accounts for the observed continuously tunable difference signal from 163.1 to 194.6nm.

Although we have mentioned examples of phase-matching behaviour that are in accord with the theory outlined in Sect.4.2, experiments have been reported in which the phase-matching characteristics have shown unexplained disagreement, e.g., FERGUSON and ARTHURS [4.47], TAYLOR [4.48]. Discrepancies can arise if the concentrations of metal vapour and buffer gas are nonuniform within the vapour column. Care is also needed to ensure that the two constituents are properly mixed. Thus, whereas KUNG et al. [1.20] at first reported observation of peak third-harmonic from Xe:Ar with $\mathcal{N}_{Xe}:\mathcal{N}_{Ar} = 1:430$, the same authors later reported, KUNG et al. [4.49] that the gases had

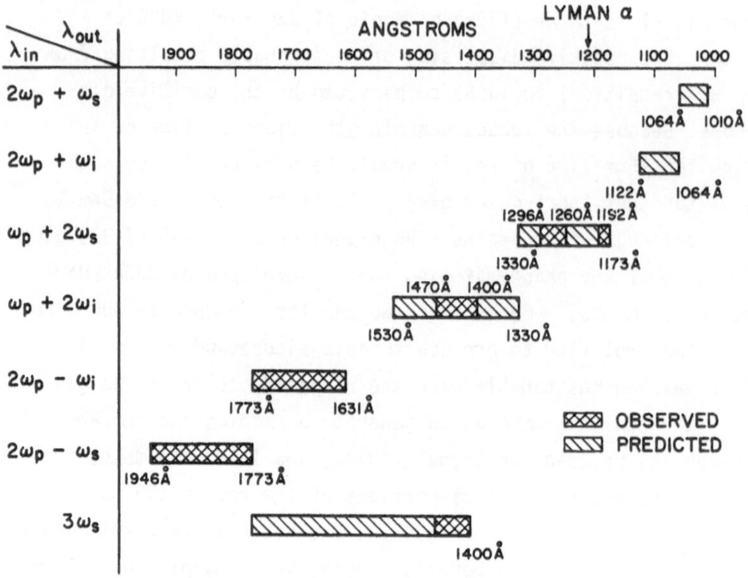

Fig.4.23. Summary of vuv tuning ranges obtained by KUNG [4.37] using frequency tripling, sum- and difference-frequency generation in Xe gas. ω_p, pump frequency at 2660Å, ω_s, signal frequency tunable from 5320 to 4200Å; ω_i, idler frequency tunable from 7200 to 5320Å. (After [4.37])

apparently been incompleted mixed; the correct phase-matching ratio was subsequently found to be $\mathcal{N}_{Xe}:N_{Ar}$ = 1:10.5.

4.4.2 Heat-Pipe Ovens

In the first experiments of YOUNG et al. [1.16] the alkali vapour (Rb) was contained in a Pyrex cell; different pressures of Xe were placed in the cells before sealing. After several hours operation above about 300°C, the Pyrex became yellow and opaque to uv radiation. YOUNG et al. suggested that this problem could be solved by using a heat-pipe oven of the type described by VIDAL and COOPER [4.50]. This suggestion has been borne out and heat-pipe ovens are now widely used as a means of producing a column of metal vapour at a uniform density and temperature (and pressures of several tens of torr), without contamination or fogging of the cell windows. We give a brief description of the principles of the heat-pipe oven here and also describe some of the more sophisticated variants that have been used. For a more detailed account of heat-pipe principles, the reader is referred to the description by VIDAL and COOPER [4.50].

Figure 4.24 shows the essential features of a "double-ended" heat-pipe oven. The main body of the oven is made of a material with which the metal vapour does not react rapidly; e.g., stainless steel is suitable for the alkali vapours. Inside this metal tube, there is a wick that consists of perhaps several turns of fine-mesh stainless-steel gauze. Initially, a charge of metal is placed in the centre of the tube and a buffer gas is introduced into the oven, at a pressure that corresponds to the metal vapour pressure at which it is desired to operate. The central region is then heated; the metal melts and wets the wick; as the heating is increased, the column of metal vapour expands to fill the tube. The cooling coils, placed between the heater and the windows, ensure a barrier of cool buffer gas between the hot-vapour region and the window. Metal vapour condenses on reaching this cool buffer and returns through the wick, by capillary action, to the central heated region. Because the windows are kept cool they can be o-ring sealed in the usual way, and they are not subject to contamination by the metal vapour. Construction details of a heat pipe of this type, used for Cs vapour, have been given by COTTER and HANNA [4.51]. One of the most important features of a heat-pipe oven is that it produces a uniform density of vapour over the length of the vapour column. For nonparametric processes, such as stimulated Raman scattering, which do not require phase matching, this homogeneity is not important; it is, however, vital when phase matching is achieved by mixing two constituents in proportions that may be very different, for example, when alkali vapours and inert gases are used. In practice, the necessary high degree of homogeneity, which may involve a temperature uniformity of better than $1^{o}C$, may require the use of a concentric heat-pipe oven [4.52]. Figure 4.25 shows an example of such an oven used by BLOOM et al. [1.17] to produce uniform mixtures of Na:Xe and Rb:Xe. The cell consists of an inner stainless-steel tube that contains a wick, with windows at each end, surrounded by an outer concentric tube that also contains a wick. Both tubes are loaded with alkali metal; they are connected to separate gas manifolds so that the buffer-gas pressure in the outer tube is set at the pressure required for the metal vapour in the inner tube. The heater supply is adjusted to produce a long column of boiling metal in the outer tube, which in turn produces an isothermal zone in the inner tube. Because the inner tube contains the same alkali as the outer, the partial pressure of the alkali vapour in the inner tube is the same as the total pressure in the outer tube. The buffer-gas pressure in the inner tube is then adjusted to the desired partial pressure for phase matching. PUELL et al. [4.46], using this type of concentric heat pipe claim that they

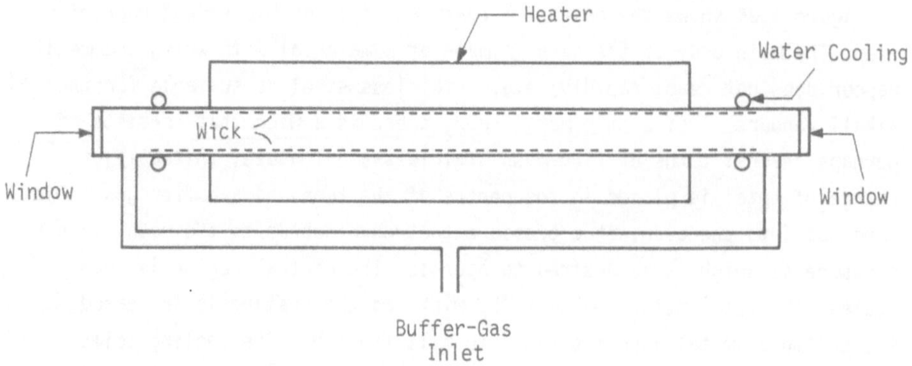

Fig.4.24. Heat-pipe oven

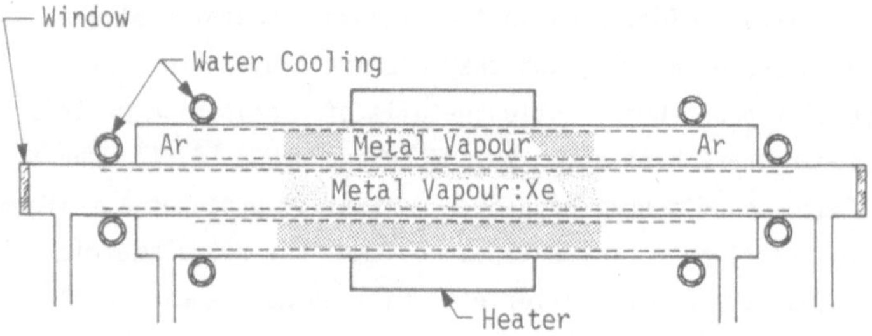

Fig.4.25. Schematic of concentric heat-pipe designed to produce a uniform mixture of metal vapours. (After [1.17])

could easily measure p_{buffer}/p_{alkali} to an accuracy of better than 1%. With their concentric heat pipe, BLOOM et al. [1.17] were able to operate at a total pressure (alkali + buffer) up to 1200 torr. At higher pressures, condensation problems (formation of a "fog") were encountered in the temperature-transition region at the end of the vapour column. Because of these experimental difficulties, the maximum third-harmonic conversion efficiency (1.06μm → 354.7nm) was limited to 10%. In an attempt to avoid the problems associated with higher pressures, BLOOM et al. [1.18] used a second metal vapour (Mg) as the buffer to provide the necessary positive dispersion. Because the Mg resonance line (285.2nm) is at a much longer wavelength than inert-gas resonance lines, the necessary ratio of buffer to alkali vapour pressure is much reduced compared with that for phase matching with inert gas. In their experiment, BLOOM et al. used Na as the nonlinear medium for THG of 1.06μm radiation and found that a ratio $\mathcal{N}_{Mg}:\mathcal{N}_{Na} = 2:1$ was required for phase matching. The heat pipe used to produce this two-metal mixture (see Fig.4.26) was of the type described by VIDAL and HESSEL [4.53]. It was similar to the concentric heat pipe shown in Fig.4.25 except that there was no wick in the central 50 cm length of the inner tube. The outer heat pipe, containing Na, provided the isothermal region in the inner pipe; the Mg that occupied the central region of the inner tube had a pressure that was determined by the temperature of the Na, which was in turn determined by the He buffer pressure in the outer tube. The inner tube also contained He buffer; the Na in the wick end-sections of this tube operated in the heat-pipe mode with a partial pressure equal to the He pressure. In the central region that was without a wick, the Na partial pressure was such that the combined pressure of Na and Mg equalled the He pressure. By independently varying the buffer pressure in the inner and outer tubes the Mg:Na ratio could be varied. In this experiment, the best conversion efficiency to 354.7nm was 3.75%, somewhat less than for the Rb:Xe case. A problem with the mixed-metal scheme was found to be poor homogeneity of the mixture. This was probably due to rapid depletion of Mg from the centre of the heat-pipe oven, because Mg with its high vapour pressure at its melting point showed a tendency to condense as a solid rather than to flow back to the centre of the oven. If these problems of homogeneity and metal mixing can be overcome, this type of phase matching scheme could be promising, because it should permit operation with a much higher vapour density than for the metal:inert-gas mixtures.

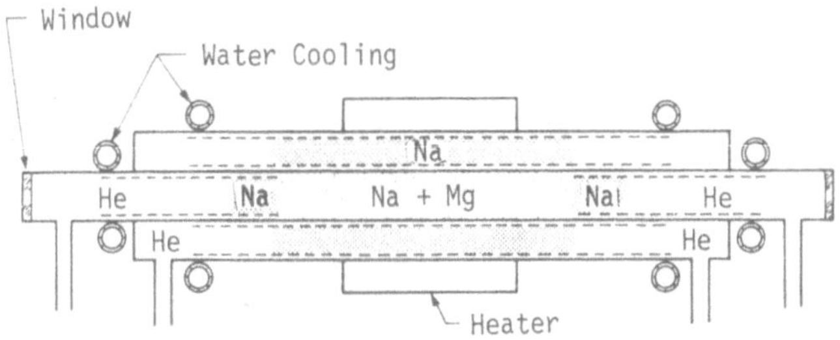

Fig.4.26. Schematic of a special concentric heat pipe, designed to produce a mixture of metal vapours. (After [1.18])

4.4.3 Phase Matching Without a Buffer Gas

More recently, BJORKLUND et al. [4.54,55] have demonstrated a technique for phase matching of two-photon resonant four-wave-mixing processes that avoids the need for a buffer gas to provide the compensating dispersion. The principle of the method is to choose the interacting frequencies around a region of anomalous dispersion, a technique first proposed by ARMSTRONG et al. [1.69]. Because only a single vapour constituent is needed, the problems of maintaining a highly uniform gas mixture and of additional absorption due to pressure broadening are thereby avoided.

BJORKLUND et al. have shown that this technique can be applied to sum- and difference-frequency generation and that it achieves phase matching over surprisingly broad ranges of the spectrum, from the far infrared to the vacuum ultraviolet. We briefly describe the results that have been reported.

To obtain a large nonlinear susceptibility, two input frequencies ω_1 and ω_2 were chosen so that $\omega_1 + \omega_2$ was exactly resonant with a two-photon transition between the ground state (energy $\hbar\Omega_g$) and a state ($\hbar\Omega_f$) of the same parity. A third input frequency ω_3 was provided so that generation of $\omega_4 = \omega_1 + \omega_2 + \omega_3$ occurred. Phase matching requires that $\eta_4\omega_4 = \eta_1\omega_1 + \eta_2\omega_2 + \eta_3\omega_3$; the only significant contribution to the refractive index for each of the four waves was assumed to come from an allowed transition between the ground state and a single intermediate state of opposite parity. This assumption of a simple three-level model turns out to be good for many two-photon resonant four-wave mixing processes, although see Sect.6.3. To generate a given ω_4, clearly ω_3 is determined (because $\omega_1 + \omega_2$ is fixed, equal to Ω_{fg}); to achieve phase matching ω_2 must be adjusted to a particular value; hence, ω_1 is also determined.

In their first experiment, BJORKLUND et al. [4.54] chose a different approach; they fixed ω_2 (it was the 1.06μm output of a Nd:YAG laser) and adjusted ω_1, which was provided by a dye laser pumped by the second harmonic of the Nd:YAG laser, around 612nm, so that $(\omega_1 + \omega_2) = \Omega_{4s-3s}$ in Na vapour. To achieve phase matching, they then had to adjust ω_3 to a particular value ($\sim$570nm); ω_4 corresponded to 231nm. When ω_3 was tuned around 570nm, the sum-frequency signal at ω_4 was significant only over a small range around the exact phase matching point (see Fig.4.27). The width of this phase matching curve was in good agreement with the theoretical value and also confirmed the prediction that it was not necessary to have a highly uniform concentration of Na atoms in the oven. The demands on the vapour-oven design are, therefore, greatly relaxed. For a few kilowatts input power in each of the three input beams, the maximum power generated at 231nm was 0.02W.

In their second experiment, BJORKLUND et al. chose to generate a particular frequency ω_4 in strontium vapour, this being a frequency for which a strong enhancement occurred owing to an autoionising resonance. Because ω_4

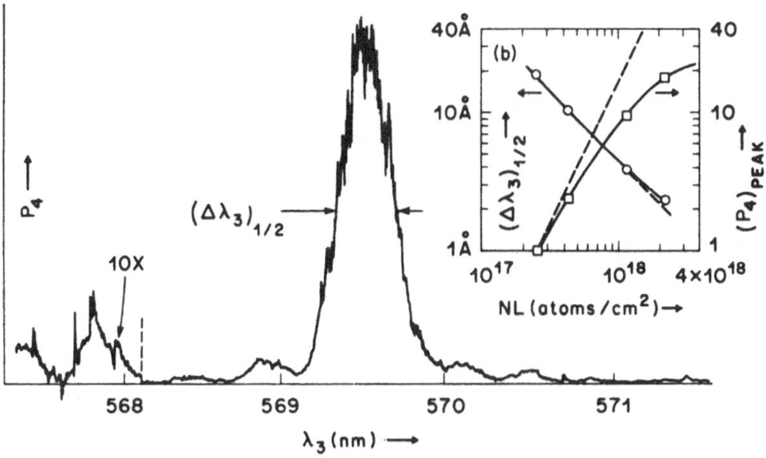

Fig.4.27. Generated power at λ_4 versus λ_3 with λ_1 and λ_2 constant and NL = 1.15 × 10^{18}cm^{-2}. (After [4.54])

was thus determined, all three input frequencies had to be adjusted to particular values; again, $\omega_1 + \omega_2$ had to satisfy the two-photon-resonance condition (for the $5s^2\,{}^1S_0 \rightarrow 5s7s\,{}^1S_0$ transition of Sr); ω_3 had then to be adjusted so that $(\omega_1 + \omega_2) + \omega_3 = \omega_4$, where ω_4 corresponded to either 169.7nm or 171.2nm, depending on the autoionising resonance chosen. Given ω_4, then ω_1 (and hence ω_2) had to be adjusted to achieve phase matching. The three adjustable frequencies were provided by three dye lasers that were pumped by either the second or third-harmonic of a single Nd:YAG laser. The best result obtained was a power of 3.2W at 171.2nm, with the input powers; P_1 = 39kW at λ_1 = 651.7nm, P_2 = 800W at λ_2 = 432.9nm and P_3 = 3.6kW at λ_3 = 501.2nm. This represented a very considerable increase of efficiency over the original sum-frequency-generation experiments in Sr [1.22]; BJORKLUND et al. [4.55] suggested that with their improved phase matching technique, it might be possible to generate useful cw powers in the vuv. This has since been demonstrated (FREEMAN et al. [4.56]) using the same resonances as in the above pulsed experiment, with ω_1 and ω_3 provided by fixed frequencies from an Ar^+ laser, and ω_2 from a cw dye laser. Rather than use a tunable ω_3 to take advantage of the autoionising resonance, a 5T magnetic field was applied to the vapour to tune one of the magnetic sublevels of the autoionising level (see also [4.57,61]). For inputs of ~ 1W, a generated vuv output of 5 × 10^{-11}W at 170nm was obtained. This figure was, however, much lower (by ~ 10^3) than would be expected from a simple extrapolation from the behaviour under pulsed conditions.

4.4.4 Group-Velocity Matching

We conclude this section on phase matching techniques by making a few comments on the special considerations that apply when the interacting frequencies are presented in the form of mode-locked pulses. Because these pulses involve a significant spread of frequencies, it is necessary to ensure that phase matching is achieved for all of the frequencies within this spread. The phase matching condition then becomes a condition for matching the group velocities of the interacting pulses, or equivalently a requirement that the pulses should remain overlapping as they propagate through the medium. We will derive this rather simple result here; we will also show that, for typical experimental conditions of mode-locked pulse length and length of nonlinear medium, the group-velocity-matching condition can be readily met.

We consider a transform-limited pulse of duration Δt, and bandwidth $\Delta\omega$, where $\Delta\omega\Delta t \simeq 1$, and centre frequency ω_1. We require, for efficient third-harmonic generation, that the coherence length for harmonic generation from any fundamental frequencies in the range $\omega_1 - \Delta\omega/2$ to $\omega_1 + \Delta\omega/2$ (the corresponding third-harmonic frequencies range from $3\omega_1 - 3\Delta\omega/2$ to $3\omega_1 + 3\Delta\omega/2$), be greater than the length of the medium, i.e., $|\Delta k|L \leq \pi$, where

$$|\Delta k| = \left|\left[k_3 + \left(\frac{dk}{d\omega}\right)_{3\omega_1} \cdot \frac{3\Delta\omega}{2}\right] - 3\left[k_1 + \left(\frac{dk}{d\omega}\right)_{\omega_1} \cdot \frac{\Delta\omega}{2}\right]\right| . \tag{4.35}$$

By phase matching exactly for the centre frequencies, i.e., $k_3 = 3k_1$, the condition becomes

$$\frac{3\Delta\omega L}{2}\left|\left[\left(\frac{d\omega}{dk}\right)^{-1}_{3\omega_1} - \left(\frac{d\omega}{dk}\right)^{-1}_{\omega_1}\right]\right| \leq \pi \tag{4.36}$$

or

$$\left|\left[\left(\frac{d\omega}{dk}\right)^{-1}_{3\omega_1} - \left(\frac{d\omega}{dk}\right)^{-1}_{\omega_1}\right]\right| \lesssim \Delta t/L . \tag{4.37}$$

This same condition is arrived at by requiring that a third-harmonic pulse, with group velocity $(d\omega/dk)_{3\omega_1}$, should not get out of step with the fundamental pulse [group velocity $(d\omega/dk)\omega_1$] by more than the pulse duration Δt in traversing a length L of medium. Thus, both ways of looking at the problem are seen to be equivalent. Finally, (4.37) can be re-expressed as

$$\eta_1\eta_3\Delta\nu_g/c^2 \lesssim \Delta t/L \quad , \tag{4.38}$$

where

$$\Delta\nu_g = \left| \left(\frac{d\omega}{dk}\right)_{3\omega_1} - \left(\frac{d\omega}{dk}\right)_{\omega_1} \right| ,$$

and the approximation $d\omega/dk \simeq c/\eta_{\omega_2}$ has been made. MILES and HARRIS [1.15] have calculated the quantity $\eta_1\eta_3\Delta\nu_g/c^2$ for phase matched third harmonic generation from 1.06μm and 0.694μm in the alkalis (using Xe buffer gas). They find, for example, that with 50 cm of Rb vapour at 10^{17} atoms cm^{-3} the minimum duration for the incident 1.06μm pulse is as short as 0.17ps.

4.5 Review of Experimental Results

In the previous paragraphs, we have referred to a number of experimental results in order to illustrate some of the general principles of uv harmonic generation and sum-frequency generation. Table 4.4 shows a fairly complete summary of results achieved up to the beginning of 1979; we list the nonlinear media involved, the type of laser used and the wavelengths generated.

In the first experiments of YOUNG et al. [1.16] a Q-switched Nd:YAG laser of 100kW power was used to generate the third harmonic in Rb vapour. Because neither the fundamental nor the harmonic were close to an atomic resonance, the susceptibility was too small to allow high conversion efficiency with the power available from such a laser. On the other hand, the generated harmonic power was found to be in good agreement with the calculated power. OHASHI et al. [4.18], using a more powerful Q-switched Nd:YAG laser, (up to 6MW) also found good agreement for lower incident powers; the harmonic power varied as the cube of the fundamental power (see Fig.4.28). Above about 1.5MW, however, they found that, in Rb:Xe, the harmonic power began to saturate. In fact, saturation behaviour of one sort or another is the common feature of various experiments in which a high conversion efficiency has been sought. Following their initial work, YOUNG et al. used a much higher power Nd:YAG laser, the higher power being achieved by mode locking the laser and thus condensing the available energy into a much shorter pulse. This ultimately led to a best efficiency of 10% for conversion from 1.064μm to 354.7nm although, here again, they found that some saturation process was limiting the harmonic power. An alternative to using high power is to increase the

Table 4.4. Summary of results (up to early 1979) for uv harmonic- and sum-frequency generation

Nonlinear medium	Nonlinear process	Incident radiation	Generated radiation	Remarks	Reference
Rb	$\omega_1+\omega_1+\omega_1$	ω_1; 1.06μm, 100kW, Q-switched NdYAG	354.7nm; 0.1mW	Xe buffer gas	[1.16]
Na, Rb	$\omega_1+\omega_1+\omega_1$	ω_1; 1.06μm, 300MW, 30ps, mode-locked NdYAG	power conversion to 354.7nm; 0.027 efficiency in Na, 0.1 eff in Rb	Xe buffer gas, concentric heat-pipe oven	[1.17]
Na	$\omega_1+\omega_1+\omega_1$	ω_1; 1.06μm, 300MW, 30ps, mode-locked NaYAG	354.7nm; 0.037 eff	Mg vapour buffer, concentric heat-pipe oven	[1.18]
Rb	$\omega_1+\omega_1+\omega_1$	ω_1; 1.06μm, up to 200MW, either 7ps or 300ps, mode-locked NdYAG	354.7nm 0.028 best eff	Xe buffer gas	[4.46]
Rb, Na	$\omega_1+\omega_1+\omega_1$	ω_1; 1.06μm, 6MW Q-switched NdYAG	354.7nm; 100W (in Rb) and 230W (in Na)	Xe buffer gas	[4.18]
Cd	$\omega_1+\omega_1+\omega_1$	ω_1; 530nm	177.3nm; 7kw, 10^{-4}eff	Ar buffer gas	[1.19]
	$\omega_1+\omega_1+\omega_1$	ω_1; 354.7nm	118.2nm; 10^{-7}eff		
	$\omega_1+\omega_1+\omega_2$	ω_1; 1.06 m; ω_2; 354.7nm (all: 50ps mode-locked NdYAG and its 2nd and 3rd harmonics, 2-20MW)	152.0nm; 10^{-6}eff from ω_1		
Na	$\omega_1+\omega_1+\omega_2$	ω_1; 685.6nm, 3kW from OPO ω_2; 10.6μm-9.26μm (step tunable), cw CO_2 laser, ~5mW	332.1nm-330.5nm, step tunable. Best eff 16.2 (1620%!). From IR (ω_2) to UV photon conversion eff 0.58.	$\omega_1+\omega_1$, two-photon resonant; possible application as IR up converter	[1.24]
Cs	$\omega_1+\omega_1+\omega_1$	ω_1; 694.3nm, 1MW Q-switched ruby laser	231.4nm, 5μW (for 100kW of ω_1)	$\omega_1+\omega_1$, two-photon resonant.	[1,23, 2.67]
Na	$\omega_1+\omega_1+\omega_1$	ω_1; 602nm, 50MW, 2ps, mode-locked rhodamine 6G	~ 200nm, 50W	Xe buffer gas $\omega_1+\omega_1$, two-photon resonant.	[4.48]

Na	$\omega_1+\omega_1+\omega_1$	ω_1; ~ 600nm, up to 100MW, 12ps, mode-locked rhodamine 6G	~ 200nm, 0.08 best eff (with 50kW of ω_1)	Xe buffer gas, $\omega_1+\omega_1$, two-photon resonant.	[4.58]
Tl	$\omega_1+\omega_1+\omega_1$	ω_1; 585.32, 50kW (0.3μs) from dye laser	195.1nm, 10^{-10} eff	$\omega_1+\omega_1$, two-photon resonant.	[4.32]
Sr	$\omega_1+\omega_1+\omega_2$	ω_1,ω_2; 15-100kW, 5ns, from dyes pumped by N_2 laser	157.8-195.7nm, best eff (from ω_2) 3×10^{-5} (measured with ω_2 input attenuated to 1.6W)	$\omega_1+\omega_1$, two-photon resonant	[1.22, 4.23]
Mg	$\omega_1+\omega_1+\omega_2$	ω_1; 430.88nm, ω_2 tunable — 50kW dye lasers pumped by N_2 laser	140-160nm, up to 2×10^{-3} eff	He buffer gas, $\omega_1+\omega_1$, two-photon resonant.	[4.59]
Mg	$\omega_1+\omega_1+\omega_2$	ω_1; 380.3nm; ω_2; 340nm, tunable dye lasers pumped by KrF — 10kW, 10ns	121.0-129.0, up to 2×10^9 photons per pulse (~ 10^8 at 121.6nm, Lyman α)	He buffer gas $\omega_1+\omega_1$, two-photon resonant	[4.60]
Ca	$\omega_1+\omega_1+\omega_1$	ω_1 ~600nm, 250MW, 3ps, mode-locked rhodamine 6G	~ 200nm, ~ 10^5W	Xe buffer gas, concentric heat-pipe oven, $\omega_1+\omega_1$ two-photon resonant.	[4.47]
Ca	$\omega_1+\omega_1+\omega_2$	ω_1; 915.3nm, 1MW, 60ns, (dye laser pumped by mode-locked ruby) ω_2; 694.3nm, 10MW, 60ns, mode-locked ruby	275.8nm	$\omega_1+\omega_1$, two-photon resonant.	[2.105]
Hg	$\omega_1+\omega_1+\omega_2$	ω_1; 312.85nm, 15ps, 1MW (second harmonic of dye, amplified by NdYAG-pumped optical parametric amplifier) ω_2; 520.5nm, 15ps, 1MW from another optical parametric generator pumped by same mode-locked NdYAG	120.28nm, 300W	$\omega_1+\omega_1$, two-photon resonant. 120.28nm is half the frequency of the $1s^1S_0-2s^1S_0$ transition in He	[4.11]
Hg	$\omega_1+\omega_1+\omega_1$	ω_1; 268.8nm (4 harmonic of Nd glass, mode-locked)	89.6nm, 10^{-6}-10^{-9}eff	two-photon resonant, using Nd-glass tunability	[4.29]

Table 4.4 (continued)

Eu	$\omega_1+\omega_1+\omega_2$	ω_1; 576.68nm ω_2; 520.13nm	185.5nm	$\omega_1+\omega_2$, two-photon resonant, ω_1 resonant with weak single-photon transition	[4.22]
Yb	$\omega_1+\omega_1+\omega_1$	ω_1; 582.07nm	194nm	$\omega_1+\omega_1$, two-photon resonant.	[4.22]
Ca II	$\omega_1+\omega_1+\omega_1$	ω_1; 383.25nm, dye laser pumped by N_2 laser	127.8nm	$\omega_1+\omega_1$, two-photon resonant with a transition of CaII	[4.62]
Na	$\omega_1+\omega_2+\omega_3$	ω_2; 1.06 m Q switched NdYAG ω_1; 612nm, ω_3; 570nm } dyes pumped by SH of NdYAG } Few kW 10ns	231nm, up to 20mW	$\omega_1+\omega_2$ two-photon resonant; phase matching without buffer	[4.54]
Sr	$\omega_1+\omega_2+\omega_3$	ω_1; 651.7nm, 39kW ω_2; 432.9nm, 800W ω_3; 501.2nm, 3.6kW } dypes pumped by SH and TH of NdYAG	171.2nm, 3.2W Similar results also achieved at 169.7nm, with different ω_1, ω_2,ω_3	$\omega_1+\omega_2$, two-photon resonant. $\omega_1+\omega_2+\omega_3$ resonant with autoionising level. Phase matching without buffer	[4.55, 4.57]
Sr	$\omega_1+\omega_2+\omega_3$	ω_1; 476.5nm, (Ar^+laser),0.425W ω_2; 572.4nm, (cw Rh6G),0.66W ω_3; 488.0nm, (Ar^+laser),0.85W	169.7nm, 5 10^{-11}W	cw, $\omega_1+\omega_2$, two-photon resonant. Magnetic field to tune autoionising level	[4.56]
Sr	$\omega_1+\omega_1+\omega_1$	ω_1; 576nm, 40kw, flash-pumped rhodamine 6G	191.97nm, 2×10^{-5}eff		[4.27]
Xe	$\omega_1+\omega_1+\omega_1$	ω_1; 354.7nm, 13MW, 25ps, third harmonic of mode-locked NdYAG	118.2nm, 10^{-3}eff	Ar buffer gas	[1.20, 4.49]
Xe	$\omega_p+\omega_p+\omega_s$ $\omega_p+\omega_p+\omega_i$ $\omega_p+\omega_s+\omega_s$	ω_p, 266nm, 30ps, 20MW, fourth harmonic of mode-locked NdYAG. ω_s, 420-530nm; ω_i, 530-720nm, respectively signal and idler	Tunable over portions of range 118 to 147 nm and continuously over 163.1 to 194.6mn. Typically 1W.		[4.37]

	$\omega_p+\omega_i+\omega_i$ $\omega_p+\omega_p-\omega_i$ $\omega_p+\omega_p-\omega_s$ $\omega_s+\omega_s+\omega_s$	of parametric generator pumped by fourth harmonic of NdYAG. 30 ps, 1MW			
Ar	$\omega_1+\omega_1+\omega_1$	ω_1, 170.9nm, 1MW, 10ns, from Xe_2 laser	57.0nm	$\omega_1+\omega_1$, two-photon resonant.	[4.38]
He, Ne	$\omega_1+\omega_1+\omega_1+\omega_1+\omega_1$	ω_1, 266.1nm, 330MW, 30ps, fourth harmonic of NdYAG	53.2nm, 10^{-5}-10^{-6}eff		[4.33]
He	$7\times\omega_1$	ω_1, 266.1nm as above	38nm, 10^{-7} to 10^{-8}eff		[4.34]
He, Ne	various fifth-order and third-order processes	ω_{IR}; 1.06μm ω_{green}; 532nm ω_{UV}; 266.1nm } fundamental and harmonics of mode-locked NdYAG	59.1nm ($5\omega_{UV}$) 62.6nm ($4\omega_{UV}+\omega_{green}$) 71.0nm ($4\omega_{UV}+\omega_{IR}$) 76.0nm ($4\omega_{UV}-\omega_{IR}$) 118.2nm ($2\omega_{UV}+\omega_{IR}$) 106.4nm ($2\omega_{UV}+\omega_{green}$) 88.7nm ($3\omega_{UV}$)		[4.12, 4.39, 4.63]
Na	$\omega_1+\omega_1+\omega_1+\omega_1+\omega_1$	ω_1; 1.06μm, 250MW, 10ps, mode-locked Nd glass	212nm, 150W	Xe buffer gas	[4.40]
	$\omega_1+\omega_1+3\omega_1$	ω_1 as above; $3\omega_1$ generated in the Na vapour	212nm, 15W		
	$\omega_1+\omega_1+3\omega_1$	ω_1 as above; $3\omega_1$, generated in KDP crystals	212nm, 7kW		
Na	$\omega_1+\omega_1+\omega_1+\omega_1+\omega_1$	ω_1 as above	212mn, 4kW	Xe buffer gas	[4.31]
NO	$\omega_1+\omega_1+\omega_2$	ω_1, ω_2 } dye lasers ~450nm, pumped by N_2 laser, ~20kW	130-150nm, up to 1mW (10^7photons/pulse)	$\omega_1+\omega_2$ tuned through several two-photon resonances	[4.64]
Kr	$\omega_1+\omega_1+\omega_1$	ω_1; 364.8nm, 10MW, 15ns, dye laser	~ 121.6nm ~ 60W	Lyman-alpha radiation	[4.35]
Kr	$\omega_1+\omega_1+\omega_1$	ω_1; 337.1nm, 1MW, 3ns N_2 laser	112.4nm, up to 10^{-6}eff	4-5 atm. optimum Kr pressure	[436]
		ω_1; 335-350nm, up to 1MW dye laser pumped by KrF laser	111.6-116.5nm, 3.10-5eff		

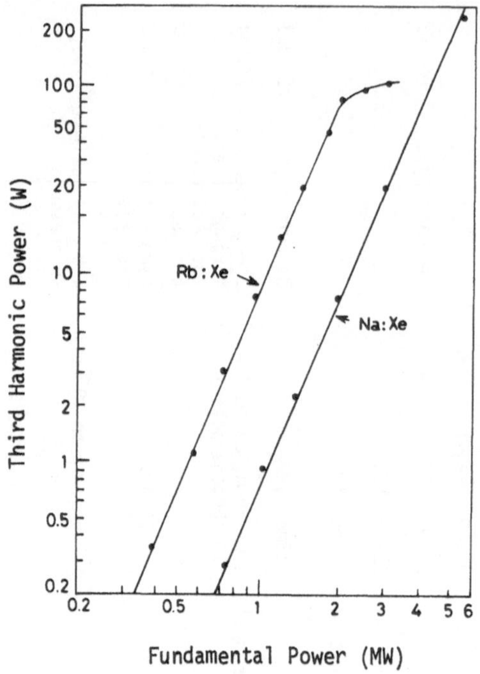

Fig.4.28. Variation of the third-harmonic power with fundamental input power for Rb:Xe and Na:Xe systems. The input beam was tightly focussed with a confocal parameter of 1cm and Rb and Na vapour pressures were 3.8 torr and 0.5 torr respectively. Note the saturation of peak output at P_1 = 1.5MW in the Rb:Xe system. (After [4.18])

susceptibility by tuning to a two-photon resonance. This allows significant harmonic power to be generated even with dye lasers of a few kilowatts power, [1.22]. However, the saturation processes also appear to become more severe; so far, apart from a notable exception [1.24], the efficiencies are well down. More recently, a number of experiments have been tried using a combination of two-photon resonance and high power (e.g., with a mode-locked dye laser, [4.47,48]). Despite this apparently favourable combination, the efficiencies have been rather disappointing and probably indicate an even more severe saturation problem. HSU et al. [4.11] obtained some more promising results in this way, however, and were able to generate significant power at 120.28nm. This wavelength was chosen because the corresponding photon energy is half that of the He $1s^1S_0 \rightarrow 2s^1S_0$ two-photon transition. High-power generation at 120.28nm is, therefore, a necessary preliminary for two-photon-resonant sum-frequency generation in He, which would allow generation of much shorter wavelengths.

Because saturation processes play such an important role in the quest for efficient harmonic generation and in other nonlinear processes [4.65] the next section, Sect.4.6, will be devoted to a discussion of various possible mechanisms. Any treatment of this topic must be somewhat qualitative and rough because many of the saturation mechanisms may be simultaneously at work. Also, the experimental results obtained for some saturation processes, such as multi-photon ionisation, are not entirely in accord with theoretically calculated ionisation rates. In many experiments, the saturation mechanisms have not yet been conclusively identified. Finally, at very high incident laser powers attempts to discuss saturation processes in terms of rates, as derived from perturbation theory, are suspect because the perturbation approach itself is of dubious validity.

4.6 Saturation and Limiting Processes

4.6.1 Single-Photon Absorption

In their original proposal for third-harmonic generation in metal vapours, HARRIS and MILES [1.14] considered the main saturation mechanisms to be either single-photon absorption of fundamental or harmonic by the pressure-broadened resonance transitions, or by the continuum. Two situations were considered, viz. where the incident pulse was either much longer or much shorter than the decay time of the absorbing level. The former situation is characterised by a saturation intensity I_{sat}, defined as the intensity that reduces the population difference of the absorbing transition by 50%. This can be calculated by considering the intensity-dependent-susceptibility formula given in (2.81), which can be written as

$$\chi(-\omega;\omega;I_\omega) = \chi^{(1)}(-\omega;\omega)/(1 + I_\omega/I_{sat}) \quad , \tag{4.39}$$

where I_{sat} may be written in terms of the absorption cross section σ and the decay time T_1 of the upper level of the transition,

$$I_{sat} = \frac{\hbar\omega}{2\sigma T_1} \quad ; \quad \sigma = \frac{\omega}{c}\,[\chi^{(1)}(-\omega;\omega)]'' \quad . \tag{4.40}$$

The usual linear susceptibility $\chi^{(1)}(-\omega;\omega)$ is given by

$$\chi^{(1)}(-\omega;\omega) = N\mu_{ab}^2/\hbar\varepsilon_0(\Delta - i\Gamma)$$

for the two-level atom. The intensity dependence contained in the factor $(1+I_\omega/I_{sat})^{-1}$ in (4.39) is attributable to the reduction of population difference N in the linear-susceptibility expression. For pulses that are longer than the dephasing time T_2', the rate-equation limit of (2.79) may be used; for a pulse shorter than T_1, the saturation is characterised by a saturation energy flux (energy/area), J_{sat}/A, given by

$$\frac{J_{sat}}{A} = I_{sat}T_1 = \frac{\hbar\omega}{2\sigma} \quad , \tag{4.41}$$

where J_{sat} is the energy that reduces the population difference by 50%. The criterion that the population should be reduced by 50% is only a crude estimate of the saturation limit. This reduction of ground-state population reduces the third-harmonic susceptibility and thus the harmonic efficiency. (This separate treatment of single-photon absorption and THG is adequate for the approximate saturation criterion we seek here; in a more-rigorous treatment of THG in intense fields, the effect of single-photon absorption would appear quite naturally.) However, the efficiency of the harmonic-generation process is much more sensitive to the change of refractive index and hence of phase matching that accompanies such population changes. The effect of population changes on the refractive index follows directly from (4.39) (for the two-level atom). We shall see that the saturation limit imposed by this need to avoid breaking of phase matching may easily be an order of magnitude lower than the crude criterion given in the foregoing.

The advantage of using short pulses is evident, because if a pulse of intensity I and duration τ contains the saturation-energy flux J_{sat}/A (i.e., $I\tau = J_{sat}/A$) then from (4.41), it follows that I can be made greater in the ratio (T_1/τ) than I_{sat}; so, the harmonic efficiency will be increased.

The value of σ appropriate to a given harmonic-generation experiment was estimated by HARRIS and MILES [1.14] on the basis of the absorption that would occur in the wings of a pressure-broadened (lorentzian) line; the linewidth was calculated from pressure-broadening data of CHEN and TAKEO [4.66] for inert-gas broadening of alkali resonance transitions. Details of the calculation are given in MILES and HARRIS [1.15]; from these, it appeared that saturation intensities would be comparable to the intensities required for efficient harmonic generation. This suggested the need for short (< few ns) pulses. If the saturation intensity for breaking phase matching is considered, the need for short pulses is even more apparent. However, in practice it has been found experimentally, by use of both picosecond pulses and nanosecond pulses that the saturation is less severe than suggested by these calculations. Thus OHASHI et al. [4.18] found that, for third-harmonic generation (of 0.3547μm radiation) in the Rb:Xe system, the saturation energy flux was $\sim$ 33J cm^{-2} rather than the 0.85J cm^{-2} which they calculated would lead to breaking of the phase matching condition. Similarly, BLOOM et al. [1.17] found no saturation at energy densities five or six times greater than those calculated for tripling of 1.06 μm in Na:Xe and Rb:Xe. The reason for these discrepancies lies in the fact that, for large detunings from the pressure-broadened resonance lines, the assumed lorentzian line shape based on the impact approximation is not valid and the quasi-static approximation must be used instead [4.67-49]. Thus, it appears that the initial estimates of restrictions due to single-photon absorption were unduly pessimistic. On the other hand, BLOOM et al. [1.17] point out that, for pulses that exceed 1ns duration, the allowable energy densities may still be seriously limited by avalanche breakdown.

4.6.2 Breaking of Phase Matching and Self Focussing

As pointed out in the foregoing, the change of population due to single-photon (or two-photon) absorption leads not only to a reduction of susceptibilities but also, more seriously, to a change of refractive index. This can lead to breaking of the phase matching condition and to focussing or defocussing of the beam as a result of the non-uniform refractive-index profile. We now examine the magnitudes of these effects.

With an incident energy flux J/A much less than the saturation energy flux J_{sat}/A, the fractional change $\delta\mathcal{N}/\mathcal{N}$ of atomic population of the ground state is given by

$$\delta\mathcal{N}/\mathcal{N} = J/J_{sat} \quad . \tag{4.42}$$

A similar result applies in the steady state [see (4.39)]. Suppose the alkali vapour alone contributes a phase mismatch $(\Delta k)_{alk}$, i.e., it has a coherence length $L_c = |\Delta\pi/(\Delta k)_{alk}|$. This is normally matched by an equal and opposite Δk provided by the buffer gas. If we assume that only the alkali undergoes saturation, then the permissible change $\delta(\Delta k)_{alk}$ is

$$|\delta(\Delta k)_{alk}L| \leq \pi \quad \text{i.e.,} \quad |\delta\eta_1 - \delta\eta_3| \leq \lambda_1/6L \quad , \tag{4.43}$$

where L is the actual length of vapour column; therefore,

$$\left|\frac{\delta(\Delta k)_{alk}}{(\Delta k)_{alk}}\right| = \frac{L_c}{L} \quad . \tag{4.44}$$

But, if we neglect the contribution to refractive index due to atoms in excited states, then

$$\left|\frac{\delta(\Delta k)_{alk}}{(\Delta k)_{alk}}\right| = \frac{\delta\mathcal{N}}{\mathcal{N}} \tag{4.45}$$

Hence

$$\frac{J_{phase\text{-}match}}{A} = \frac{L_c}{L}\,\frac{J_{sat}}{A} \quad , \tag{4.46}$$

where $J_{phase\text{-}match}/A$ is the incident energy flux required to break phase matching. Thus, $J_{phase\text{-}match}/A$ may be much less then J_{sat}/A.

The change of population does not occur uniformly across the incident beam because typically, the intensity is not uniform but decreases toward the beam edge. This radially varying refractive-index profile acts as a lens-like medium, which will either focus or defocus the incident beam, depending on the sign of refractive-index change. With the usual situation of the incident frequency having a refractive index greater than one, which decreases on becoming saturated, the result will be a defocussing, because the medium acts as a negative lens. MILES and HARRIS [1.15] made a rough calculation of the tolerance of the phase-matched process to this defocussing effect and found that it was roughly the same as the tolerance on breaking phase matching, i.e., the energy should be limited to $\sim(L_c/L)(J_{sat}/A)$.

The foregoing discussion of population and refractive-index changes has been expressed in terms of a two-level atom. This simple model is, unfortunately, not always appropriate, as for example in nonresonant situations where the multi-level nature of the atom must be taken into account. Although it is to be expected that certain features of the two-level system, such as saturation intensity, still have a meaning, it is not possible to express this in such a simple form. We must return to the full power-series expansion of the nonlinear polarisation. The relation between the latter approach and the simple formula (4.39) of the two-level atom, was discussed at length in Sect.2.6.

To find the relationship between the change of refractive index $\delta\eta\omega$ and the incident intensity, we express the component of polarisation at frequency ω in terms of the incident fields [from (2.11)]

$$P_\omega = \varepsilon_0\chi^{(1)}(-\omega;\omega)E_\omega + \frac{3}{4}\varepsilon_0\chi^{(3)}(-\omega;\omega-\omega_1\omega)|E_\omega|^2E_\omega + \dots \quad . \tag{4.47}$$

Thus, the dielectric constant is

$$\varepsilon_\omega = \frac{\varepsilon_0E_\omega+P_\omega}{\varepsilon_0E_\omega} = 1 + \chi^{(1)}(-\omega;\omega) + \frac{3}{4}\chi^{(3)}(-\omega;\omega,-\omega,\omega)|E_\omega|^2 + \dots \quad .$$

Therefore

$$\delta\eta_\omega \equiv (\varepsilon_\omega)^{\frac{1}{2}} - [1 + \chi^{(1)}(-\omega;\omega)]^{\frac{1}{2}} \simeq \frac{3}{8}\cdot\frac{\chi^{(3)}(-\omega;\omega,-\omega,\omega)|E_\omega|^2}{[1+\chi^{(1)}(-\omega;\omega)]^{\frac{1}{2}}} \tag{4.48}$$

$$= \frac{3}{4}\cdot\frac{\chi^{(3)}(-\omega;\omega,-\omega,\omega)I_\omega}{\varepsilon_0 c n_0^2} , \tag{4.49}$$

where n_0 is the unperturbed refractive index at frequency ω. Equations (4.48,49) apply even if the susceptibilities are complex, in which case the refractive indices are also complex. However, in using (4.49) to describe self-focussing behaviour, we shall assume that all quantities are real. Moreover, when the susceptibilities are real (i.e., the frequencies are sufficiently removed from resonance that the adiabatic conditions (Sect.2.6) are satisfied), then the field quantities E_ω may be allowed to be time dependent. At high intensities, it is necessary to include higher-order terms, such as $\chi^{(5)}(-\omega;\omega,-\omega,\omega,-\omega,\omega)|E_\omega|^4E$ in (4.47) to (4.49). Also, if the third-harmonic field grows to a significant level then there will be an optical Kerr-effect contribution $\chi^{(3)}(-\omega;3\omega,-3\omega,\omega)|E_{3\omega}|^2$ to $\delta\eta\omega$. Similarly, there are contributions $\chi^{(3)}(-3\omega;3\omega,-3\omega,3\omega)|E_{3\omega}|^2$ and $\chi^{(3)}(-3\omega;\omega,-\omega,3\omega)|E_\omega|^2$ to $\delta\eta_{3\omega}$.

The various susceptibilities, such as $\chi^{(3)}(-\omega;\omega,-\omega,\omega)$ can be calculated in the same way as was used to calculate χ_{THG} (see Sect.2.6.5). MILES and HARRIS calculated values for all the alkalis and for incident wavelengths of 1.06μm or 0.6943μm (their values refer to the susceptibility per atom). Combining (4.49) and (4.43) gives the fundamental intensity for which phase matching is broken; a typical value, for tripling 1.06μm in Rb, given by MILES and HARRIS, is 2×10^{10} Wcm^{-2} for 10^{16} atoms cm^{-3}. This gives a rather rough estimate of the limiting intensity. A detailed set of experimental measurements and theoretical calculations by PUELL et al. [4.46] are consistent with this value. Figure 4.29 shows a plot of measured third-harmonic intensity (0.3547μm) against fundamental intensity, for tripling in phase-matched Rb:Xe as obtained by PUELL et al. At low intensities, the results follow the expected cubic relation, $I_{3\omega} \propto I_\omega^3$, shown by the dotted line. At higher intensities, however, the harmonic increases less rapidly, and the fit is much better with the broken line, which resulted from a calculation in which the third-order $[\chi^{(3)}(-\omega;\omega,-\omega,\omega)]$ intensity-dependent refractive-index effects were included. The solid line is the result of a calculation in which the fifth-order $[\chi^{(5)}(-\omega;\omega,-\omega,\omega,-\omega,\omega)]$ intensity-dependent refractive index is also included. The need to include these higher-order terms is more apparent in Fig.4.30 which shows the results of PUELL et.al. for the higher intensities available from a laser with a much shorter pulse (7ps as opposed to 300ps). The conclusion drawn by PUELL et.al. is that the departure of their experimental data from the $I_{3\omega} \propto I_\omega^3$ behaviour is due to these intensity-dependent changes in the phase matching. PUELL and VIDAL [4.70] have since considered how the effects of intensity-dependent refractive-index changes can be minimised, and suggest that the Na:Xe system should be much better than the Rb:Xe system. A somewhat different Kerr nonlinearity, $\chi^{(3)}(-3\omega;3\omega,-\omega,\omega)$, has been identified by ZYCH and YOUNG [4.71] as responsible for limiting the efficiency of 354.7nm → 118.2nm conversion in Xe.

The possibility of self focussing due to the intensity-dependent refractive index was also considered by PUELL et al. If we express (4.49) in terms of refractive indices, we can write

$$\eta_\omega = \eta_0 + \eta_2|E_\omega|^2 \quad , \tag{4.50}$$

where (see also Sects.2.6.4,5)

$$\eta_2 = \frac{3}{8}\frac{\chi^{(3)}(-\omega;\omega,-\omega,\omega)}{\eta_0} \quad . \tag{4.51}$$

Then it can be shown that there is a critical power P_{cr} at which the focussing due to the nonlinear refractive-index coefficient η_2 (assumed positive,

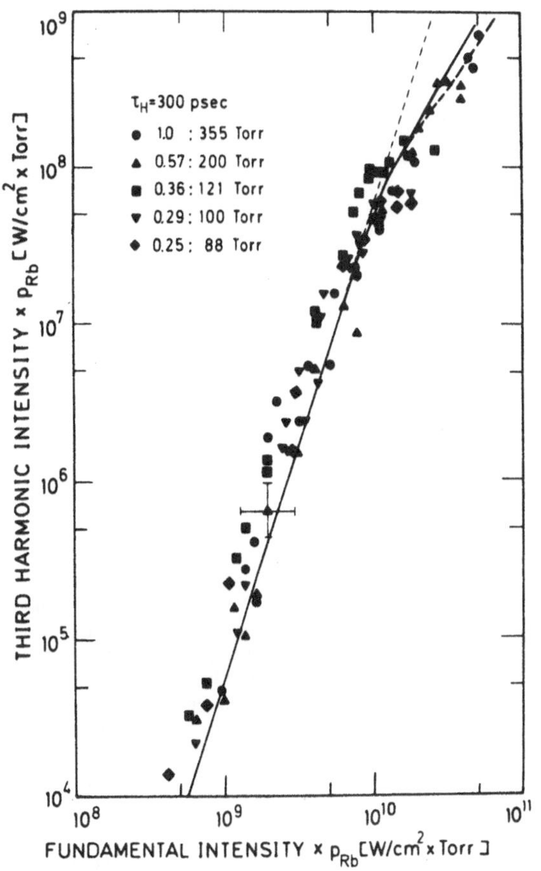

Fig.4.29. Plot of third-harmonic intensity times Rb vapour pressure versus input intensity times Rb vapour pressure (pulse duration 300ps). Theoretical curves are given for the small-signal limit (dashed line), for the third-order calculation (broken line) and the fifth-order calculation (solid line). (After [4.46])

as is indeed the case for 1.06μm in Rb vapour) overcomes the diffraction spread [4.72-74]. P_{cr} is given by

$$P_{cr} = \frac{(1.22\lambda)^2 c\pi\varepsilon_0}{64 n_2} \ . \tag{4.52}$$

For an incident power $P > P_{cr}$, a collimated beam with beam spot size w_0 becomes sharply focussed after a distance z_F, for which the approximate analytic expression is

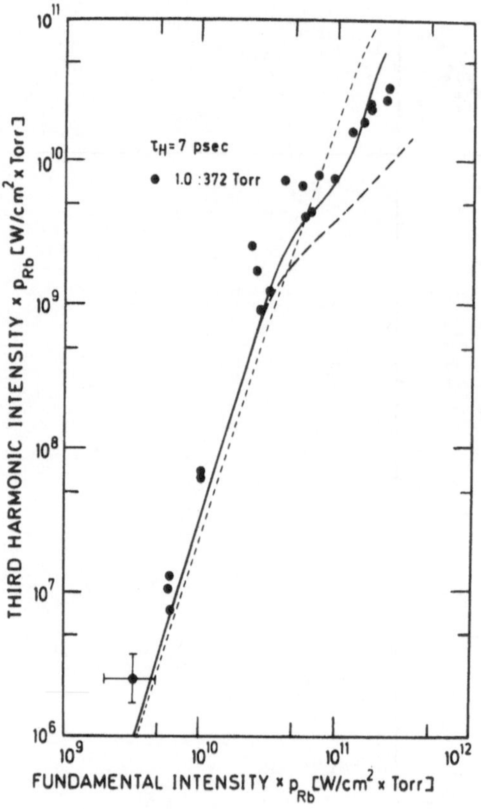

Fig.4.30. Plot of third-harmonic intensity times Rb vapour pressure versus input intensity times Rb vapour pressure (pulse duration 7ps). Theoretical curves are given for the small-signal limit (dashed line), for the third-order calculation (broken line) and the fifth-order calculation (solid line). (After [4.46])

$$z_F \simeq \frac{\pi w_0^2}{\lambda\left[\left(\frac{P}{P_{cr}}\right)^{\frac{1}{2}}-1\right]} \quad . \tag{4.53}$$

If the incident beam is converging or diverging with radius of curvature R (positive or negative, respectively) then this focussing distance is modified and becomes

$$z_{FR} = \frac{z_F R}{R+z_F} \quad . \tag{4.54}$$

In fact PUELL et al. saw no evidence of such focussing at the highest power available (6×10^7W) when 300ps pulses were used. This was in accord with the predictions of (4.52,53) for these particular experimental conditions. However, using 7ps pulses with powers up to 2.5×10^8W, the threshold power for self focussing, [4.6×10^7W using (4.52)] was exceeded; but again, observations of the beam emerging from the vapour showed no sign of focussing. PUELL et al. explained this negative finding by including the effect of the next higher-order term in (4.47) (i.e., $5/8\,\chi^{(5)}(-\omega;\omega,-\omega,\omega,-\omega,\omega)$ $|E_\omega|^4E_\omega$) in which case (4.50) becomes

$$n_\omega = n_o + n_2|E_\omega|^2 + n_4|E_\omega|^4 \quad . \tag{4.55}$$

Because n_4 is negative for 1.06µm in Rb vapour, the additional term reduces the tendency to self focussing; using the calculated value of n_4, PUELL et al. found that the terms $n_2|E|^2$ and $n_4|E|^4$ should cancel at an intensity of 1.5×10^{11}Wcm^{-2}. This approached the maximum intensity used. At higher intensities, the $n_4|E|^4$ term would dominate, thus leading to defocussing; and PUELL et al. offer some indirect evidence for this. Working at higher rubidium pressure (~ 3 torr, as opposed to the 1 torr of PUELL et al.), BLOOM et al. [1.17] did directly observe the occurrence of self focussing at ~ 10^{10}W cm^{-2} incident intensity. Thus, self focussing effects may be important limiting factors under some conditions; equations (4.47-55) are useful in providing a rough estimate of the conditions under which it could occur. The threshold for self focussing can vary significantly, however, for different vapours and different incident wavelengths.

Our discussion of self-focussing effects is necessarily rather brief; it can hardly do justice to the large amount of work, both theoretical and experimental, that has been performed on this subject. However, we will mention some further aspects of self focussing that have relevance to topics covered in other chapters of this book. First, the susceptibility that describes self focussing can, under some conditions, become resonantly large as, for example, when the incident frequency ω is near a two-photon resonance. An example of such a resonant behaviour is observed when 1.06µm radiation passes through Cs vapour; two photons of 1.06µm are close to resonance with the $6s^2S_{1/2} \rightarrow 7s^2S_{1/2}$ transition (~ 260cm^{-1} off resonance). The value of n_2 is negative; it therefore leads to defocussing. LEHMBERG et al. [2.68, 4.75,76] examined this case in some detail, both experimentally and theoretically. They suggest that this defocussing behaviour might be used to compensate the self focussing that occurs in high-power Nd:glass lasers (due to the positive n_2 of the glass) and thus permit higher useful output power from such a laser.

Another example of resonantly enhanced focussing or defocussing is that produced by the real part of the Raman susceptibility $\chi^{(3)}(-\omega_s;\omega_p,-\omega_p,\omega_s)$. The imaginary part leads to stimulated Raman gain, but the real part represents a change of refractive index at frequency ω_s, due to the presence of the pump field E_p (Sect.3.3). The latter is, of course, the optical Kerr effect, enhanced by the Raman resonance; the effect can, therefore, be large. OWYOUNG [4.77] has experimentally verified this focussing behaviour and its dispersion by injecting a dye-laser beam (tunable around the Stokes frequency) into a cell of benzene that was pumped so as to give Raman gain.

4.6.3 Two-Photon Absorption

In Sects.4.6.1,2, we discussed how single-photon absorption can produce saturation by reducing the ground-level population, thus leading to breaking of the phase-matching condition and self focussing. Equations (4.40,41) which express the criterion for saturation, may be rewritten as

$$W^{(1)}\tau = 1/2 \quad , \tag{4.56}$$

where $W^{(1)}$ is the single-photon transition rate and τ is the incident pulse length Δt, or the decay time T_1 of the upper level, whichever is shorter. Under conditions of near resonance with a two-photon transition, the two-photon absorption rate can be large, and this, rather than single-photon absorption, may become an important saturation mechanism. The criterion for saturation, (4.56), then becomes

$$W^{(2)}\tau = 1/2 \quad , \tag{4.57}$$

where $W^{(2)}$ is the two-photon transition rate and τ again refers to the shorter of Δt or T_1 (the time for an atom to return from the upper (two-photon) level back to the ground level). Radiation trapping can have a significant effect on T_1. In fact, (4.57) like (4.56) provides only a crude estimate of the saturation limit. A more realistic criterion, based on the condition for breaking phase matching, can be obtained in exactly the same way as discussed in Sect.4.6.2.

The derivation of the two-photon transition rate $W^{(2)}$ follows the same lines as the Raman transition-rate calculation, which led to (3.23). For the case of absorption of two photons that have, in general, different energies $\hbar\omega_1$, $\hbar\omega_2$ and that are linearly polarised in the z direction, the transition rate from level g to level f, $W^{(2)}$, is

$$W^{(2)} = \frac{\Gamma}{2\varepsilon_0^2\hbar^4c^2[(\Omega_{fg}-\omega_1-\omega_2)^2+\Gamma^2]}\left|\sum_i\left(\frac{\mu_{fi}\mu_{ig}}{\Omega_{ig}-\omega_2}+\frac{\mu_{fi}\mu_{ig}}{\Omega_{ig}-\omega_1}\right)\right|^2 I_1I_2 \quad , \tag{4.58}$$

where the μ are matrix elements of the z component of the dipole-moment operator.

This expression for $W^{(2)}$ contains factors that also appear in the expression for the power generated at a sum frequency ω_4, $(\omega_1 + \omega_2 + \omega_3 = \omega_4)$, under two-photon resonance conditions, $\Omega_{fg} - \omega_1 - \omega_2 \simeq 0$. Therefore, the conditions under which sum-frequency generation is optimised, i.e., the exact two-photon resonance and high incident intensities, also lead to the two-photon absorption being maximised. The question that we now examine, is whether two-photon absorption will, in practice, prevent efficient sum-frequency generation.

Following HARRIS and BLOOM [4.78], we simplify the discussion by considering just four levels, 0, 1, 2, 3, with allowed transitions $0 \leftrightarrow 1$, $1 \leftrightarrow 2$, $2 \leftrightarrow 3$, $3 \leftrightarrow 0$, having dipole matrix elements μ_{01}, μ_{12}, μ_{23}, μ_{30} (see Fig.4.31). The two-photon resonance Ω_{20} is achieved by the photons ω_1 and ω_2, i.e., $\omega_1 + \omega_2 \simeq \Omega_{20}$. We will assume that $|\Omega_{10} - \omega_1| \ll |\Omega_{10} - \omega_2|$; then, taking the dominant term in the susceptibility $\chi^{(3)}(-\omega_4;\omega_1\omega_2\omega_3)$, we find the polarisation at the sum frequency,

$$P_{\omega_4} = \frac{N\mu_{01}\mu_{12}\mu_{23}\mu_{30}E_1E_2E_3}{4\hbar^3\Delta\omega_1(\Delta\omega_2 - i\Gamma)\Delta\omega_3} \quad , \tag{4.59}$$

where $\Delta\omega_1 = \Omega_{10} - \omega_1$, $\Delta\omega_2 = \Omega_{20} - (\omega_1 + \omega_2)$, $\Delta\omega_3 = \Omega_{30} - (\omega_1 + \omega_2 + \omega_3)$ and 2Γ is the line width (FWHM) of the two-photon transition $0 \leftrightarrow 2$. For plane waves, the intensity at frequency ω_4 generated in a single coherence length L_c is given by (3.14), i.e.,

$$I_4 = \frac{\omega_4^2}{2\pi^2 c\varepsilon_0 n_4} |P_{\omega_4}|^2 L_c^2 \quad . \tag{4.60}$$

This expression is also approximately correct for the optimised tight-focussing case, as discussed in Sect.4.2.

The coherence length is given by

$$L_c = \frac{c\pi}{|\omega_1\Delta n_1 + \omega_2\Delta n_2 + \omega_3\Delta n_3 - \omega_4\Delta n_4|} \quad , \tag{4.61}$$

where $\Delta n_i = n_i - 1$, n_i being the refractive index at frequency ω_i. We assume, for simplicity, that the dominant contribution in the denominator of (4.61) is the $\omega_4\Delta n_4$ term and that the main contribution to Δn_4 comes from the transition $0 \to 3$, with which ω_4 is close to resonance. The expression for L_c then becomes

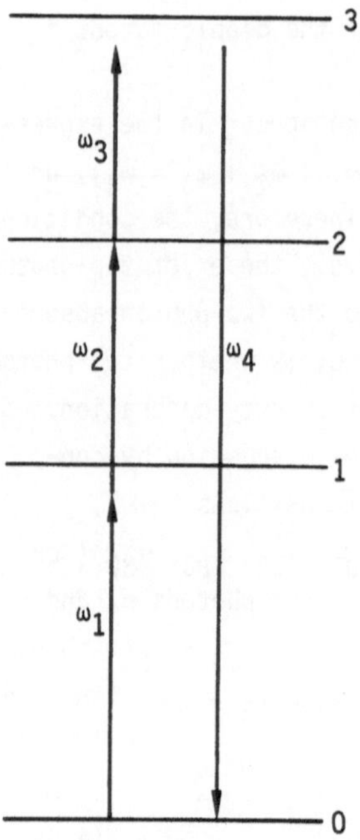

Fig.4.31. Energy-level scheme for two-photon resonant sum-frequency generation, $\omega_4 = \omega_1 + \omega_2 + \omega_3$

$$L_c = \frac{c\pi}{\omega_4|n_4-1|} = \frac{2\pi\varepsilon_0 c\hbar|\Delta\omega_3|}{N\omega_4\mu_{03}^2} . \tag{4.62}$$

Substitution of (4.59,62) into (4.60) gives an expression for the photon-conversion efficiency $\mathscr{E}$ of sum-frequency generation,

$$\mathscr{E} = \frac{I_4/\omega_4}{I_3/\omega_3} = \frac{\omega_3\mu_{01}^2\mu_{12}^2\mu_{23}^2 I_1 I_2}{\omega_4\hbar^4 n_1 n_2 n_3 n_4 c^2\varepsilon_0^2\mu_{03}^2\Delta\omega_1^2(\Delta\omega_2^2+\Gamma^2)} . \tag{4.63}$$

The maximum efficiency is obtained by substituting into (4.63) the maximum permissible value of I_1I_2, i.e., the value of I_1I_2 that makes $W^{(2)}\tau = 1/2$, obtained from (4.58),

$$(I_1I_2)_{max} = \frac{\varepsilon_0^2\hbar^4c^2[\Delta\omega_2^2+\Gamma^2]\Delta\omega_1^2}{\mu_{01}^2\mu_{12}^2\Gamma\tau} . \tag{4.64}$$

The final result is a maximum photon-conversion efficiency,

$$\mathscr{E} = \frac{\omega_3 \mu_{23}^2 T_2}{\omega_4 \mu_{03}^2 \tau \eta_1 \eta_2 \eta_3 \eta_4} \quad , \tag{4.65}$$

where $T_2 = \Gamma^{-1}$ is the dephasing time of level 2 and τ is the shorter of T_1 and the pulse duration Δt. Note that the detuning $\Delta\omega_2$ does not appear explicitly in this expression, although a larger detuning would of course require a higher incident intensity to produce this maximum conversion efficiency. Also, it should be noted that the maximum efficiency is determined by a ratio of matrix elements rather than by their absolute values, so high conversion efficiency is possible even with small matrix elements, although this would, again, require high incident intensities.

Equation (4.65), as it stands, applies to the case of a plane-wave interaction over a length $L = L_c$ or where tight focussing is used, according to the prescription $b\Delta k = -2$ (see Sect.4.2) and the medium is not phase matched by a buffer. If such tight focussing is impractical, then a buffer gas may be added to increase the coherence length and thus permit looser focussing. Although this reduces $W^{(2)}$ it also reduces the amount of absorption needed to break phase-matching, in accordance with (4.46). In practice, this means that when a phase matching buffer is used, the increase of maximum efficiency will be approximately linearly dependent on L/L_c, where L is the cell length and L_c is the non-phase-matched coherence length.

We have considered here the limitations imposed by two-photon absorption that leads to a change of population. However, particularly in a long sample, a significant limiting factor may be the reduction of incident intensity by two-photon absorption. Such a limitation has been considered by POPOV and TIMOFEEV [4.30] and STAPPAERTS [4.65]. Experimental evidence for this behaviour has been reported by BJORKLUND et al. [4.57].

4.6.4 Multiphoton Ionisation

Multiphoton ionisation and its effects can limit the efficiency of a desired nonlinear process in a number of ways. First there is the reduction of intensity of the incident light, due to absorption. The associated reduction of ground-state population may lead to self focussing and breaking of phase matching, as already discussed. The presence of a large population of photoelectrons and ions can produce large broadening and shifts of the levels of the remaining neutral atoms, via the Stark effect, and thus reduce the susceptibility of a highly resonant nonlinear process. Under more extreme conditions, the initial photoelectrons can gain energy by inverse bremsstrahlung absorption and then by an avalanche ionisation process finally form an opaque plasma. This rich variety of effects poses formidable problems for any attempt at a quantitative analysis of a real experimental situation. Further complicating features are (i) the presence (in alkali vapours) of a significant population of molecules that can also undergo multi-photon transitions, often more efficiently than the atoms, and (ii) photon-correlation effects. MEASURES [4.79] has also suggested that superelastic collisions of photoelectrons with excited atoms can lead to dramatically enhanced rates of photoionisation, by virtue of the elevated electron temperature, and also because the excited atoms in effect constitute a reservoir of atoms with reduced ionisation potentials.

Given these complications, it is not surprising that comparisons between experimentally measured ionisation rates and available theoretical calculations have in many cases shown poor agreement. We will therefore, confine ourselves here to quoting a few published figures (from the very extensive literature), which may at least provide a rough guide to the magnitude of effects that should be anticipated. For a review of this subject, the reader is referred to the articles by LAMBROPOULOS and LAMBROPOULOS [4.80] and GREY-MORGAN [4.81].

BEBB [4.16] calculated the two-photon ionisation rates for all of the alkalis, as functions of photon energy. Figure 4.32 shows a typical result, for Cs, in which the ordinate is W/F^2, where W is the photoionisation rate per atom, and F is the photon flux (photons $cm^{-2}s^{-1}$). To see the implications of these numbers, we consider a photon flux of 10^{27} photons $cm^{-2}s^{-1}$ of 2eV photons (i.e., $3 \times 10^8 Wcm^{-2}$). This, it is predicted from Fig.4.32, will produce an ionisation rate $\sim (10^{-48}) \times (10^{27})^2 = 10^6 s^{-1}$. Thus, in a 10ns pulse at the stated intensity, some 1% of the atoms would undergo two-photon ionisation. As can be seen from the figure, a much greater rate is predicted when the incident-photon energy is close to a resonance transition. This suggests that virtually complete ionisation could occur under the conditions of intense irradiation by resonant radiation used in a number of frequency-conversion experiments. LAMBROPOULOS and TEAGUE [4.82] also report calculated two-photon ionisation rates in caesium for a range of Ar laser wavelengths that have been used by GRANNEMAN and VAN DER WIEL [4.83] in experimental measurements. These measurements cover the range of wavelengths spanning the deep minimum on the low-frequency side of the 6s-7p transition in cesium. The observed rates were found to be two or three orders of magnitude greater than the theoretical values. This discrepancy spurred TEAGUE et al. [4.84] to make new calculations using a variety of available information on matrix-element values. However, they found that the disagreement persists; therefore, it remains a puzzle and a salutary warning against placing too much reliance on theoretical values (see also [4.85]).

GEORGES et al. [2.60,4.86] have incorporated the effects of multiphoton ionisation in a theoretical treatment of third-harmonic generation under two-photon resonant conditions. Their general treatment is applied to give numerical predictions to a particular experiment, that of WARD and

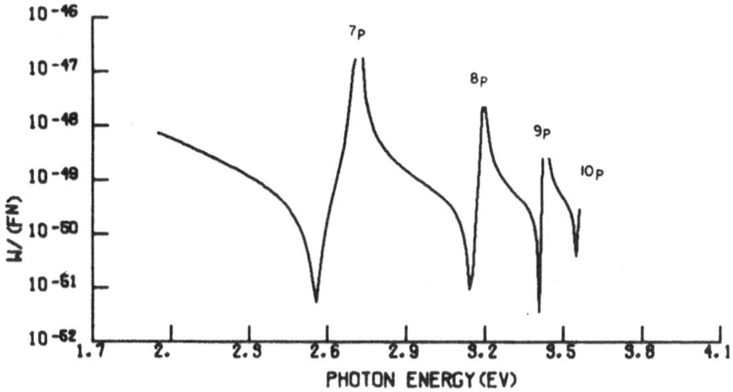

Fig.4.32. Two-photon ionisation rate W/F^2 versus incident photon energy, for Cs. W is the photoionisation rate per atom and F is the photon flux (photon $cm^{-2}s^{-1}$). (After [4.16])

SMITH [2.67], in which the third harmonic of ruby-laser radiation was generated in cesium vapour. This system offers a fortuitous coincidence between the ruby wavelengths and the two-photon transition $6s\,^2S_{1/2} \rightarrow 9d\,^2D_{3/2}$. WARD and SMITH offered an explanation of their observed third-harmonic saturation behaviour in terms of power broadening of the two-photon transition and population of the two-photon resonant state. WANG and DAVIS [4.32], in a companion paper, proposed the same saturation mechanism to explain third-harmonic generation behaviour in thallium vapour. In analysing the experimental data of WARD and SMITH, GEORGES et al. also took into account the laser-induced ac Stark shift, the loss of atoms due to ionisation, the change of phase matching due to population redistribution, and ionisation broadening of the two-photon resonant level (i.e., broadened because ionisation from that level reduced its lifetime). Their results indicate that these effects make substantial contributions in a quantitative analysis of the data. So far, however, rather few experimental data are available that can be subjected to close numerical scrutiny of this sort; there is a need for further careful experimental measurements using lasers of well-defined spatial and frequency characteristics.

To conclude this section on multiphoton ionisation, we briefly consider the question of Stark broadening of neutral atomic levels by the photoelectrons themselves. As we have seen, the theoretical data on photoionisation predicts that under conditions typical to many nonlinear-optics experiments in vapours, e.g., resonance irradiation by a tightly focussed dye-laser beam, a significant fraction of the atoms should be ionised. In fact, experimental measurements by LUCARTORTO and McILRATH [4.87] showed that it is possible to ionise completely a 10 torr-cm column of Na vapour using a 1MW dye-laser beam. The same authors [4.42] also report similar results for Li. Thus, photoelectron densities of the order of $10^{22}m^{-3}$ seem to be produced quite readily; it will be shown below that such densities can lead to very considerable Stark broadening of the atomic levels.

If we assume that the perturbing field that acts on an atom is due to a single charge Ze at a distance R, then the quadratic Stark shift ΔE is of the form

$$\Delta E = C_4/R^4 \quad . \tag{4.66}$$

The explicit form of C_4 is found by perturbation theory; thus, the energy shift ΔE of the unperturbed state $|\alpha JM_J\rangle$ is given by (see e.g., [2.65])

$$\Delta E_{\alpha JM_J} = \left(\frac{Ze}{4\pi\varepsilon_0 R}\right)^2 \sum_{\alpha' J'} \frac{|\langle\alpha JM_J|ez|\alpha' J' M_J\rangle|^2}{E_{\alpha J}-E_{\alpha' J'}} \quad . \tag{4.67}$$

This expression can usually be greatly simplified, because it is often the case that one term in the summation dominates. Thus, in considering the shift of ns levels in an alkali atom the dominant contribution is from the p level that has the same principal quantum number. If spin-orbit splitting is neglected, this gives

$$C_4 \simeq \left(\frac{Ze}{4\pi\varepsilon_0}\right)^2 \frac{|\langle ns|ez|np\rangle|^2}{E_{ns}-E_{np}} \quad . \tag{4.68}$$

HINDMARSH [4.68] showed that, within the impact approximation, the Stark-broadened line width γ (HWHM, in rad s^{-1}) is given by

$$\gamma = 11.4 \left(\frac{C_4}{\hbar}\right)^{2/3} (\bar{v})^{1/3} N \quad , \tag{4.69}$$

where N is the number density of the perturbing charged particles, whose mean speed is $\bar{v}$. Furthermore (see e.g., GRIEM [4.69], HINDMARSH [4.68], and SOBELMAN [2.65]), the impact approximation holds for energy shifts ΔE that satisfy

$$|\Delta E| \ll (\hbar\bar{v})^{4/3}(C_4)^{-1/3} \quad . \tag{4.70}$$

If we consider the particular case of cesium, then, using (4.68), we find that $C_4 \sim 10^{-55}\,\mathrm{Jm}^4$ for the 7s level and $\sim 8 \times 10^{-55}\,\mathrm{Jm}^4$ for the 8s level. Thus, for electron velocities of $5 \times 10^5\,\mathrm{ms}^{-1}$, (4.70) shows that the impact approximation for electron broadening is valid for shifts of several hundred cm^{-1}. For ions, with their lower velocities, the impact approximation is valid for shifts of only up to 1 cm^{-1} or so. Equation (4.69) also shows that the dominant broadening is that due to electrons, on account of their higher velocity. When these values of C_4 are inserted with typical values of $N = 10^{22}\,\mathrm{m}^{-3}$ and $\bar{v} = 5 \times 10^5\,\mathrm{ms}^{-1}$, a line width (FWHM) of $\sim 1.5\,\mathrm{cm}^{-1}$ is obtained for the 7s level of Cs and $\sim 5\,\mathrm{cm}^{-1}$ for the 8s level. These rough estimates are enough to indicate that Stark broadening by photoelectrons can be a significant factor, particularly when high-lying levels are involved, since the matrix elements in (4.68) are then very large and the frequency denominators are small. For more careful calculations of these Stark-broadened linewidths, the reader is referred to the book by GRIEM [4.69], which contains tabulations of calculated linewidths for a number of elements, including Li, Na, K, Cs, Mg, and Ca.

4.6.5 The ac Stark Effect

The level shift induced by an optical-frequency field, or ac Stark effect as it is widely known, has already been introduced in Sect.2.6.3. The possibility of such a shift has long been known and was observed many years ago [2.43]. However, the effect has attracted much more interest recently, because it is rather prominent in high-resolution spectroscopic techniques, such as Doppler-free two-photon absorption. A qualitative notion of how this shift can affect a nonlinear process such as two-photon resonant THG can be seen as follows. We consider a three-level atom, with electric-dipole allowed transitions $1 \leftrightarrow 2$ and $2 \leftrightarrow 3$. The atom is irradiated with an optical frequency ω, adjusted so that $2\hbar\omega = \Omega_{31}$. This would provide resonance enhancement for generation of 3ω. However, as the intensity of the incident frequency is increased, so the atomic levels undergo significant ac Stark shifts and exact resonance is destroyed, thus reducing the susceptibility for harmonic generation. We now examine this notion in more detail.

As shown in Sect.2.6.3, the energy shift $\hbar\Delta\Omega_n$ of an atomic level n induced by an optical field $E = E_0 \cos\omega t$ (assumed for simplicity to be linearly polarised in the z direction) is given by

$$\hbar\Delta\Omega_n = -\frac{1}{2\hbar}\sum_m \left\{\frac{1}{\Omega_{mn}-\omega} + \frac{1}{\Omega_{mn}+\omega}\right\}|\mu_{mn}|^2\overline{E^2} \quad , \tag{4.71}$$

where $\overline{E^2} = \frac{1}{2}E_0^2$; to avoid confusion concerning signs, we reiterate that $\Omega_{mn} = (E_m - E_n)/\hbar$. In neglecting damping factors in the denominators, we assume that frequency detunings are much greater than the natural widths of the levels. Damping factors can, however, be added in the usual way, as in Sect.2.6.3. For static fields (i.e., $\omega \to 0$), (4.71) reduces to the same form as (4.67), where the static field in (4.67) is that produced at a distance R from a charge Ze, viz. $Ze/4\pi\varepsilon_0 R^2$; (4.71) is valid only when $|\mu_{mn}|^2\overline{E^2} \ll \hbar^2(\Omega_{mn} \pm \omega)^2$. If this inequality is not satisfied, then (4.71) is replaced by a more complicated expression [see the discussion following (2.61)]. This situation is analogous to the transition from quadratic to linear Stark effect in the case of a strong dc field, or more familiarly, the transition from the weak field Zeeman effect to the strong field (Paschen-Back) effect for strong magnetic field.

If for simplicity, we consider the case of a two-level atom whose unperturbed levels have energies $\hbar\Omega_2$, $\hbar\Omega_1$, then (4.71) reduces to [see (2.61)]

$$\hbar\Delta\Omega_2 = -\hbar\Delta\Omega_1 \simeq |\mu_{12}|^2\overline{E^2}/2\hbar(\Omega_{21} - \omega) \quad . \tag{4.72}$$

Thus, if ω is less than the unperturbed transition frequency Ω_{21}, level 1 (the lower level, see Fig.4.33) moves down, and level 2 moves up by the same amount. The reverse applies if $\omega > \Omega_{21}$, thus the behaviour is always a repulsion of the transition frequency away from the incident frequency. When a three-level atom is considered (two-photon transition $1 \to 3$, with an intermediate level 2) it can be seen that with an appropriate choice of incident frequencies ω_1, ω_2 (see Fig.4.34) the Stark shift of levels 1 and 3 can actually bring the two-photon transition into resonance with $\omega_1 + \omega_2$. This possibility led to the suggestion by GRISCHKOWSKY and LOY [2.79], subsequently demonstrated by LOY [2.81], that a population inversion of level 3 with respect to level 1 might be achieved by a two-photon adiabatic rapid passage (ARP), as discussed in Sect.2.6.7. ARP can be achieved in single-photon transitions by applying a dc Stark field or by using a frequency sweep of the incident radiation (see Sect.5.5.2). The unique feature of this two-photon ARP is that the necessary sweep of two-photon transition frequency relative to the field frequencies is self-induced and can be produced by pulse sources at a fixed frequency.

An approximate expression for the shift of the two-photon transition frequency is now derived. Frequency ω_1 is assumed to be detuned by an amount Δ_1 below the transition $1 \rightarrow 2$, and $\omega_1 + \omega_2$ detuned by Δ_2 below the two-photon transition $1 \rightarrow 3$ (see Fig.4.34). We also assume $\Delta_2 \ll \Delta_1$. Thus $\Omega_{21} - \omega_1 = \Delta_1$, $\Delta_{31} - (\omega_1 + \omega_2) = \Delta_2$; therefore, $\Omega_{32} - \omega_2 \simeq - \Delta_1$. When the levels are

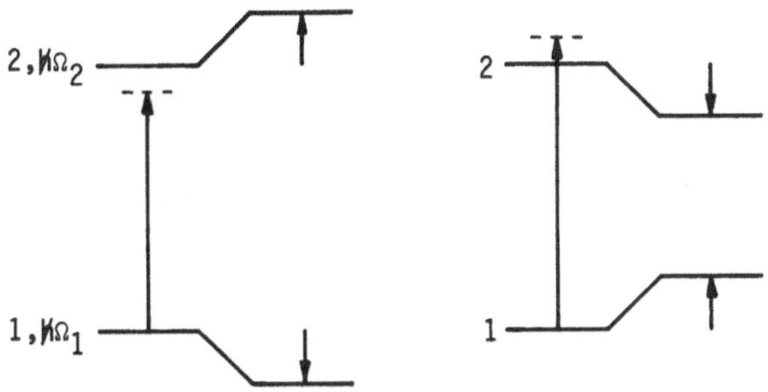

Fig.4.33. ac Stark shifts produced by field with $\omega < \Omega_{21}$ and $\omega > \Omega_{21}$

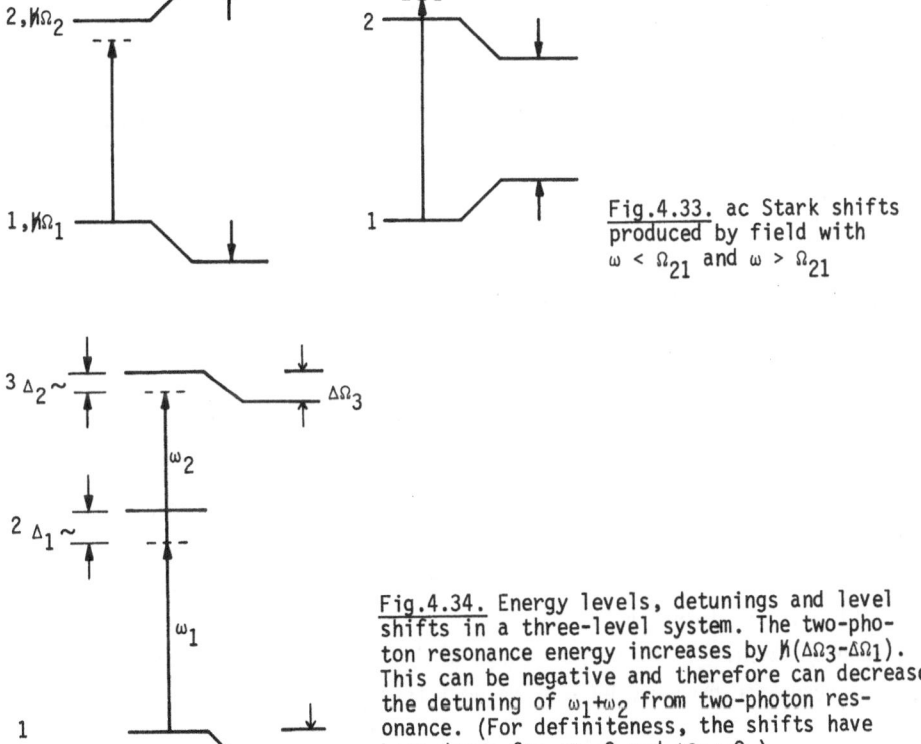

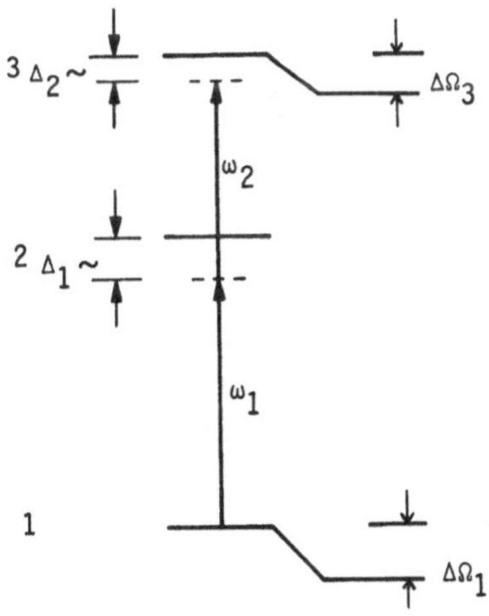

Fig.4.34. Energy levels, detunings and level shifts in a three-level system. The two-photon resonance energy increases by $\hbar(\Delta\Omega_3 - \Delta\Omega_1)$. This can be negative and therefore can decrease the detuning of $\omega_1 + \omega_2$ from two-photon resonance. (For definiteness, the shifts have been drawn for $\Delta\Omega_2 < 0$ and $\Delta\Omega_1 > 0$.)

ac Stark shifted, the detuning from the two-photon resonance is

$$\Omega_{31} + \Delta\Omega_3 - \Delta\Omega_1 - (\omega_1 + \omega_2) = \Delta_2 + (\Delta\Omega_3 - \Delta\Omega_1) \quad ; \tag{4.73}$$

with the help of equation (4.72) (generalised to include two incident frequencies) this is found to be

$$\Delta_2 + \frac{1}{2\hbar^2\Delta_1}\left\{|\mu_{12}|^2\overline{E_1^2} - |\mu_{23}|^2\overline{E_2^2}\right\} \quad . \tag{4.74}$$

In deriving this expression, we assumed that the incident frequency ω_1 is the dominant agent in shifting level 1 and similarly frequency ω_2 in shifting level 3. It can be seen that the detuning in (4.74) can increase or decrease with the application of the fields at ω_1, ω_2 depending on the relative intensities $\overline{E_1^2}$, $\overline{E_2^2}$ and on the relative signs of Δ_1, Δ_2. A numerical estimate of typical shifts will be given in Sect.5.4.2.

The presence of the two incident frequencies ω_1, ω_2 will produce two-photon transitions at the same time as shifting the levels; LIAO and BJORKHOLM [2.51] pointed out that there is a simple relationship (for the three-level atom) between the two-photon transition rate $W^{(2)}$ and the level shifts $\hbar\Delta\Omega_1$, $\hbar\Delta\Omega_3$,

$$W^{(2)}(1 \rightarrow 3) \propto -\Delta\Omega_1\Delta\Omega_3 \quad . \tag{4.75}$$

This relation follows from comparing equation (4.72) with the expression for $W^{(2)}$ [see (4.58)],

$$W^{(2)} \propto \left|\sum_m \left(\frac{\mu_{3m}\mu_{m1}}{\Omega_{m1}-\omega_1} + \frac{\mu_{3m}\mu_{m1}}{\Omega_{m1}-\omega_2}\right)\right|^2 \overline{E_1^2}\,\overline{E_2^2} \quad . \tag{4.76}$$

The sum over intermediate states $|m\rangle$ reduces to a single term for the case of a three-level atom. From (4.75), it is seen that the energy-level shifts are large whenever the two-photon transition rate is large. However, it is possible, if the level shifts $\Delta\Omega_1$, $\Delta\Omega_3$ are significantly different, to have a large shift $\hbar(\Delta\Omega_3 - \Delta\Omega_1)$ even though $W^{(2)}$ is small. Thus, given a two-photon resonance produced by two incident frequencies, there is some scope for varying the extent of any effects induced by level shifts.

We have restricted ourselves in this section to a discussion of level shifts and two-photon absorption within the framework of a perturbation-theory analysis. Clearly, however, such an approach will be inadequate at high intensities and small detunings. Recently, there has been a great amount of theoretical work aimed at nonperturbative and transient analysis of the interaction between radiation and (for example) a three-level atom; this was reviewed in Sect.2.6. Such interrelated phenomena as level shifts, power broadening, and two-photon absorption are handled together rather than in the separate and ad hoc fashion we have just given. This type of treatment leads to some interesting conclusions; for example, if sum-frequency generation $\omega_1 + \omega_1 + \omega_2 \rightarrow \omega_3$ (with $\omega_1 + \omega_1$ two-photon resonant) occurs in a transient regime, then the highest generation efficiency may occur when the pulse of ω_2 radiation is not synchronised with the pulse of ω_1 radiation. This effect has been observed experimentally [2.106], and theoretical discussions have been given by NEW [4.88], and ELGIN and NEW [4.89].

Clearly these nonperturbative treatments are of considerable relevance to resonant nonlinear interactions, particularly in connection with limiting and saturation. However, when the perturbation approach is abandoned, the analysis generally becomes much more complicated and many of the explicit formulae obtained by perturbation theory have to be replaced by numerical calculations.. For this reason, it is likely that perturbation approaches will continue to be widely used even when it is appreciated that their validity is questionable.

5. Stimulated Electronic Raman Scattering

5.1 Background Material

It was pointed out in Chap.1 that stimulated Raman scattering is one of the earliest observed and most widely studied of nonlinear optical processes. The subject has been extensively reviewed, and our introduction is, therefore, a brief one (see also Sects.2.3,3.3). For detailed introductions the reader may wish to consult some of the following reviews; [2.14, 90,100,102,5.1-6]. The greatest emphasis in the work covered by these reviews (which is by no means a complete list) has been on stimulated vibrational and rotational Raman scattering in molecular liquids and gases. In addition, there has been much interest in stimulated polariton scattering in crystals [5.7,8] and stimulated spin-flip Raman scattering in semiconductor crystals [1.8,5.9]. These processes can be used to generate tunable far-ir and medium-ir radiation. Thus a rich variety of stimulated Raman scattering processes have been observed and exploited. In this section, we review the results obtained from stimulated Raman scattering in atomic vapours, a process that we describe as stimulated electronic Raman scattering (SERS), because the energy levels involved are those of isolated atoms and hence are purely electronic.

Stimulated electronic Raman scattering was first observed by ROKNI and YATSIV [1.33,34] and SOROKIN et al. [1.32].

The experiment by ROKNI and YATSIV, in particular, serves as a useful introduction to some of the important features of SERS. They studied SERS in potassium vapour, using a Q-switched ruby laser as the pump. The energy-level scheme is shown in Fig.5.1; the Raman transition observed was $4p_{3/2} \rightarrow 5p_{3/2}$ (i.e., atoms initially in the $4p_{3/2}$ level were excited to $5p_{3/2}$). The energy of this transition is 11677 cm^{-1}, and so the pump beam ($\omega_p \sim 14398$ cm^{-1}) generated radiation at the Stokes frequency $\omega_s \sim 2721$ cm^{-1} (3.7μm). This particular energy level scheme is favourable because the ruby-laser frequency is within ~ 10 cm^{-1} of the allowed single-photon transition $4p_{3/2}$ - 6s, thereby greatly enhancing the Raman polarisability. However, with

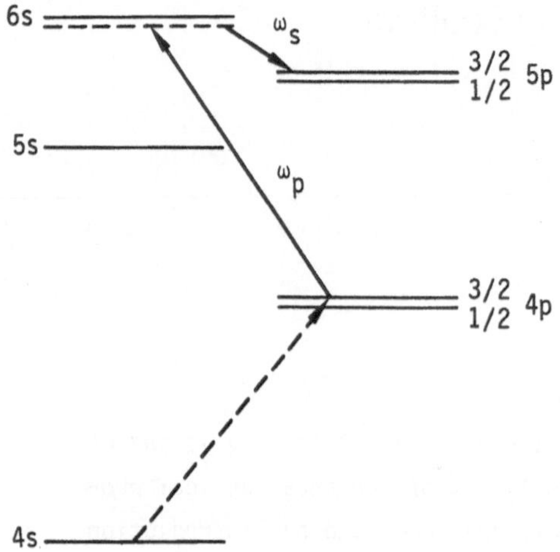

Fig.5.1. Energy-level diagram of potassium, showing the SERS scheme used by ROKNI and YATSIV [1.33,34]. Optical pumping (broken line) to the $4p_{3/2}$ level is followed by SERS on the $4p_{3/2}$ - $5p_{3/2}$ Raman transition, resonantly enhanced by the 6s level

this level scheme the initial Raman level ($4p_{3/2}$) is an excited level with negligible thermal population. It was, therefore, necessary first to prepare a significant population of atoms in this state. This can be achieved in a number of ways; various possible methods are described briefly in Sect.5.2.1. The simple and effective method used by ROKNI and YATSIV was to frequency shift some of the output from the ruby laser into near coincidence with the 4s - $4p_{3/2}$ D-line transition by means of stimulated (vibrational) Raman scattering in a suitably chosen liquid, and thereby optically pump the $4p_{3/2}$ level. Experimentally this simply required the liquid cell to be placed between the ruby laser and the potassium vapour cell. Four different Raman-scattering liquids were used with success; for example, nitrobenzene (Raman shift: 1345 cm^{-1}) shifted the ruby-laser frequency to within $\sim$ 10 cm^{-1} of the D-line. At the potassium vapour pressures used there was sufficient absorption in the wings of the D-line to produce a substantial population of atoms in the excited $4p_{3/2}$ state. The remaining ruby-laser light, which was not shifted in the liquid cell, could then produce SERS from the excited potassium atoms.

In this experiment, ROKNI and YATSIV demonstrated the ability to tune the generated Stokes frequency. By varying the temperature of the ruby-laser rod, the pump frequency was shifted by up to 3 cm^{-1}; it was observed that the frequency of the SERS emission changed by the same amount. Subsequent experiments, using the output of widely tunable dye lasers as the pump, have shown Stokes tuning ranges of several hundred cm^{-1}.

Before going on to discuss experimental results more generally, we first give some order-of-magnitude estimates of Raman gain, showing the effect of resonance enhancement and comparing the results with typical figures for vibrational Raman scattering in molecules. The starting point for a calculation of Raman gain is the Raman-susceptibility expression, from (2.19) and Table 2.2,

$$\chi_R \equiv \chi^{(3)}(-\omega_s;\omega_p-\omega_p\omega_s) = \frac{N/6\hbar\varepsilon_0}{\Omega_{fg}+\omega_s-\omega_p+i\Gamma}\,\overline{|\alpha_R|^2}\ , \tag{5.1a}$$

where the Raman polarisability is

$$\alpha_R = \frac{1}{\hbar}\sum_i\left(\frac{\underline{\varepsilon}_s^*\cdot\underline{Q}_{fi}\underline{\varepsilon}_p\cdot\underline{Q}_{ig}}{\Omega_{ig}-\omega_p}+\frac{\underline{\varepsilon}_p\cdot\underline{Q}_{fi}\underline{\varepsilon}_s^*\cdot\underline{Q}_{ig}}{\Omega_{ig}+\omega_s}\right)\ . \tag{5.2}$$

As discussed in Sect.2.5, N is the effective atomic density for the Raman transition $g \rightarrow f$; the orientation average denoted by a bar in (5.1a) may be effected by summing over the degeneracy of the initial level; a sum over the degeneracy of the final level is also implied. For most descriptive purposes, it is convenient to have at hand the simplified form of (5.1a) that results when the pump and Stokes polarisation vectors are linearly polarised and parallel [cf. the simplification of χ_{THG} in passing from (2.18) to (4.1)],

$$\chi_R = \frac{N/6\hbar^3\varepsilon_0}{\Omega_{fg}+\omega_s-\omega_p+i\Gamma}\left|\sum_i \mu_{fi}\mu_{ig}\left(\frac{1}{\Omega_{ig}-\omega_p}+\frac{1}{\Omega_{ig}+\omega_s}\right)\right|^2\ . \tag{5.1b}$$

[In the interest of simplicity, the orientation average has been omitted from (5.1b), but it must be borne in mind when attempting numerical calculations.]

The frequency denominators in (5.1) show that by tuning the pump frequency close to one of the intermediate levels ($\omega_p \simeq$ some Ω_{ig}), the Raman susceptibility can be greatly increased. Thus, in the experiment by ROKNI and YATSIV [1.33,34] described earlier, a large gain for SERS was ensured by selecting an energy-level scheme in which the ruby-laser (pump) frequency was nearly resonant with an appropriate intermediate level. By using a dye laser as a tunable pump source, many more suitable intermediate resonances can be found, rather than relying on chance coincidences. This ability to enhance resonantly the susceptibility by a suitable choice of pump frequency leads to SERS gain coefficients in atomic vapours that are comparable to the highest SRS gains in liquids and high-pressure molecular gases, despite the very much lower number densities. The following example will demonstrate this point.

In Sect.3.3 (3.19), the exponential gain coefficient G_R was derived in terms of the Raman susceptibility χ_R,

$$G_R \equiv g_R I_{po} = (-3\omega_s/\varepsilon_o c^2 n_s n_p)\chi''_R I_{po} \quad , \tag{5.3}$$

where I_{po} is the incident pump intensity and χ''_R is the imaginary part of χ_R. For typical vapours and gases the refractive indices n_s and n_p can be set equal to unity. For molecular vibrational-rotational Raman scattering, the gain coefficient is more often expressed in terms of the Raman-scattering cross section $d\sigma_R/d\Omega$, because the latter can be determined from spontaneous Raman-scattering data [5.3,10]. From (2.84,85), the ratio of the number of Stokes photons scattered to the incident pump flux is $d\sigma_R/d\Omega = [\omega_s^3\omega_p/(4\pi\varepsilon_o c^2)^2]|\alpha_R|^2$, [2.19], whence (5.1,3) give

$$g_R = \frac{8\pi^2c^2}{\hbar n_s n_p \omega_s^2 \omega_p} \frac{N\Gamma}{(\Omega_{fg}+\omega_s-\omega_p)^2+\Gamma^2} \frac{d\sigma_R}{d\Omega} , \tag{5.4a}$$

which becomes, on exact Raman resonance,

$$g_R = \frac{8\pi^2c^2}{\hbar n_s n_p \omega_s^2 \omega_p} \frac{N}{\Gamma} \frac{d\sigma_R}{d\Omega} \quad , \tag{5.4b}$$

from which it can be seen that the Raman gain is inversely proportional to the linewidth Γ. In deriving (5.4) we assumed that the pump is monochromatic (or at least that the pump linewidth is much less than Γ); the effect of a nonmonochromatic pump is discussed in Sect.5.3.3. The calculation of the Raman gain based on (5.4) requires knowledge of the value of Γ; often, uncertainty of this value makes the over-all calculation of the gain subject to considerable error, even if $d\sigma_R/d\Omega$ is accurately known.

For the case of SERS in atoms, direct experimental data on Raman-scattering cross-sections is generally not available; however, the relevant data on matrix elements often exists, in which case G_R can be calculated from (5.3), with χ''_R given by (5.1). A knowledge of Γ is again required; this is more of a problem, because there is evidence, from a number of experiments, that considerable line broadening occurs in SERS. These points are discussed further in Sects.5.3,4, together with detailed calculations of the nonlinear susceptibilities and gain coefficients for the alkali metals.

However, it is instructive at this stage to compare the typical gain coefficients found in stimulated vibrational-rotational Raman scattering with those of SERS. The gain factor $g_R \equiv G_R/I_{po}$ (given here in units of cm GW^{-1}, where I_{po} is the incident pump intensity in GW cm^{-2}) has been measured for a number of Raman-scattering liquids and gases ([5.3], and references therein). Examples of measurements made with a ruby laser (694nm) are 16 ± 5 for liquid nitrogen (Raman shift: 2326 cm^{-1}), 24 for CS_2 (656 cm^{-1} shift), 2.8 for benzene (992 cm^{-1} shift), and 1.5 for hydrogen gas above 20 atmospheres (4155 cm^{-1} shift). As an example of SERS, we can use the simplified expressions (5.5,6) for the gain (Sect.5.3.1) to estimate the gain coefficient in the experiment of ROKNI and YATSIV [1.33,34] in potassium, described earlier (Fig.5.1). A reasonable simplifying assumption is that, because the ruby-laser frequency is only $\sim$ 10 cm^{-1} away from exact resonance with the $4p_{3/2}$ - 6s transition, the 6s level is the intermediate level that makes the dominant contribution. The potassium-vapour temperature was 300°C; on the assumption that about 10% of the potassium atoms were excited from the ground state to the $4p_{3/2}$ level, and by use of (4.33,34), the $4p_{3/2}$ population density is estimated to be $\sim 5 \times 10^{20} m^{-3}$. (This is a relatively low value, because for the many SERS experiments in which the ground state is the initial Raman level, and with vapour pressures of about 10 torr, typically the initial-state number densities are $\sim 10^{23} m^{-3}$.) Making the pessimistic assumption that Γ is as large as 4 cm^{-1}, then from (5.1,2) we arrive at an estimate for the gain factor of $g_R \sim 10$ cm GW^{-1}, which is comparable to the gain in liquid nitrogen. By use of a power density of $\sim$ 250 MW cm^{-2} (easily achieved with a Q-switched ruby laser) the corresponding small-signal gain would be $\sim e^{50}$ over a vapour-path length of 20 cms. This calculated gain is well in excess of the e^{30} gain (see Sect.3.3) needed to produce an intense SERS Stokes signal in a single pass through the vapour. However, it should be noted that, in this example, the pump-laser frequency was only 10 cm^{-1} from resonance with the intermediate level, whereas the SERS gain is inversely proportional to the detuning squared [(5.6), Sect.5.3]. On the other hand, the Raman gains quoted for molecular systems were measured under conditions in which the pump frequency is far removed from intermediate resonances; the gain is then a slowly varying function of the pump frequency. These examples demonstrate the great advantage to be gained from exploiting local resonances in atomic vapours, although this necessarily implies that the tuning ranges are more limited in extent.

Before leaving this brief discussion of resonance enhancement, we note that there are two types of intermediate-state resonance. These correspond to the two frequency denominators in (5.2), and are illustrated in Fig.5.2a. The first frequency denominator is resonant when $\omega_p \simeq \Omega_{ig}$, where i is an intermediate state that lies above the initial Raman level g. This is the most commonly used resonance, and is familiar from intuitive pictures of Raman scattering. The second frequency denominator in (5.2) is resonant when $\omega_s \simeq -\Omega_{i'g}$ so that i' must be below the initial Raman level, which in turn means that g must be an excited state. (This type of resonance has its analogues in any process, as has been pointed out in Sect.2.4). DUCUING et al. [1.61] have considered resonances of this second type in connection with a scheme for far-infrared generation by SRS on pure-rotational transitions in HF and HCl; see Sect.7.6. Such resonances could also be of importance in SERS experiments in which the atoms are first prepared in an excited state g.

So far, we have considered only Stokes generation, i.e., where $\omega_s < \omega_p$. When the generated frequency is greater than ω_p the term anti-Stokes is used, and we replace the symbol ω_s by ω_{as}, although there is no change in the actual process; thus $\Omega_{fg} + \omega_{as} - \omega_p \simeq 0$ still applies. However, it is now necessary that the initial level g lie above the final level f, i.e., $\Omega_{fg} < 0$ (thus if $\Omega_{fg} = -\Omega$, where Ω is positive, then $\omega_{as} \simeq \omega_p + \Omega$); therefore, g must always be an excited state, so that, for atoms, some preparation of the initial level will be needed for anti-Stokes generation.[1] Again there are two types of intermediate-state resonance; for comparison with the case of Stokes scattering, these are shown in Fig.5.2b.

Both denominators in (5.1) always contribute to Raman scattering but, for the level schemes we have been considering, resonance enhancement of the Raman gain is obtained by making one of the denominators very small, and therefore dominant. For the majority of spectroscopic applications of spontaneous Raman scattering in molecules there is no such enhancement, and the denominators contribute almost equally; this is also true of some infrared-generation schemes in molecules (Chap.7).

For the most part, we shall confine our attention to the first type of resonance enhancement, in which $\omega_p \simeq \Omega_{ig}$, and where g is the ground level; this is by far the most common arrangement. Anti-Stokes SERS will be briefly discussed in Sect.5.5. Naturally many practical details change when anti-Stokes SERS and/or the second type of resonance are considered, but the

---

[1]It should be noted that we are not considering here the parametric generation of ω_{as} by "coherent" anti-Stokes Raman scattering (CARS); see Chap.6.

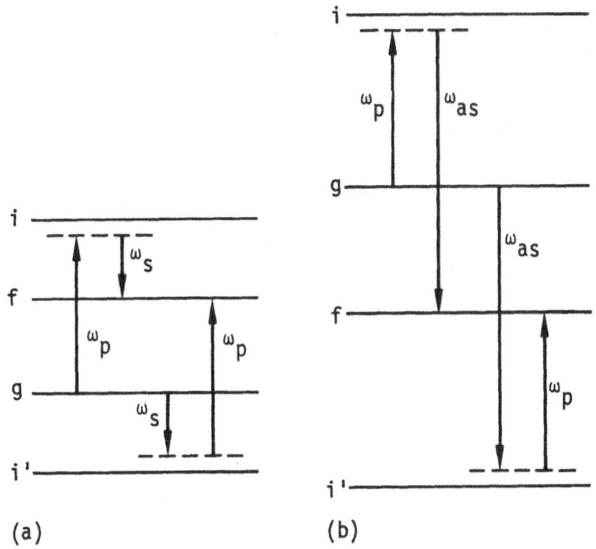

Fig.5.2a and b. Resonance enhancement in SRS. Both Stokes (ω_s) and anti-Stokes (ω_{as}) schemes are drawn for a fixed pump frequency ω_p. (a) *Stokes generation* (The final Raman level f lies above the initial level g.) The Raman susceptibility may be enhanced by tuning ω_p close to resonance with an intermediate state i lying above g, or by tuning ω_p so that ω_s is close to resonance with an intermediate state i' lying below g. In the latter case, g must be a populated excited state, but in the first case g may be any state, including the ground state. (b) *Anti-Stokes generation* (f lies below g) As for (a), but with ω_s replaced by ω_{as}. Notice that g must always be a populated excited state

principles of SERS as applied to the resonance $\omega_p \simeq \Omega_{ig}$ hold true for these other variations.

Finally, before discussing, in the next subsection, the application of SERS to tunable-infrared generation, a brief mention should be made of the recent application of SERS as a method of shifting the output from ultra-violet excimer lasers to longer wavelengths. Noble-gas halide excimer lasers (or more correctly, exciplex lasers), both electric-discharge pumped [5.11, 12] and e-beam pumped [5.13,14], are rapidly gaining importance as convenient and efficient high-energy sources for the uv and vuv regions. For example, discharge-pumped KrF lasers, which are no larger than a 10mJ nitrogen laser, readily give pulsed-output energies in excess of 200mJ at 248nm. (Other excimer and exciplex laser wavelengths, all in nm, are Ar_2 126, Kr_2 146, Xe_2 172, F_2 157, Cl_2 259, Br_2 292, I_2 342, ArF 193, KrF 248, XeF 351, ArCl 175, KrCl 222, XeCl 308, XeBr 282.) If the output of these lasers could be shifted efficiently to longer wavelengths, the range of possible applications, particularly for photochemistry and isotope separation, could be greatly extended. DJEU and BURNHAM [1.41] used SERS in barium vapour to convert the 351nm output from a XeF electric discharge laser to Stokes radiation at 585nm with an 80% photon-conversion efficiency. This very high efficiency was obtained by taking special precautions to ensure good spatial-beam quality from the excimer laser, and also to narrow the frequency of the emission, although this was achieved at the expense of laser output energy, which was only 25μJ. More recently, COTTER and ZAPKA [1.42] also used SERS in barium vapour to shift the output of a XeCl laser, although an energy level scheme different from that of DJEU and BURNHAM was employed. In this case, the excimer-laser output at 308nm was converted to blue-green Stokes radiation at 475nm. A photon conversion efficiency of up to 20% and blue output energies of greater than 5mJ were obtained, even though the excimer laser beam was of poor spatial quality, with an angular divergence more than 25 times greater than the diffraction limit, and although no attempt was made to narrow the laser linewidth of 0.5nm. BURNHAM and DJEU [1.43] recently reported using SERS transitions in Ba, Tl, Bi, and Pb vapours to shift the XeCl laser output to the blue visible region; in the case of Pb, a photon-conversion efficiency of 60% was achieved, with outputs up to 20mJ. Because the nonlinear medium can be readily made larger, these results indicate that SERS has considerable potential as a method of frequency shifting the output of very powerful excimer lasers. Thus, the development of high-energy visible laser sources with high over-all efficiencies, in the range 0.1 - 1% in the case of discharge-pumped lasers, can be foreseen.

5.2 Tunable Infrared Generation

5.2.1 Experimental Techniques and Results

The simplest scheme for generation of tunable infrared radiation by means of SERS is illustrated by the energy-level diagram in Fig.5.3, where as a specific example, the atom has been assumed to be an alkali. Intense pulsed dye-laser light is used as the pump to excite an electronic Raman transition between the atomic ground (g) and excited (f) states, both of which are nondegenerate s states. In this way, Stokes radiation is produced at the frequency $\omega_s = \omega_p - \Omega_{fg}$. The Raman shifts, Ω_{fg}, are typically 20,000-30,000 cm^{-1} in the alkalis. Thus dye lasers that work in the blue and near-uv region can be used directly to generate tunable medium-infrared radiation (1-20μm). The

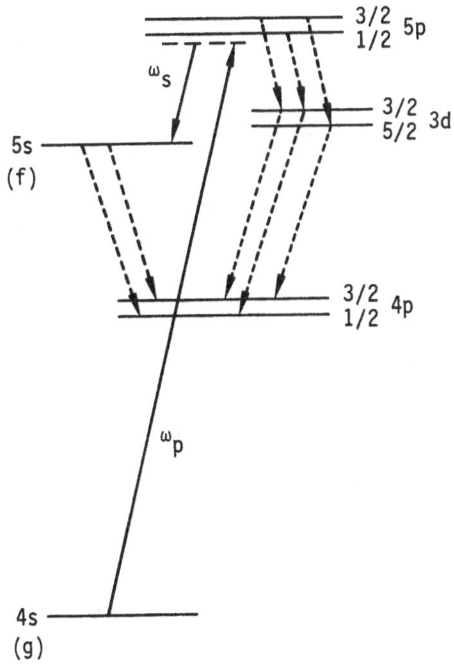

Fig.5.3. Energy-level diagram of potassium, showing the 4s-5s SERS transition. The tunable Stokes output ω_s is accompanied by various amplified spontaneous emissions (broken lines)

wavelength region in which the dye laser is operated is chosen to be near to appropriate single-photon resonances. In the alkali metals these are the principal-series resonance lines. In Sect.5.3.1, it is shown that this produces a large Raman gain for Stokes scattering to the excited s level that has the same principal quantum number as the resonant intermediate p level. With dye-laser powers of a few tens of kilowatts, infrared tuning ranges of several hundreds of cm^{-1} have been readily obtained in alkali vapours, with peak photon-conversion efficiencies as high as 50%.

In practical terms, the SERS method is very straightforward; the pulsed dye-laser beam is simply focussed into a heated cell that contains the alkali vapour, and the infrared radiation emerges in a narrow collimated beam collinear with the incident laser beam. A convenient type of vapour cell, which is most often used, is the heat-pipe oven. One of the important features of heat-pipe ovens is that it is possible to maintain a very uniform vapour density over great lengths; as described in Sect.4.4.2, this is necessary where phase matching is achieved by mixing with another vapour or gas. For SERS, there is no phase-matching condition to be satisfied; the main advantage of using a heat pipe in this case is that the cell can be operated at relatively high vapour pressures (typically 10 torr) over long periods of time (up to several weeks with some alkali metals) without the usual problems caused by the vapourised material continually depositing near (or on) the cold end windows. At the output of the vapour cell, it is usually necessary to have some appropriate optical filter to discriminate between the SERS Stokes output and other radiation that emerges from the atomic vapour. Figure 5.4a shows a typical arrangement.

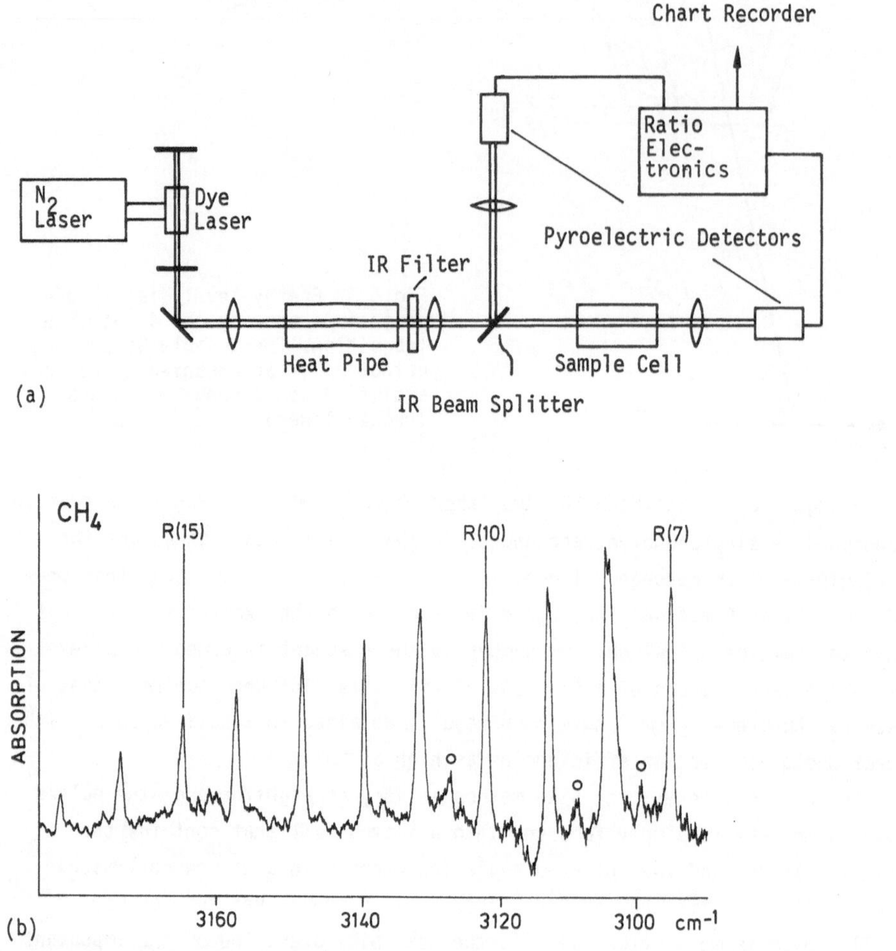

Fig.5.4. (a) Layout of the tunable infrared source based on SERS, showing the dual-beam spectrometer arrangement used for taking infrared absorption spectra (after [1.37]). (b) Absorption spectrum of North Sea gas (100 torr, 15°C, 7 cm length) showing a part of the Ch_4 R-branch. (After [4.5])

SOROKIN et al. [1.35] were the first to observe a widely tunable infrared output produced by SERS. They used potassium vapour; the energy-level scheme was that shown in Fig.5.3. By using a dye laser tunable near the 4s-5p principal resonance lines (~ 404nm), intense tunable infrared radiation was observed. The Raman shift was 21027 cm^{-1} (corresponding to the energy difference between the 5s and 4s levels). Consequently, the infrared radiation was generated in the region of 3700 cm^{-1} (2.7μm). SOROKIN et al. used a nitrogen-laser-pumped dye laser of ~ 1kW output power; the beam was focussed into a 30 cm column of potassium vapour at a pressure of 10 torr. In this way, an infrared tuning range of ~ 220 cm^{-1} between 2.63μm and 2.79μm was obtained. They used this infrared emission as one of the driving waves in a four-wave parametric-mixing process to generate a further tunable infrared output; this work is described in Sect.6.2.

COTTER et al. [1.37] repeated this experiment of SOROKIN et al. in potassium except that a higher dye-laser power was used (20-30kW) in a nearly diffraction-limited beam. The infrared output was found to be tunable from 2.56μm to 3.5μm, a range of ~ 1000 cm^{-1}. This was limited at the short-wavelength end by the dye-laser tuning range. Infrared output powers in excess of 1kW were obtained over the central 250 cm^{-1} of the range, apart from two gaps when the dye laser was tuned within a few cm^{-1} of the 4s-5p doublet absorption lines. Throughout the 1000 cm^{-1} tuning range as defined, the output power was in excess of 100mW. To test the usefulness of this tunable source for infrared spectroscopic applications, and also to measure (indirectly) the output linewidth, the generated radiation was used to record absorption spectra of CO_2, CH_4, and H_2O vapour in the region of 2.7μm. Figure 5.4a shows the dual-beam-ratiometer arrangement that was used and Fig.5.4b shows a typical spectrum obtained with it [4.51]. Operation of this infrared source proved to be very simple and reliable.

However, a somewhat disappointing result of these measurements was that the resolution in the infrared absorption spectra was only ~ 0.4 cm^{-1}. This was despite the fact that the dye-laser linewidth was 0.1-0.2 cm^{-1} and the calculated Doppler width of the 4s-5s Raman transition was even less, ~ 0.06 cm^{-1}. The generated infrared output had been expected to have a linewidth closer to that of the dye laser, or possibly somewhat narrower because of gain narrowing. This observation was the first indication that some significant line-broadening processes could be occurring; as will be described later, this result has since been confirmed on a number of occasions by more-direct methods of measurement.

The 6s-7s SERS transition in cesium has also been studied in detail, by COTTER and HANNA [4.51] and KUNG and ITZKAN [5.15]. Using a nitrogen laser-pumped dye laser of 20kW peak power, COTTER and HANNA obtained results similar to those found earlier using potassium; the output power was in excess of 10W over a continuous 860 cm^{-1} tuning range around 3μm. The peak output power was 1.5kW, representing a photon-conversion efficiency of 50%. The performance of this SERS source and its dependence on various experimental parameters, such as vapour pressure and dye-laser focussing, were examined in detail. The best output power and tuning range were obtained at 10 torr cesium-vapour pressure and with the dye-laser beam focussed confocally over the 25 cm vapour column (i.e., focussed at the centre of the vapour so that the confocal parameter was approximately equal to the column length). It was found that these optimum values are mainly dictated by the need to minimise the absorption loss due to Cs_2 dimers (see Sect.5.4.1).

Intense tunable infrared radiation generated by SERS in barium vapour has been reported by CARLSTEN and DUNN [1.36]. The Raman transition was from the $6s^2\ ^1S_0$ singlet ground state to the metastable $6s5d^3D_{1,2}$ levels in the triplet system, giving Raman shifts of 9033 cm^{-1} and 9215 cm^{-1}, respectively. Resonance enhancement was obtained by tuning the input dye laser in the region of the 791.1nm intercombination line between the ground state and $6s6p^3P_1$ level. Using the 250mJ output from a ruby-laser-pumped dye laser, CARLSTEN and DUNN obtained an output that was tunable over a range of 130 cm^{-1}, around 2.9μm. A notable feature of their experiment was the very high output energies that were obtained; up to 30mJ (1MW) with as much as 40% photon-conversion efficiency. Figure 5.5 shows the generated infrared energy as a function of the dye-laser frequency for three different barium vapour pressures. The tuning profiles shown are typical of those found in SERS. Especially at the higher pressure, the profile shows a distinct dip when the dye laser is tuned exactly to the intermediate resonance. This reduction of output is due to resonance absorption of the dye-laser light as well as to other resonantly enhanced competing processes (Sect.5.4.1). Either side of the resonance frequency there are output maxima, and then the infrared output decreases gradually as the dye laser is tuned further from resonance. In the course of their work, CARLSTEN and DUNN observed saturation of the generated infrared output due both to depletion of the pump-beam intensity and to reduction of the atomic-ground-state population that results from excitation by SERS. A rate-equation analysis of these effects (described in Sect.5.4.3) also leads to a plausible explanation for the observed tuning profiles.

In the alkali metals, the frequency interval between ns and np levels becomes smaller for successively higher principal quantum number n. This means that by using Raman transitions from the ground state to higher-lying final ns levels, and using the correspondingly higher intermediate resonance levels np, longer infrared Stokes wavelengths can be generated. A

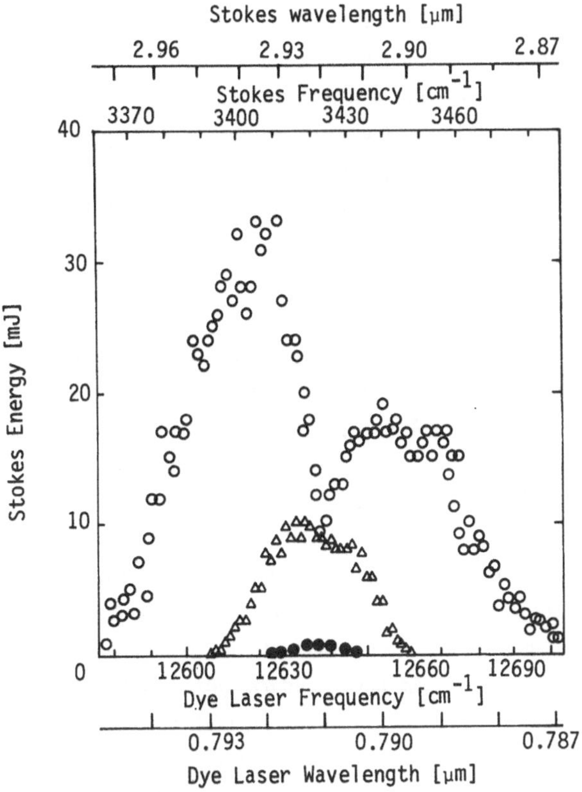

Fig.5.5. Tuning profiles of the Stokes signal generated in barium vapour. For • the vapour pressure was 0.02 torr, for Δ, 0.2 torr; for o, 2.0 torr. (After [1.36])

disadvantage, however, is that the pump frequencies that are required extend further into the uv. Also, the lighter the alkali atom, the higher are its energy levels. Most SERS experiments have, therefore, been carried out on the heavier alkalis, because they offer a number of possible transitions from the ground state that are accessible by use of dye lasers pumped by a nitrogen laser (337nm) or the second harmonic of a ruby laser (347nm). In particular, cesium provides SERS transitions to the first three excited ns states, all of which can be resonantly enhanced by tuning these dye lasers close to the corresponding intermediate np levels. Thus, SERS outputs centred around ~ 3μm on the 6s-7s transition, ~ 7μm on the 6s-8s transition and ~ 13μm on the 6s-9s transition are obtained. A major stimulus to the experimental investigation of these transitions [1.39] was that calculations of the SERS gain and threshold (Sect.5.3), based on the assumption that the Raman transition linewidth Γ is Doppler-limited, indicated that complete coverage of the range 2-20μm should be possible by use of a diffraction-limited dye-laser beam of ~ 30kW power and 0.1 cm^{-1} bandwidth. The Stokes frequencies that would be generated by use of the SERS transitions 6s → 7s, 6s → 8s and 6s → 9s would then correspond to infrared wavelengths in the ranges 2-5μm, 5-10μm and 9-20μm, respectively, and the required corresponding dye-laser wavelengths are 425-487nm, 380-395nm and 357-365nm. The experiments were carried out using a dye-laser oscillator-amplifier system pumped by the 200mJ second-harmonic output from a ruby laser. This dye laser, which used a selection of blue and uv dyes, gave an output power in the range 250-750kW. The beam was focussed with a 50-cm lens into a 40 cm cesium heat-pipe oven with vapour pressures of 3-30 torr; the generated SERS output was detected after passing it through semiconductor filters and a grating monochromator. In this way, infrared radiation continuously tunable over the ranges 2.5-4.75μm, 5.67-8.65μm, and 11.65-15μm was obtained, with peak output powers of 25kW, 7kW, and 2kW, respectively.

The discrepancy between the predicted and observed tuning ranges in this experiment is due, at least in part, to the incorrect assumption that the Raman linewidth Γ remains Doppler limited under the conditions of the high laser intensities that were used. Various effects that can contribute to line broadening are discussed in Sect.5.4.2. An idea of the extent of this line broadening is indicated by some preliminary measurements of the infrared output linewidth made by scanning the output with an f/10 grating monochromator with 0.15 cm^{-1} resolution, WYATT [5.16]. On the 6s-9s Raman transition, Stokes bandwidths of 0.7-0.9 cm^{-1} were obtained at cesium-vapour pressures of 5 and 30 torr, with a dye-laser bandwidth of 0.15 cm^{-1} and power of 250kW. At the same time, the measured bandwidth of the fixed-frequency amplified spontaneous emission (ASE) output (see Sect.5.2.2) that corresponds to the transition $9s_{1/2}$ - $9p_{3/2}$ was instrument limited at 0.15 cm^{-1}. Reducing the dye-laser power to ~ 4kW decreased the measured Stokes bandwidth to 0.3 - 0.4 cm^{-1}. For the 6s-8s transition, WYATT measured the SERS linewidth as a function of pump-beam focussing. A factor of four increase from 0.4 cm^{-1} to 1.65 cm^{-1} was observed when the pump intensity was increased by four orders of magnitude from ~ 2MW cm^{-2} to ~ 10GW cm^{-2}.

To make a more intensive study of the dependence of the infrared bandwidth on the various experimental parameters, a nitrogen-laser-pumped dye laser was used [4.51]. Although it provided a lower power than the ruby-pumped laser it had the advantage of a higher repetition rate. For this work, the 6s-7s cesium Raman transition was used. The Stokes linewidth was again monitored by scanning with an infrared monochromator, with resolution 0.1 cm^{-1}. The dye-laser output was 100-120μJ in 7ns pulses (14-17kW), and with 0.07-0.1 cm^{-1} bandwidth. Over a wide range of vapour pressures, laser-beam focussing conditions and frequencies, the measured linewidth of the generated infrared fell in the range 0.25-0.55 cm^{-1}. Within this range, the linewidth was observed to increase as the vapour pressure was increased from 1 to 30 torr, and also as the dye laser was tuned closer to the intermediate 7p levels; the total variation amounted to no more than a factor of two in each case. Also, increasing the peak dye-laser intensity in the vapour over two orders of magnitude caused the Stokes bandwidth to increase by about a factor of two.

Under all conditions of measurement, the SERS output always occurred in a single well-defined spectral line with no obvious structure. Since the hyperfine splitting of the cesium ground state is $\sim$ 0.3 cm^{-1} the Stokes output might have been expected to have been observed as two lines with frequencies separated by this amount, which would have been easily resolved experimentally. The Boltzmann factor can be neglected, because the hyperfine splitting is much less than k_BT, and ground-state atoms, therefore, occupy the two hyperfine levels with populations proportional to their degeneracies; g = 7 for the F = 3 level and g = 9 for F = 4. YURATICH [2.21] found that the Raman gains for transitions from the initial levels F = 3 and F = 4 are simply in the ratio of their degeneracies, i.e., 7:9, when the final-state hyperfine structure is neglected. If the output Stokes intensity were limited by pump saturation, then the Stokes output on the hyperfine transition with the highest gain (i.e., from $6s^2S_{1/2}F = 4$) would be expected to effectively suppress that on the lower-gain transition. However, there is evidence that, over part of the tuning range, the Stokes intensity is limited by atomic depletion, in which case it would be expected that both Stokes lines should be present with comparable magnitude. This was not observed experimentally; the negative result has not been accounted for.

The last two or three years have seen important technical developments in high-power, uv dye lasers. This will allow still higher Raman levels to be used for the generation of longer infrared wavelengths. An early example of this possibility was the generation of 20μm SERS in Cs (6s → 10s) [5.96], using the second harmonic of a mode-locked ruby laser as pump (347nm). One approach to tunable uv generation is to use a pulsed short-wavelength laser, such as the fourth harmonic of Nd:YAG (266 nm) or a discharge excimer laser (notably KrF 248nm and XeCl 308nm) to pump uv-lasing dyes. Output powers in excess of 1MW are now available from commercial dye lasers. The short-wavelength limit attained with this approach currently stands at 322nm, the edge of the lasing band of the dye *p*-terphenyl [5.17-20,97]. A second approach is to frequency double the output from dye lasers in the visible.

By pumping efficient dyes (such as rhodamine 6G) using the frequency-doubled output from an unstable-resonator Nd:YAG laser, BYER and HERBST [5.21] obtained frequency-doubled dye-laser outputs as high as 16mJ (4MW) at wavelengths down to 275nm. These short-wavelength pump sources permit the study of Raman transitions to atomic Rydberg states, with the possibility of generating radiation over wide ranges of the far infrared. WYNNE and SOROKIN [5.22] have observed SERS in potassium by using a frequency-doubled dye laser as pump; however, the power was too low to permit any significant tuning. The possibility of using transitions between Rydberg levels for optical-frequency-conversion processes has been discussed by LAU et al. [5.23]. The over-all efficiency of such processes is likely to be low, however, because the ratio of the generated Stokes photon energy to the pump energy, ω_S/ω_p, is very small.

An alternative approach to the generation of longer infrared wavelengths by SERS is to use Raman transitions from previously populated excited states. In the alkali metals, transitions from the first excited p states are of particular interest. As will be described, it is relatively easy to populate these particular levels in a number of ways; also the effective lifetimes are very long (on the order of microseconds) because of radiation trapping. In fact, as mentioned earlier, the first demonstration of SERS by ROKNI and YATSIV [1.33,34] (Fig.5.1) and also by SOROKIN et al. [1.32] involved transitions from the excited $4p_{3/2}$ level of potassium, which was populated by direct optical pumping from the ground state. Recently, KUNG and ITZKAN [5.24] have also produced SERS from the 4p levels of potassium; they generated 8.5μm and 16μm radiation by use of the 4p-6p and 4p-7p Raman transitions. Their method for preparing the initial excited-state population is shown in Fig.5.6. A dye laser tuned close to the 4s-5p single-photon transitions was used to prime the vapour by producing simultaneous SERS on the 4s-5s transition and amplified spontaneous emission (ASE) on the 5p-3d transitions. This serves to populate the 5s and 3d levels; the resulting population inversion between these levels and the lower 4p levels causes further ASE on the 5s-4p and 3d-4p transitions, and leads ultimately to population of the 4p levels. (This interpretation of the priming sequence is based on our own observations of SERS and ASE in potassium vapour, described in Sect.5.2.2; it differs somewhat from the explanation given by KUNG and ITZKAN. They propose that the 3d levels are populated via SERS from the ground state, but we have been unable to observe this SERS process and believe that ASE on the 5p-3d transition is the dominant mechanism.) At the

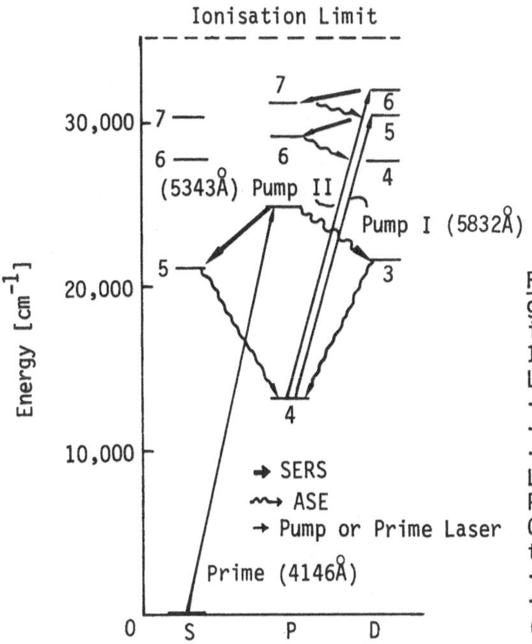

Fig.5.6. Potassium energy-level diagram, showing the priming and pumping systems used for 8.5µm and 16µm generation. Priming sequence: Laser - 404.7nm ($4s-5p_{1/2}$); Raman - 2.7µm ($5p_{1/2}$ - $5s_{1/2}$); ASE - 3.2µm ($5p_{1/2}$ - $3d_{3/2}$); Cascade - (5s-4p, 3d-4p). Pump system I: Laser - 583.2nm ($4p_{3/2}$ - $5d_{5/2}$); Raman - 8.5µm ($5d_{5/2}$ - $6p_{3/2}$); Cascade - 6.2µm (6p-4d). Pump system II: Laser - 534.3nm ($4p_{1/2}$ -$6d_{3/2}$); Raman - 16.0µm ($6d_{3/2}$ -$7p_{1/2}$); Cascade - 11.4µm (7p-5d). (After [5.24])

vapour pressures used (1-10 torr, typically), the large excess ground-state population results in radiation trapping of the growing 4p population. The trapped lifetime depends on the rate of radiation leakage from the wings of the emission lines, which is determined by the homogeneously broadened linewidth. By use of formulae due to HOLSTEIN [5.25,26], the radiatively trapped lifetimes for the lowest p levels in the alkalis are predicted to be on the order of tens of microseconds at vapour pressures of a few torr. In practice, the lifetimes may be somewhat shorter, being determined mainly by quenching through collisions between the alkali atoms and buffer-gas molecules [5.27].

This method of populating the 4p levels was used by KUNG and ITZKAN [5.24] in preference to direct optical pumping, because the required wavelength for the priming laser was more convenient in practice (~ 400nm instead of ~ 770nm). Using 1kW pulses from both the prime and pump lasers, they obtained a Stokes output on the $4p_{1/2}$ - $7p_{1/2}$ Raman transition of ~ 10mW at 16µm, tunable over a 2-4 cm^{-1} range. By introducing a variable delay between the two lasers, they were able to estimate the lifetime of the populations in the 4p levels, as indicated by the reduction of the Stokes output as the delay was increased. They thus estimated a 4p lifetime of

3.3μs when partial pressures of 3 torr K and 1.5 torr He were used, falling to 1μs when the He pressure was increased to 31 torr.

The same Raman transition in potassium ($4p_{1/2}$ - $7p_{1/2}$) was also used by GRISCHKOWSKY et al. {[5.28], footnote (12)} to generate 16μm radiation. However, they employed an alternative method for populating the $4p_{1/2}$ level; they operated the potassium-vapour cell as a heat-pipe discharge tube [5.29]. This device is similar to the normal type of heat-pipe oven (Fig.4.24), except that a ceramic envelope is used in place of the more usual stainless-steel tube and the wick is split into two halves. This construction allows a voltage to be sustained between the two parts of the wick. A glow discharge can be maintained, resulting in continuous pumping of the $4p_{1/2}$ atomic level. With the discharge tube operating at a potassium-vapour pressure of 1 torr (total number density ~ $1.6 \times 10^{22} m^{-3}$), a current of 30mA, and at a temperature of 390°C, GRISCHKOWSKY et.al. measured a number density of ~ $6 \times 10^{17} m^{-3}$ for atoms in the $4p_{1/2}$ excited state.

The use of yet another method of producing a population in the 4p levels of potassium was reported earlier by GRISCHKOWSKY et al. [5.30]. This method involved using an intense flash-lamp pulse to photodissociate K_2 dimers (see Sect.5.4.1), yielding excited atoms in the 4p levels. However, compared to this photodissociation technique, the discharge tube method has the advantage that the 4p levels are pumped continuously. Also, there is not the technical difficulty of admitting the flash-lamp light to the vapour, and there are no synchronisation requirements.

Like KUNG and ITZKAN (Fig.5.6), GRISCHKOWSKY et al. [5.28] used a nitrogen-laser-pumped dye laser to drive the $4p_{1/2}$ - $7p_{1/2}$ SERS transition, resonantly enhanced via the 6d levels, resulting in the generation of radiation tunable in the region of 16μm. A closely related approach that they also studied involved direct optical pumping of the $6d_{3/2}$ level, with subsequent amplified spontaneous emission on the ~ 16μm $6d_{3/2}$ - $7p_{1/2}$ transition. (The transition frequency may be dc Stark tuned.) The optical pumping was with a flash-lamp-pumped dye laser; it resulted in 16μm output pulses of about 1μs duration. The number of 16μm photons generated (equivalent to pulse energies of up to 1μJ) were as much as ten times greater than the number of $4p_{1/2}$ atoms present in the beam path at the start of the pumping pulse. This indicates that, because of rapid cascading of the atoms from the upper final level $7p_{1/2}$ back down to the 4p levels via a sequence of spontaneous and amplified spontaneous emission steps, the atoms can be recycled many times during the pumping pulse.

Apart from the additional technical complexity, an obvious disadvantage of using excited metastable states as the initial level for SERS, compared to ground-state systems, is the much-lower initial population number densities, which results in reduced gain. However, as we have just suggested, there is the possibility that the atoms can be recirculated many times during the pumping pulse; this represents a significant advantage. When the atomic ground state is used on the other hand, radiative trapping in the first excited p levels prevents recirculation back to the ground state; this results in the possibility of early saturation of the generated signal. This saturation process and its influence on the tuning profile in SERS generation is discussed further in Sect.5.4.3.

COTTER and HANNA [4.51] discussed the relative merits of using SERS to generate tunable infrared radiation for spectroscopic applications, compared with other available types of sources. The SERS method has advantages of economy, simplicity of construction and operation, and scalability, although the bandwidth of the infrared output, 0.3 - 0.5 cm^{-1} typically, is considerably broader than from some other tunable infrared sources. PETERSON et al. [5.32]report molecular-energy-transfer studies using a tunable SERS source.

BETHUNE et al. [5.31] have recently demonstrated that SERS can generate broadband radiation (they achieved 400 cm^{-1}) for use in single-shot ir spectroscopy. That stimulated Raman scattering can be produced as readily using a wide-band pump source (bandwidth much greater than the spontaneous Raman linewidth) is not obvious from an elementary treatment of Raman scattering. However, as will be discussed further in Sect.5.3.3, an analysis of Raman scattering by a stochastic (incoherent broad-band) pump leads to the conclusion that under these conditions the threshold is independent of pump linewidth. COTTER et al. [1.37] obtained some evidence for this in experiments on SERS in potassium vapour; it was noticed that increasing the dye-laser linewidth by more than an order of magnitude, from 0.1 to 2 cm^{-1} produced a negligible difference in the ir tuning range [5.16]. Indeed, a strong SERS output could be obtained even when the dye laser was operated as an ASE source. BETHUNE et al. [5.31] used this effect to create wide-band infrared radiation and obtained single-shot infrared spectra of absorbing species. The ir beam transmitted by the sample was then up-converted to the visible region by use of a further nonlinear-mixing stage; the infrared absorption spectra thus obtained could be recorded photographically. The nanosecond time resolution obtained with this technique should prove valuable for recording spectra of transient chemical species.

5.2.2 Amplified Spontaneous Emission

In the majority of experiments on SERS using alkali vapours, the Stokes output is accompanied by a number of fixed-frequency emissions whose wavelengths correspond to single-photon atomic transitions. As has just been discussed, in some SERS schemes these amplified spontaneous-emission (ASE) processes can play a useful role in assisting the rapid recirculation of atoms and thus prevent saturation. On the other hand, some of these ASE signals are frequently comparable in pulse energy to the Stokes output itself; for some applications, these extraneous signals can be a nuisance, particularly when their wavelengths are close to the desired tunable Stokes signal.

In our SERS experiments [1.37] using the 4s-5s transition in potassium (Fig.5.3), strong ASE signals were observed on the 5p-3d transitions (3.14μm and 3.16μm), amongst others. These signals are greatest when the dye laser is tuned very close to the 5p levels; this is consistent with the idea that these levels are populated by direct optical pumping from the ground state. We have observed, however, that these ASE signals remain quite strong even when the dye laser is detuned from the 5p levels by as much as several hundred cm^{-1}; therefore, some other mechanism for populating the 5p levels must be involved. SOROKIN and LANKARD [5.33] observed these emissions when potassium vapour was excited by a ruby laser, and suggested that they were the result of two-photon absorption by K_2 molecules (see Sect.5.4.1) followed by dissociation leaving K atoms in the 5p state. It may indeed be possible that molecular multi-photon absorption and subsequent dissociation produces atoms in more-highly excited states, and that these atoms then lose their energy by a series of rapidly cascading ASE steps. This explanation is consistent with the observation, in the case of caesium, that when a Stokes output was generated on the 6s-7s transition, simultaneous ASE outputs occurred on all transitions between adjacent s and p states from at least as high as 10s down to 6p, with the exception of 7p-7s [5.34].

For SERS from the ground state of the alkali atoms, the Stokes output is always observed only on the Raman transition to the ns state (where n is the principal quantum number of the resonant intermediate p level), and not to the adjacent (n-2)d level, or to any other s or d levels. This is borne out by the nonlinear-susceptibility calculations presented in Sect.5.3.1, where it is shown that the Raman transition to the ns state has the greatest cross section. However, the same argument holds for ASE from the np levels, because the calculated cross sections for ASE on the np → ns transitions are greater than for the corresponding np → (n-2)d transitions, or other possible single-photon transitions. Nevertheless, when SERS was produced on the 4s-5s Raman transition in potassium, which is resonantly enhanced by tuning near the 5p levels (see Fig.5.3), strong ASE outputs were observed on the 5p-3d transitions. On the other hand, the emissions that correspond to the theoretically more probable 5p-5s transitions were more than three orders of magnitude weaker, at least when the pump frequency was tuned more than $\sim$ 1-2 cm^{-1} from the 5p levels so that the ASE could be resolved from the SERS signal. Similar anomalous behaviour is observed in the case of Rb; strong ASE occurs on the 6p-4d transitions, and on the 7p-5d transitions in the case of Cs. WYNNE and SOROKIN [5.22] observed ASE to both p and d levels when they optically pumped higher-lying p levels (up to 16p) in potassium.

The ASE signals on the transitions from the final Raman level, ns → (n-1)p, also exhibit an unexplained behaviour. Because SERS involves excitation of ground-state atoms to the final Raman level, this process tends to produce a population inversion between this level and lower-lying p levels. This inversion can rapidly reach the point at which strong ASE can occur. In the case of SERS using the 4s-5s transition in potassium (Fig.5.3), a strong fixed-frequency emission at around 1.25µm, corresponding to 5s-4p, is observed. However, this emission could be observed even when the dye laser was tuned outside the range of pump frequencies for which SERS was detectable; therefore, some other (as yet unknown) mechanism for populating the 5s level must play a role. A possibility is that K_2 dimers photodissociate, leaving some atoms in the 5s excited state. However, it is clear that much further work will be required to unravel these complexities.

From a practical viewpoint, it is necessary in many spectroscopic applications to remove these spurious ASE signals from the generated Stokes output. In some cases, this must be done by placing a tunable filter, such as an infrared monochromator, after the heat-pipe oven. This has the disadvantage that, when the dye frequency is scanned, the filter and the dye laser must be tuned synchronously. In other cases, however, the wavelength separation between the Stokes and ASE signals is sufficiently great that a simpler arrangement, using dielectric filters, can be used (as for example in Fig.5.4a).

Finally, we note that these ASE signals, together with the pump and Stokes waves, can take part in a number of possible parametric-mixing processes, resulting in a profusion of fixed-frequency and tunable outputs

(Sect.5.4.1, see also [5.35]). However, all of these outputs are generally very weak compared to the original Stokes and ASE signals.

5.3 The Calculation of SERS Gain

5.3.1 Plane-Wave Gain Coefficient

The expression (5.1b) for the Raman susceptibility reduces, in the case of a three-level system (i.e., a single intermediate level), to

$$\chi_R = \frac{N\mu_{fi}^2\mu_{ig}^2}{6\hbar^3\varepsilon_0(\Omega_{fg}+\omega_s-\omega_p+i\Gamma)}\left(\frac{1}{\Omega_{ig}-\omega_p}+\frac{1}{\Omega_{ig}+\omega_s}\right)^2 . \tag{5.5}$$

As noted at the end of Sect.5.1, by suitable tuning of ω_p, either of the frequency denominators in (5.5) can be made small, depending on the position of the intermediate state. We confine our attention to the case in which i lies above g, and $\omega_p \simeq \Omega_{ig}$ (Fig.5.2a), so that only the first denominator need be retained. (The analysis is the same if the other resonance is employed.) This shows that the susceptibility varies with pump detuning from the intermediate level as $(\Omega_{ig} - \omega_p)^{-2}$. Substitution of this approximate form of χ_R into (5.3) shows that the peak Raman gain (i.e., when $\Omega_{fg} + \omega_s - \Omega_p = 0$) is

$$g_R = N\omega_s\mu_{fi}^2\mu_{ig}^2/2\varepsilon_0^2c^2\hbar^3\Gamma(\Omega_{ig} - \omega_p)^2 \tag{5.6a}$$

(we have set $\eta_s = \eta_p = 1$). This formula was used for the illustrative calculation of SERS gain in Sect.5.1. As shown in Sect.2.8, for a three-level system, the Raman polarisability may be simply expressed in terms of oscillator strengths. For the linearly polarised geometry assumed here, and a nondegenerate ground state (e.g., ns in an alkali atom or $(ns)^2\ {}^1S_0$ in an alkaline earth), (2.111) for $|\alpha_R|^2$ with (5.1,3) shows that

$$g_R = \frac{2\pi^2 r_e^2 c^2}{\hbar}\ \frac{N\omega_s}{\Gamma}\ \frac{f_{fi}f_{gi}}{\Omega_{if}\Omega_{ig}}\ \frac{1}{(\Omega_{ig}-\omega_p)^2}\quad (J_f = 0\ ;\ \times 2 \text{ if } J_f = 2)\ . \tag{5.6b}$$

(The classical electron radius $r_e \simeq 2.82 \times 10^{-15}$m.)

In practice, these simple expressions, although they provide useful estimates of the gain, are often valid only over limited ranges of pump frequencies. Thus, if we consider the example of an intermediate ^2P term in an alkali vapour system, this can be treated as a single level (i.e., its spin-orbit splitting neglected) only when the pump detuning is substantially greater than the spin-orbit splitting. In the heavier alkali metals, these splittings can be very large (for example, the separation of the $7p^2P_{1/2,3/2}$ levels in cesium is $\sim 180\ \mathrm{cm}^{-1}$); therefore, it is necessary to take account of the individual contributions from these levels over a major proportion of the observed SERS tuning range on the 6s-7s transition. On the other hand, if the pump frequency is tuned far enough away from the main intermediate level, then contributions from other levels can begin to be important.

These restrictions may be lifted by applying the theory sketched in Sect.2.8. It was shown there that the quantity $\overline{|\alpha_R|^2}$, which enters the Raman susceptibility (5.1), may be written as a linear combination of angular factors (depending on the pump and Stokes polarisation vectors) and reduced-matrix elements of the Raman-polarisability operator. Each term in this combination is well characterised by a spherical tensor rank, K. It is an easy matter to use the expression (2.107) in (5.1,3) and thereby write the more general form of Raman gain. As an illustration of this, and because we will use it later to demonstrate the interplay between the effects of spin-orbit splitting and polarisation geometry, we give the result for n_gs - n_fs scattering here [2.111];

$$g_R = (N\omega_s/2\varepsilon_0^2c^2\hbar\Gamma)\overline{|\alpha_R|^2} \quad , \tag{5.7a}$$

where

$$\begin{aligned}\overline{|\alpha_R|^2} = \frac{e^4}{3\hbar^2}\left|\sum_{n_i}\left[\frac{2}{3}\,\Phi(n_ip\tfrac{3}{2}0) + \frac{1}{3}\,\Phi(n_ip\tfrac{1}{2}0)\right]\right|^2 \Theta_R(0)\\ + \frac{2e^4}{81\hbar^2}\left|\sum_{n_i}\left[\Phi(n_ip\tfrac{3}{2}1) - \Phi(n_ip\tfrac{1}{2}1)\right]\right|^2 \Theta_R^{(1)}\end{aligned} \tag{5.7b}$$

and

$$\Phi(n_ipJK) = \langle n_fs|r|n_ip\rangle\langle n_ip|r|n_gs\rangle\left(\frac{1}{\Omega_{ig}-\omega_p} + \frac{(-1)^K}{\Omega_{ig}+\omega_s}\right) \quad . \tag{5.7c}$$

The two parts of (5.7b) are termed symmetric and antisymmetric, respectively. It turns out that the denominator $\Omega_{ig} + \omega_s$ in Φ makes a negligible contribution for all calculations in the alkali atoms when the initial state is the ground state. Using this fact, and ignoring any J dependence of the radial matrix elements, (5.7b) may be simplified to

$$\overline{|\alpha_R|^2} = \frac{e^4}{9\hbar^2}\left|\sum_{n_i} \langle n_f s|r|n_i p\rangle\langle n_i p|r|n_g s\rangle \left(\frac{2/3}{\Omega_{i3/2}-\omega_p} + \frac{1/3}{\Omega_{i1/2}-\omega_p}\right)\right|^2 |\underline{\varepsilon}_s^* \cdot \underline{\varepsilon}_p|^2$$

$$+ \frac{e^4}{81\hbar^2}\left|\sum_{n_i} \langle n_f s|r|n_i p\rangle\langle n_i p|r|n_g s\rangle \left(\frac{1}{\Omega_{i3/2}-\omega_p} - \frac{1}{\Omega_{i1/2}-\omega_p}\right)\right|^2 |\underline{\varepsilon}_s^* \times \underline{\varepsilon}_p|^2 \; . \quad (5.8)$$

The notation $\Omega_{i3/2}$ has been used for the transition frequency $\Omega_{n_i p_{3/2} - n_g s_{1/2}}$, etc. Also, the angular expressions have been taken from Sect.2.8; the dot and cross have their usual vector meanings. Before proceeding, we should emphasise that the calculations reported later in this section were all made with the full expression (5.7b), but for most quantitative and all qualitative purposes (5.8) is adequate. One important quantitative exception to (5.8) occurs for cesium, for which the radial matrix elements show significant J dependence; again, the actual calculations incorporated this.

To make the connection between (5.6a) and (5.8), consider what happens as ω_p is tuned well away from $\Omega_{i3/2}$ and $\Omega_{i1/2}$. The denominators $\Omega_{i3/2} - \omega_p$ and $\Omega_{i1/2} - \omega_p$ both tend to $\Omega_i - \omega_p$ where Ω_i is an average frequency for the doublet. The second part of (5.8), therefore, vanishes for sufficiently large detuning, leaving

$$\overline{|\alpha_R|^2} = (e^4/9\hbar^2)\left|\sum_{n_i} \langle n_f s|r|n_i p\rangle\langle n_i p|r|n_g s\rangle/(\Omega_i - \omega_p)\right|^2 \quad (5.9)$$

for $\underline{\varepsilon}_s$, $\underline{\varepsilon}_p$ of the same polarisation. If the sum over n_i is restricted to the dominant intermediate state, and the radial matrix elements are related to the z matrix elements through (2.109), then (5.9) and (5.7a) together reduce to (5.6). Later in this section we will discuss in detail the behaviour of (5.8) when ω_p is close to a doublet.

The Raman susceptibility and gain for a large number of possible SERS transitions in the different alkali metals have been calculated according to (5.7) and its companion formula for $f = n_f d\,^2D_{3/2,5/2}$. Published values for the transition frequencies and electric dipole matrix elements or oscillator strengths were used. Some of the sources of data have been referred to in Sects.2.9,4.3,3.2; a further list (mainly for cesium) will be given on page 213. In practice, the summation over intermediate levels is rapidly convergent; typically, only the three or four most prominent intermediate states are required to keep the truncation error smaller than the error due to uncertain parameters. Also, the contribution from the continuum can usually be neglected, as can be seen from an application of the Thomas-Kuhn sum rule.

Some general conclusions about SERS from ground-state alkali atoms may be drawn from these calculations. Firstly, the nonlinear susceptibility χ_R is greater for transitions to excited n_fs levels than for transitions to n_fd levels (for the same Raman linewidth Γ). This is because the matrix elements that connect a given p state to excited s states are larger than the corresponding elements to d states. (An obvious exception to this, however, occurs when the pump frequency ω_p is below the first excited s level but above the lowest d level. In this case, the d level is the only possible final Raman level, a situation that occurs in rubidium and cesium. HODGSON [5.36] observed the generation of 2.4μm radiation by SERS to the $5d_{5/2}$ level when cesium vapour was pumped by the 532nm second harmonic of a Nd:YAG laser.) A second conclusion is that when the Raman susceptibility is enhanced by tuning ω_p in the region of a principal-series doublet term, $n_i{}^2P$, the frequency denominators and products of matrix elements in (5.7) ensure that the greatest gain is for the Raman transition to the excited n_fs state that has the same principal quantum number, i.e., n_fp. Therefore, transitions of this type are expected to dominate, which result has been confirmed experimentally. To our knowledge the experiment by HODGSON referred to above is the only observation of an s-d SERS transition in alkali atoms. As noted in Sect.8.2, however, COTTER and YURATICH [5.98] have recently observed tunable 4s-4p SERS in potassium vapour, using resonantly enhanced multipole transitions.

A question that arises in the course of these calculations is the appropriate choice of polarisation geometry. In experiments on two-photon absorption or ionisation, processes closely related to SERS, there is usually no ambiguity because the light beams (or beam) are injected into the medium and hence can have well defined polarisations. In SERS, however, the Stokes radiation is generated within the medium; consequently, because of the high exponential gain, the Stokes output will have the polarisation that experiences the maxium gain. In the calculations, it is, therefore, important not to make premature assumptions about the polarisation of the Stokes wave, but to determine the optimum case.

Consider again the expression (5.8), which refers to $n_g s$ - $n_f s$ Raman scattering, and assume that the pump frequency is sufficiently resonant with an intermediate-state doublet that the sum over n_i may be restricted to one term. If we first take the pump and Stokes waves to be parallel and linearly polarised (as for example can be arranged in a SERS amplifier, where the Stokes wave is input), then the angular factor in the antisymmetric part vanishes identically, leaving the symmetric part;

$$\overline{|\alpha_R|^2} = \frac{e^4}{9\hbar^2} \langle n_f s|r|n_i p\rangle^2 \langle n_i p|r|n_g s\rangle^2 \left(\frac{2/3}{\Omega_{i3/2}-\omega_p} + \frac{1/3}{\Omega_{i1/2}-\omega_p}\right)^2 \quad . \tag{5.10}$$

When ω_p is above or below the doublet frequencies, then the two frequency denominators have the same sign, and coalesce as described above for (5.9). Between the doublet, however, the denominators take opposite signs; therefore, at some point (5.10) and hence the Raman gain, will vanish. [The actual point of cancellation is determined partly by the J dependence of the radial integrals, which has been ignored in (5.10).] A similar behaviour is predicted for two-photon absorption or ionisation.

In earlier calculations of the SERS susceptibilities in the case of cesium [1.39], the incorrect assumption was made that because the pump-laser beam used in the experiments was linearly polarised, the Stokes radiation would also be linearly polarised and in the same direction. Therefore, it was expected that the cancellation point just discussed would be seen as ω_p was tuned through the doublet. However, all experimental observations showed that, although the Stokes output dropped to a minimum for some frequency ω_p between the doublet, nevertheless a strong Stokes signal could always be observed. The nonexistence of the predicted cancellation was a source of confusion, especially because a similar cancellation had been clearly observed in the two-photon absorption experiments in Na vapour by BJORKHOLM and LIAO [5.37].

The explanation becomes quite clear, however, when the full expression (5.7) or (5.8) is used. The case where the pump beam is assumed linearly polarised but the Stokes wave can take a general elliptical polarisation was discussed by COTTER and HANNA [2.112]. They showed that the Raman susceptibility is maximised when the Stokes wave is also linearly polarised, in which case (5.8) reduces to $\hbar^2\overline{|\alpha_R|^2} = \mathscr{S}\cos^2\beta + \mathscr{A}\sin^2\beta$, where β is the angle between the pump and Stokes polarisation vectors; the symmetric ($\mathscr{S}$) and antisymmetric ($\mathscr{A}$) parts appear in (5.8) [or (5.7)] as the coefficients of $|\underline{\varepsilon}_s^* \cdot \underline{\varepsilon}_p|^2$ and $|\underline{\varepsilon}_s^* \times \underline{\varepsilon}_p|^2$, respectively. For example, $\mathscr{S}$ is approximately given by (5.10). The direction of β is, however, fixed by the relative magnitudes of $\mathscr{S}$ and $\mathscr{A}$. When $\mathscr{S} > \mathscr{A}$ the maximum occurs for perpendicular polarisation ($\beta = \pi/2$). Thus, by considering only $\beta = 0$ in the earlier work, the effect of the antisymmetric part and its implication for the polarisation of the generated Stokes wave was ignored. As in the derivation of (5.10), the antisymmetric part is found to have a frequency dependence $[(\Omega_{i3/2} - \omega_p)^{-1} - (\Omega_{i1/2} - \omega_p)^{-1}]^2$; this decreases very rapidly as ω_p is tuned away from the doublet. Unlike the $\mathscr{S}$ part, however, the $\mathscr{A}$ part is seen not to vanish when ω_p is tuned *between* the doublet, and, therefore, $\mathscr{S} < \mathscr{A}$ for part of the tuning range.

Figure 5.7 shows the calculated values of the $\mathscr{S}$ and $\mathscr{A}$ parts of $|\alpha_R|^2/6\hbar\varepsilon_0$ as a function of ω_p for the $6s^2S_{1/2} - 7s^2S_{1/2}$ transition in cesium, for which (5.7) was used. From (5.1), it is seen that these quantities are simply the $\mathscr{S}$ and $\mathscr{A}$ parts of the Raman susceptibility normalised to N/Γ. The data of WARNER [4.17] were used for these calculations, rather than those from other sources (such as [1.15,2.116,4.13,5.38-43]), because Warner's tabulations are the most extensive, thus allowing the calculation of susceptibilities for high-lying SERS transitions. Also, he includes spin-orbit splitting, and his calculated results are generally in reasonable agreement with measured values where they are available [5.43-50]. In the calculations for Fig.5.7, the summation was taken over the p states up to $12p^2P$.

As can be seen from Fig.5.7, over the major part of the tuning range, the usual symmetric term plays the dominant role. But for pump frequencies in the range that extends from the lower resonance to about midway through the doublet interval, the antisymmetric term becomes dominant and should, therefore, compensate for the destructive cancellation in the symmetric term. This explains why a zero in the Stokes output could not be found. Further experimental verification of the theory was obtained by observing the polaristion direction of the generated Stokes signal as a function of tuning, COTTER and HANNA [2.112]. Using an infrared grid polariser after the heat-pipe oven, they measured the polarisation ratio $\mathscr{R} = (W_x - W_y)/(W_x + W_y)$, where W_x and W_y are the energies of the detected Stokes signals when the polariser is set parallel and perpendicular to the polarisation direction of the pump beam. As shown in Fig.5.8, the Stokes polarisation direction changed from parallel to orthogonal ($\mathscr{R}$ = 1 to -1) according to whether the symmetric or antisymmetric part of the susceptibility was dominant, as predicted theoretically.

This "polarisation-flip" effect could be used to measure the doublet-intensity ratio of the infrared $n_f s$-$n_i p$ transitions, which are generally not

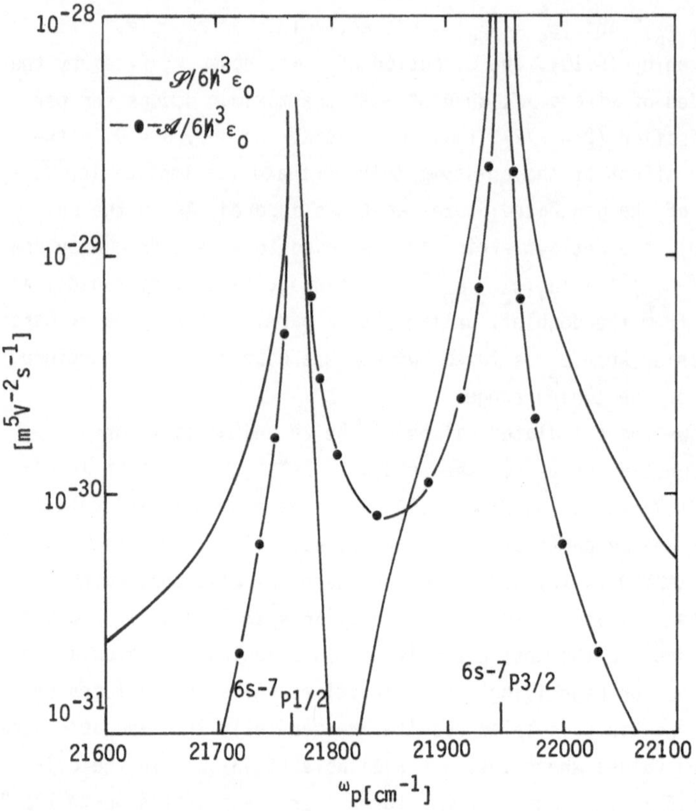

Fig.5.7. Calculated values of the symmetric ($\mathscr{S}$) and antisymmetric ($\mathscr{A}$) parts of the SERS susceptibility for 6s-7s Raman scattering in cesium. (After [2.112])

accessible to more conventional measurement techniques; this has been discussed by COTTER and HANNA [2.112], CORNEY and GARDNER [5.51], and COTTER [5.52].

In passing, we note that for two-photon ionisation (TPI) or absorption (TPA) using two beams of the same polarisation, the angular factors in the appropriate form of (5.7) or (5.8) become $|\underline{\varepsilon}\cdot\underline{\varepsilon}|^2 = 1$ and $|\underline{\varepsilon}\times\underline{\varepsilon}|^2 = 0$, so that the antisymmetric part makes no contribution. This conclusion also applies for TPI/A with a single inputwave regardless of polarisation. Moreover, for $^2S_{1/2}$ - $^2D_{3/2}$ transitions there is also an antisymmetric part, which again makes no contribution under the above conditions.

The first extensive calculations of TPI in the alkalis (linearly polarised light), including spin-orbit splitting, were reported by BEBB [4.16].

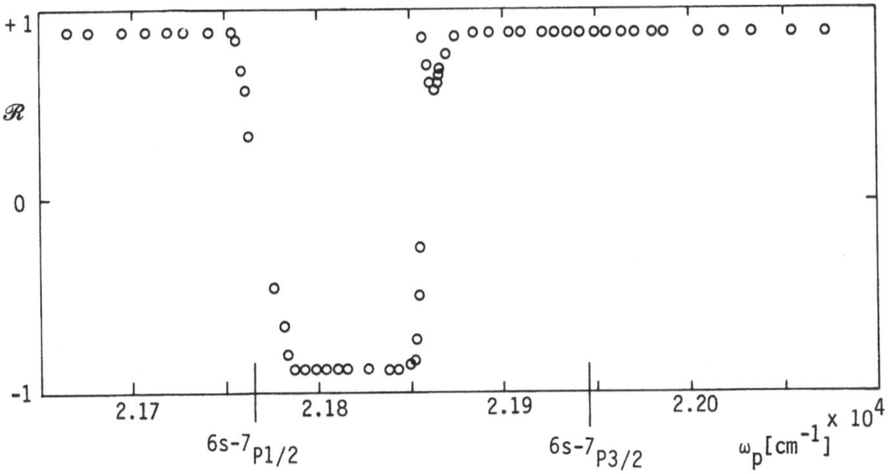

Fig.5.8. Measurements of the SERS polarisation ratio, showing the polarisation "flip" as the pump laser is tuned through the intermediate doublet. (After [2.112])

In the body of this paper, the ionisation rates were expressed in a form recognisable as being of symmetric character alone. However, he included a footnote that states that what are here called the antisymmetric parts had been overlooked. These were not incorporated into the numerical calculations, but never exceeded 8% of the published results. As the photon energy was tuned away from the intermediate-state doublets the correction became negligible. However, as we have just seen, when both photons have the same polarisation the antisymmetric part can make no contribution; therefore, some error in the analysis that led to Bebb's footnote must have occurred.

5.3.2 SERS Threshold

In SRS, some starting noise at the Stokes frequency (produced by spontaneous Raman scattering) is amplified over many orders of magnitude to a level at which the resulting beam of Stokes light can be detected. The question of estimating the power of the starting-noise signal was discussed in Sect.3.5; it was shown in Sect.3.3 that a rough criterion can be found, viz. that a gain of $\sim \exp(30)$ is needed to amplify the noise to a level comparable to the incident pump intensity. In practice, a Stokes signal may be detectable at a power level that is a few orders of magnitude less than this, but we shall continue to use exp(30) as our criterion of the gain required in order to reach "threshold", because the exponent is not sensitive to either the value of starting noise or the threshold Stokes intensity. Thus, by use of calculated values of the Raman gain, the threshold pump power can be calculated from $g_R I_{po} L = 30$, where L is the vapour column length. Moreover, from a knowledge of the frequency dependence of the Raman susceptibility the tuning range can be predicted for a given incident pump intensity (defined as the range of pump frequencies for which the Raman threshold is exceeded).

However, there are a number of reasons why the simple gain formula (5.3) is unlikely to predict accurately the threshold and tuning range. Two of these are discussed below, viz. the loss due to diffraction of the Stokes wave and the uncertain value of the linewidth Γ. A third problem, that of competing processes, is discussed in Sect.5.4.1.

The theory developed in Sect.3.3 was for plane waves, whereas in practice the pump radiation is in the form of a beam of finite diameter, which may be tightly focussed. Thus, a correct description must take account of the nonuniform intensity within the beam with a corresponding nonuniformity of the gain. Also, the propagation behaviour of the Stokes radiation may be strongly influenced by diffraction, particularly because the Stokes wavelength may be much longer than the pump wavelength. As a result of this diffraction, there can be significant loss due to spread of the generated radiation outside the small amplifying volume. Only the Stokes radiation that remains inside the narrow pumped region is amplified and can emerge as a beam that has dimensions similar to the pump beam. This guiding effect is termed gain focussing. Clearly, the diffraction loss increases as the ratio κ of the Stokes and pump frequencies, $\kappa = \omega_S/\omega_p$, becomes smaller. When visible or near-infrared lasers are used to produce SRS in liquids or solids, where the Stokes shift is not greater than a few hundred cm^{-1}, then the ratio κ is close to unity and the Stokes diffraction loss can usually be safely neglected. However, in SERS, the Stokes shifts can be very large, and the ratio of frequencies κ is generally in the range 0.001 - 0.2. This implies a substantial diffraction loss for the (infrared) Stokes wave, and can result in the SERS threshold being more than an order of a magnitude greater than is predicted by the plane-wave formula (5.3). For a given pump power, this increase of threshold implies a reduced tuning range.

The practical situation of greatest importance, which we analyse here, is of pump radiation in the form of a diffraction-limited gaussian beam ([4.7], see also Fig.4.2). The beam, travelling in the z direction, enters the vapour at $z = 0$, converges to a focus at $z = f$ (waist spot size w_{po}, confocal parameter b_p), and leaves the vapour at $z = L$. Then a spatially dependent Raman gain can be defined,

$$G_R(\xi,r) = G_{Ro}(1+\xi^2)^{-1}\exp[-2r^2/w_{po}^2(1+\xi^2)] \quad , \tag{5.11}$$

where $\xi = 2(z-f)/b_p$ and $G_{Ro} = G_R(0,0)$ is the peak Raman gain obtained from (5.3) evaluated with the on-axis pump intensity at the focus. (The optical Kerr effect is thus ignored for the moment.) The gain (5.11) may be substituted into an equation for the Stokes wave that is similar to the plane-wave formula (3.17), but with transverse derivatives retained, ie.,

$$(2ik_s)^{-1}\left(\frac{\partial^2}{\partial x^2}+\frac{\partial^2}{\partial y^2}\right)E_s + \frac{\partial E_s}{\partial z} = \frac{1}{2}G_R(\xi,r)E_s \quad .$$

[This equation may be derived directly from (3.3).] In general, numerical techniques are required in order to analyse the behaviour of the Stokes wave (and in particular, to calculate the net effective gain) as it travels through such an amplifying medium. However, a considerable simplification results if the exponential in (5.11) is approximated by $1 - 2r^2/w_{po}^2(1+\xi^2)$, so that the gain has a parabolic radial dependence. The Raman medium then appears like the generalised lens-like medium considered by KOGELNIK [5.53]; in this case, the problem has exact solutions. In particular, the generated Stokes wave is also a gaussian beam, and a closed expression for it is easily derived. This approach was used by COTTER et al. [1.38] to analyse the small-signal growth in infrared Raman generation; the principal results are described below. A check was made of the usefulness of the parabolic approximation by comparing its analytical solutions with exact computer solutions obtained for the gaussian pump beam. For typical experimental conditions the approximation appears to be good.

It is convenient to introduce the dimensionless parameter $\tilde{P}_p$, proportional to the total power of the pump-laser beam,

$$\tilde{P}_p = \frac{\omega_s}{c}G_{Ro}w_{po}^2 = G_{Ro}b_p\kappa \quad . \tag{5.12}$$

In terms of this, the generated Stokes power $P_s(L)$ at the cell exit is given by

$$P_s(L) = P_s(0)\exp[(\tilde{P}_p - 2\sqrt{\tilde{P}_p})\tan^{-1}(L/b_p)/\kappa] \quad , \tag{5.13}$$

where the beam waist has been assumed to be in the centre of the vapour region, i.e., $f = L/2$. The dependence on how tightly the pump beam is focussed is contained in the term $\tan^{-1}(L/b_p)$. As expected, tighter focussing (increasing L/b_p) leads to increased Stokes-power gain, although the behaviour of the $\tan^{-1}$ term shows that focussing much more tightly than the confocal condition $b_p = L$ has a relatively small effect. In any practical case there will, of course, be a limit to how tight the focussing should be, depending on the rate of saturation of the medium and the influence of various competing processes (see Sect.5.4).

The term $2\sqrt{\tilde{P}_p}$ in (5.13) represents the diffraction loss of the Stokes light. In particular, this shows that there is over-all net gain for the Stokes wave only if $\tilde{P}_p > 4$. If $\tilde{P}_p$ is less than this, then the loss due to diffraction experienced by the Stokes wave exceeds its gain from the SRS process, and so no Stokes generation is possible.

When radiation is generated that is to be tuned in frequency, it is clearly of interest to predict the pump power P_{pth} that would be required to reach threshold at a particular wavelength, or alternatively to estimate the Stokes tuning range for a given pump power. From (5.3,12,13), we find

$$P_{pth} = \frac{\pi c}{6\mu_0 |\chi_R''| \kappa^2 \omega_p^2} \left[1 + \left(1 + \frac{\kappa \ln[P_{sth}/P_s(0)]}{\tan^{-1}(L/b_p)}\right)^{\frac{1}{2}}\right]^2 , \tag{5.14}$$

where $P_s(L) = P_{sth}$ is the Stokes power at which detection becomes possible. In view of our earlier discussion, we can put $\ln[P_{sth}/P_s(0)] = 30$. If the Raman shift is small ($\kappa = \omega_s/\omega_p \rightarrow 1$), then (5.14) may be approximated by

$$P_{pth} \simeq \frac{\pi c}{6\mu_0 |\chi_R''| \kappa \omega_p^2} \, \frac{\ln[P_{sth}/P_s(0)]}{\tan^{-1}(L/b_p)} \quad ; \quad \kappa \rightarrow 1 \ .$$

This is the result that is usually obtained by ignoring the effects of the Stokes-wave diffraction, and implies a ω_s^{-1} threshold dependence. However, as mentioned earlier, for generating infrared radiation by SERS, κ is in the range 0.001 - 0.2, and then (5.14) tends towards a ω_s^{-2} threshold dependence, reducing as $\kappa \rightarrow 0$ to the condition $\tilde{P}_p = G_{Ro} b_p \kappa = 4$ mentioned earlier.

This analysis of the gain-focussing effect can also be extended to include the effect of the real part of the Raman susceptibility χ_R [1.38]. As mentioned in Sects.3.3,4.6.2, the real part represents a contribution to the Stokes refractive index that is proportional to the incident pump intensity (optical Kerr effect). This gives rise to pump-induced focussing and defocussing of the Stokes light, according to the sign of $\Omega_{fg} + \omega_s - \omega_p$.

[This quantity was of course implicitly taken as zero in the foregoing analysis, because that is the condition that χ_R be purely imaginary and the gain $G_R(\xi,r)$ thus real.] When $\omega_p - \omega_s$ is less than the Raman transition frequency Ω_{fg}, there is an increase of the Stokes refractive index at the centre of the beam where the incident radiation is strongest. This tends to reduce the divergence of the Stokes wavefront, and so decreases the diffraction loss. There is a corresponding increase of divergence when $\omega_p - \omega_s > \Omega_{fg}$. When this diffraction loss is taken into account, the net Stokes gain is, therefore, expected to be greatest for some $\omega_s > \omega_p - \Omega_{fg}$. The analysis indicates that this frequency-pulling effect can result in the generated Stokes radiation being shifted from the exact Raman resonance by an amount of the order of the linewidth Γ.

With the pump radiation in the form of a pulse, the centre frequency of the Stokes wave would be swept during the pulse. There would also be pulse reshaping of the Stokes wave during its propagation, due to the spread of frequencies implied by its pulse shape experiencing different gains. These two effects are therefore sources of line broadening. However, as described in Sect.5.2, experimental measurements of Stokes linewidth have revealed very considerable broadening, with linewidths up to an order of magnitude greater than that of the pump, and two orders of magnitude greater than the Doppler width of the Raman transition. Some possible causes of this line broadening are given in Sect.5.4.2.

The presence of broadening makes it difficult to decide on the value of Γ to be used in the Raman susceptibility, with a consequent uncertainty in the calculated Raman gain. Because the predicted Raman threshold is proportional to $1/\Gamma$ (in both the plane- or focussed-wave treatments), an error of one or two orders of magnitude in Γ results in a corresponding error in the calculated threshold. Certainly, large discrepancies have been found between theoretical and measured thresholds and tuning ranges, when Doppler-limited values of Γ are used in the formulae [1.39]. However, for the SERS experiments with potassium and cesium we have found that by taking values of Γ equal to the observed linewidths of the Stokes output, there is generally reasonable agreement between measured and calculated values of threshold and tuning range. This shows that the formulae are useful for predicting the performance that might be obtained with other SERS transitions, although there is no firm justification for this empirical procedure.

5.3.3 Transient and Broad-Band Pumping

Our discussion of SERS has assumed the validity of the steady-state-growth equations derived in Chap.3, yet the SERS experiments described have all been carried out with pulsed pump lasers. It is, therefore, appropriate to consider the conditions under which the steady-state equations may be applied to pulsed sources (c.f. Sect.2.6.1). In fact, this is a rather complex question, which depends on the relative time scales of pump fluctuations, medium relaxation time, group-velocity dispersion, etc. We will, therefore, confine our remarks to a few simple considerations, and will give references to where the reader can find further discussions.

An obvious steady-state requirement is that the characteristic time scale of changes of the pump intensity, τ_c, be longer than the response time associated with the Raman transition, $T_2 = 1/\Gamma$, i.e., $\tau_c \gg T_2$. Introducing the retarded time $\tau = t - z/u_p$, where u_p is the group velocity of the pump pulse, then

$$I_s(\tau,z) = I_{so} \exp[g_R I_{po}(\tau)z]$$

in the small-signal region. A further requirement is clearly that the time scale of intensity changes in the generated Stokes radiation should also be longer than T_2. This can be expressed as

$$\frac{1}{T_2} \gg \frac{1}{I_s(\tau,z)} \frac{\partial I_s(\tau,z)}{\partial \tau} = g_R \frac{\partial I_{po}(\tau)z}{\partial \tau} = [g_R I_{po}(\tau)z]/\tau_c \quad .$$

This places a more stringent requirement on the pump pulse, namely that

$$\tau_c \gg g_R I_{po}(\tau) z T_2 \equiv (GL)T_2 \quad ,$$

and arises from the fact that the temporal changes in the pump pulse are magnified by a factor GL in the Stokes wave. The gain GL can be estimated as GL = 30, for threshold, in which case we have $\tau_c \gg 0.16/\Gamma$, where Γ is in cm^{-1} and τ_c is in ns. Thus, for example, taking $\Gamma = 0.1\ cm^{-1}$ gives $\tau_c \gg 1.6$ns. For smooth (transform-limited) pulses this criterion would be well satisfied with a pulse length $\tau_p \sim 10$ns. Thus, on that basis, the steady-state description would be appropriate for the experiments just described. However, the pump sources used are usually far from transform limited; for example, a typical pump linewidth of $\sim 0.1\ cm^{-1}$ corresponds to $\tau_c \sim 0.05$ns. To assess the effects of this short τ_c properly requires a knowledge of the pulse structure. Nevertheless, it is apparent that for pump fluctuations $\tau_c \ll T_2$, and yet $\tau_p > GLT_2$, the effect of the fluctuations is likely to be averaged out to a large extent.

This is confirmed by an analysis of the behaviour of stimulated Raman scattering in a stochastically varying pump field (AKHMANOV et al. [2.98]; reprinted in [5.54] with additions). They show that the Raman-gain coefficient is then just $g_R\bar{I}_{po}(\tau)$, where $\bar{I}_{po}(\tau)$ is the pump intensity averaged over the fluctuations. This leads to the important conclusion that, under these conditions and for a given average pump intensity, the gain coefficient is *independent* of the pump linewidth $\Delta\omega_p$(HWHM).[2] As τ_c becomes shorter, the effects of group-velocity dispersion can begin to play a significant role. AKHMANOV et al. [2.98] show that these effects can be neglected only if

$$\left|\left(\frac{d\omega}{dk}\right)_p^{-1} - \left(\frac{d\omega}{dk}\right)_s^{-1}\right| < \frac{g_R\bar{I}_{po}(\tau)}{2\Delta\omega_p} \quad , \tag{5.15}$$

where the subscripts p and s distinguish the pump and Stokes group velocities $d\omega/dk$. If the pump intensity is reduced below a critical value implied by the inequality (5.15), then the group-velocity mismatch leads to a reduction of the effective Raman gain coefficient below the nondispersive value of $g_R\bar{I}_{po}(\tau)$. Thus, if the value of $g_R\bar{I}_{po}(\tau)$ needed to satisfy (5.15) is greater than $\sim 30/L$ then (5.15) imposes a more stringent threshold condition. Group-velocity dispersion can become significant in SERS when the pump is tuned close to an atomic resonance line. If there is only one intermediate resonance level, both sides of the inequality (5.15) share the same inverse-square-law dependence on resonance detuning; hence there is an appreciable range of pump frequencies around the resonance for which the SERS threshold pump intensity is *independent of detuning*. Experimental confirmation of these predictions was obtained recently by MIKHAILOV et al. [5.56] and KOROLEV et al. [5.57] for SERS in Rb vapour. These experiments show very clearly the importance of group-velocity dispersion effects under conditions similar to those used in many of the experiments described in Sect.5.2, i.e.,

---

[2] This prediction differs fundamentally from one found in the early literature (for example, [5.55]), that the effect of a finite pump linewidth $\Delta\omega_p$ is to reduce the Raman gain by the ratio $\Gamma/(\Gamma + \Delta\omega_p)$.

near-resonant SERS using dye-laser pulses that are not transform-limited. In the nondispersive regime (5.15), the SERS gain coefficient was also found to be independent of pump-laser linewidth, as predicted.

The description of stimulated Raman scattering for stochastic pump pulses is due mainly to AKHMANOV et al. [2.97,98], (see also [5.54,58,99-101], and references therein). On the other hand, SRS with transform-limited short pulses, which is often called transient SRS, has been investigated by a number of workers, e.g. [2.90,96,97,99,5.59,60] and see Sect.2.7.

For ultra-short pulses, the relaxation times of the medium no longer play a role, although group-velocity dispersion assumes more importance. Under these conditions, a number of interesting coherent phenomena such as self-induced transparency and pulse break-up occur. For examples of this type of behaviour the reader is referred to papers by BELENOV and POLUÉKTOV [5.61], POLUÉKTOV et al. [5.62], TAN-NO et al. [2.27], TAKATSUJI [2.9], ELGIN et al. [5.63].

5.4 Limiting Mechanisms

A number of mechanisms can limit the efficiency of infrared generation by SERS. First, several simultaneous competing processes can divert part of the pump energy from Stokes generation. An obvious example of this is absorption of the pump radiation by single-photon transitions to the nearly resonant intermediate levels. When the pump is detuned from resonance by more than a few cm^{-1}, however, single-photon absorption by the atoms is generally not as serious as absorption by dimers and other possible multiphoton processes. These various competing processes are discussed in Sect.5.4.1. Second, a number of possible mechanisms could have the effect of broadening the two-photon transition linewidth Γ. Although these effects are not competing processes in the above sense, because they do not actually divert energy from the pump, nevertheless they may tend to reduce the Raman gain. These line-broadening mechanisms are summarised briefly in Sect.5.4.2. Limits are imposed on the SERS output energy by both depletion of the pump and saturation of the Raman transition; these saturation processes are considered in Sect.5.4.3.

5.4.1 Competing Processes

Single- and Multiphoton Absorption by Atoms

Efficient SERS generation is obtained when the pump frequency is tuned near to resonance with an intermediate level. However, as well as enhancing the nonlinear susceptibility for SERS, a number of other processes are similarly enhanced. The simplest of these is single-photon absorption, in which atoms are excited from the initial level to the intermediate levels at the expense of the pump photons. Rather than give a general discussion (which in principle must treat the problem of resonance Raman scattering), we simply show the sort of magnitudes that may be expected for a typical SERS experiment. We consider a situation in which the pump frequency ω_p falls in the wings of the resonance line, so that the lorentzian line shape dominates the Doppler-broadened, gaussian line shape. The absorption coefficient β, given by $\beta = (\omega_p/c)\mathrm{Im}\,\chi^{(1)}(-\omega_p;\omega_p)$, can thus be written as

$$\beta = \frac{8\pi r_e c f_{gi} N\omega_p^2 \Gamma_i}{(\Omega_{ig}^2-\omega_p^2)^2} \simeq \frac{2\pi r_e c f_{gi} N\Gamma_i}{(\Omega_{ig}-\omega_p)^2} \quad , \tag{5.16}$$

where Ω_{ig} is the resonance frequency, f_{gi} the transition oscillator strength, and Γ_i is the transition linewidth (HWHM). This expression may be derived from (2.33) by use of the assumption $|\Omega_{ig} - \omega_p| >> \Gamma_i$. To calculate β, a knowledge of Γ_i is required. This linewidth includes contributions from resonance self-broadening (see Sect.5.4.2) and pressure broadening due to collisions with the buffer-gas atoms, as well as the natural width. MILES and HARRIS [1.15] tabulate values for all three of these linewidth contributions for a number of alkali resonance lines. We consider the example of the second resonance lines of potassium (4s-5p), which are used to resonantly enhance the 4s-5s SERS transition (Fig.5.3). Foreign-gas broadening can be neglected, because in a correctly operated heat pipe (see Sect.4.4.2) there is no mixing of the vapour and the buffer gas (except at the short transition regions at each end). For a vapour pressure of 10 torr, ($N = 1.35\times10^{17}$ cm^{-3}), Γ_i is ~ 0.001 cm^{-1} due to self-broadening; natural broadening is negligible. The oscillator strength is $f_{4s-5p} \simeq 0.0154$; therefore, $\beta \simeq 0.7/(\Omega_{ig} - \omega_p)^2$ cm^{-1}, (Ω_{ig},ω_p in cm^{-1}). Thus, if the pump detuning is ~ 4 cm^{-1}, then $\beta \sim 0.04$ cm^{-1}. This amount of absorption means that the dye-laser (pump) power would fall to 1/e of its initial value after passing through a 25 cm-long vapour column. However, (5.16) shows that the absorption coefficient decreases as the square of the detuning from resonance; therefore, in the example there is negligible depletion of the pump power by this mechanism, over most of the observed SERS tuning range. Also, for SERS transitions to higher-lying s levels, single-photon absorption by the resonant p levels is, successively, less for a given amount of detuning. There are similar absorption coefficients for the corresponding resonance lines in the other alkalis. On the other hand, absorption in the region of the principal resonance line will be strong over a much-wider range (because, as will be seen later, the resonance-broadened linewidth is proportional to f_{gi}/Ω_{ig}; f_{gi} is greater, and Ω_{ig} less, for the principal line); it may easily be a hundred times greater.

It can be seen from (5.6,16) that, when the pump frequency is tuned in the region of one of the intermediate levels, the SERS gain and single-photon absorption coefficients have the same inverse-square dependence on the detuning from resonance, and both are proportional to the oscillator strength f_{gi}. The consequence of this can be examined by considering the following simplified (and plane-wave) model for SERS in a region of strong absorption of the pump beam. If the attenuation of the pump by single-photon absorption is assumed to be much greater than any depletion due to SERS, then the pump intensity $I_p(z)$ decreases according to $I_p(z) = I_{po} \exp(-\beta z)$. When this is substituted for $I_p(z)$ in the growth equation for the Stokes intensity,

$$\frac{dI_s}{dz} = G_R\, e^{-\beta z} I_s \quad , \tag{5.17a}$$

the solution is obtained,

$$I_s(z) = I_{so} \exp[(G_R/\beta)(1 - e^{-\beta z})] \quad . \tag{5.17b}$$

This shows that, because of the decreasing pump intensity, the over-all Stokes gain saturates after a distance equal to a few times the absorption length β^{-1}, reaching a maximum value of $\exp(G_R/\beta)$. It is interesting to compare this result with the expression for gain, $\exp(G_R L)$, which applies when the pump-beam absorption can be neglected. Thus, in the presence of strong absorption, a high gain is obtained by making G_R/β large. From (5.6,16), it follows that $G_{R/\beta} \propto f_{fi}$, and is independent of the number density, the detuning $\Omega_{ig} - \omega_p$, and the oscillator strength f_{gi}. By contrast, if there is no pump absorption, the aim would be to achieve a large G_R; from (5.6), this is proportional to $f_{gi}f_{fi}$. This result can be a little misleading if the effect of pump absorption is ignored, because it suggests that, for a suitable SERS transition, a high value of the product $f_{gi}f_{fi}$ is needed, without regard to the individual values of f_{gi} and f_{fi}. However, for a fixed product $f_{gi}f_{fi}$ that is obtained from a high f_{gi} but a low f_{fi}, the large f_{gi} will tend to cause high pump-beam absorption; as shown in the foregoing, the Raman gain will then actually be determined by the low f_{fi}. Therefore, Raman transitions should be sought for which f_{fi} is high.

For SERS from the ground state in cesium vapour, the values of the oscillator strengths f_{fi} are $\simeq$ 1.58 (f-i = 7s-7p), 2.01 (8s-8p), 2.43(9s-9p) for the dominant intermediate states in the 6s-7s, 6s-8s and 6s-9s SERS transitions, respectively. There are similar high values for the corresponding transitions in the other alkalis.

On the other hand, the oscillator strengths can be much less favourable for some other SERS transitions in alkaline earths. For example, we have attempted to observe a Stokes output in the region of $\sim$ 6.5μm using the $(5s)^2\,{}^1S_0$ - $5s4d\,{}^1D_2$ SERS transition in strontium vapour, resonantly enhanced by tuning ω_p near to the $(5s)^2\,{}^1S_0$ - $5s5p\,{}^1P_1$ resonance line at 460.7nm. A dye-laser power of $\sim$ 250kW and strontium vapour pressures up to 10 torr were used for this experiment. Even when rather pessimistic assumptions are made about the appropriate value for the Raman transition linewidth (say, $\Gamma = 5\mathrm{cm}^{-1}$), the parameters of this experiment suggest that it should be possible to exceed the Raman threshold by at least an order of magnitude [according to the threshold formula (5.14)] when ω_p is tuned within $\sim$ 60 cm^{-1} of resonance with Ω_{ig}. However, we were not able to observe any Stokes output while the dye laser was tuned over a wide range of frequencies, from several hundred cm^{-1} from the resonance line to within $\sim$ 1 - 2 cm^{-1}. WYNNE and SOROKIN [5.102] also report having tried this experiment without success. We believe that this failure to reach SERS threshold is due to the strong absorption of the pump beam, the oscillator strengths having the values $f_{gi} \simeq 1$, $f_{fi} \simeq 0.01$ in this case.

However, the oscillator strengths are far more favourable if higher intermediate ${}^1P_1$ levels are used; this is then analogous to the alkali-atom schemes. There is also the possibility of using intercombination (singlet-triplet) lines in the alkaline earths, for which the oscillator strengths are again favourable. For example, in the experiments of CARLSTEN and DUNN [1.36] using Ba vapour (Sect.5.2.1), the pump frequency was closely resonant with the relatively weak intercombination transition $(6s)^2\ ^1S_0 - 6s6p\,^3P_1$ ($f_{gi} \simeq 0.01$), whereas the Stokes frequencies were close to resonance with the strong triplet-triplet transitions $6s5d\,^3D_{1,2} - 6s6p\,^3P_1$ ($f_{fi} \simeq 0.02$, 0.03, [5.64]). Thus, as shown in Fig.5.2, although there is a dip in the Stokes output, a strong signal could, nevertheless, be observed even when the pump frequency was tuned into exact resonance with the intermediate level. In this case, pump absorption is not severe, because of the comparatively low oscillator strength f_{gi}; the gain is determined by the product $f_{gi}f_{fi}$.

A final point concerning single-photon absorption is that, in the extreme wings of the resonance lines, even as far as several tens or hundreds of cm^{-1} from resonance, there may be slight residual absorption, which, although irrelevant from the point of view of pump-beam depletion, could populate the intermediate level sufficiently to produce significant resonance broadening of the final Raman level. Self-broadening is discussed further in Sect.5.4.2.

Multiphoton ionisation of atoms and the consequential effects of this have already been mentioned in Sect.4.6.4. For a typical SERS scheme in alkali vapours, two pump photons have enough energy to photoionise the atom. A detailed analysis of the effects of multiphoton ionisation on the SERS behaviour is probably impractical, and the use of calculated ionisation rates is of questionable value when the poor agreement between the calculated and measured rates is considered. Fortunately, some measured two-photon ionisation rates are available for cesium in the region of the 6s-7p resonances [4.83], this being the region of pump frequencies required for 6s→7s SERS. The measured values are very large; GRANNEMAN et al. [5.65] found off-resonant values of $W/F^2 \sim 10^{-47}\ cm^4s$, where W is the photoionisation rate per atom and F is the incident photon flux (photons $cm^{-2}s^{-1}$). Thus, for example, a photon flux of 10^{27} photons $cm^{-2}s^{-1}$ of 2.7eV photons (450MW cm^{-2}; 21744 cm^{-1}) is sufficient to produce a photoionisation rate of $\sim 10^7 s^{-1}$, which indicates that about 10% of the atoms are photoionised during the 10ns laser pulse. Moreover, the photoionisation rate is strongly enhanced when the incident radiation is tuned close to the 6s-7p resonance lines; GRANNEMAN et al. found that W/F^2 increased by over an order of magnitude, to $3 \times 10^{-46}\ cm^4s$ when the incident frequency was detuned from the $7p_{3/2}$ level by $\sim 20\ cm^{-1}$. Thus, it appears that a significant fraction of atoms may undergo two-photon ionisation when subjected to the pulsed intensities typically involved in an SERS experiment. However, these rough estimates also indicate that photoionisation is unlikely to provide any serious competition to SERS by depleting the pump beam.

Absorption by Dimers

In addition to absorption by alkali atoms, it is necessary to consider the effects of absorption by alkali dimers, because the dimer concentrations become quite large at vapour pressures of a few torr. They cause strong absorption over a substantial part of the visible spectrum. EVANS et al. [5.66] have collected all of the necessary thermodynamical data to allow calculation of both the monomer and dimer densities as functions of temperature. The results are given in Fig.5.9, which shows that the lightest alkali metals are more susceptible to the formation of molecules than the heavier ones. Equations (4.33,34) give the saturated vapour pressure and total number density as a function of temperature. Thus, in the case of cesium, for example, at a pressure of 10 torr (646 K, total number density $1.5 \times 10^{23} m^{-3}$) the dimer density is $\sim 2.3 \times 10^{21} m^{-3}$, i.e., about 1.5% of the total.

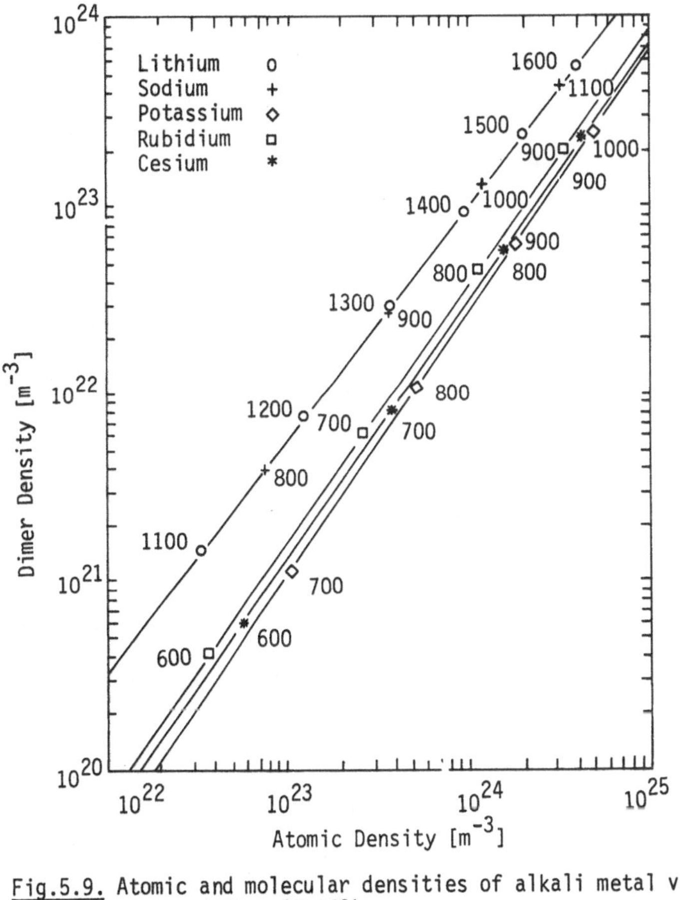

Fig.5.9. Atomic and molecular densities of alkali metal vapours as functions of temperature. (After [5.34])

The broad absorption bands in the visible spectrum due to these dimers give the different alkali vapours their characteristic colours when seen in transmission; for example, green for potassium and greenish yellow for caesium. The frequencies of the main absorption bands of alkali molecules have been known for a long time [5.67], but data on their cross sections are far from complete. LAPP and HARRIS [5.68] have, however, studied the absorption of K_2 molecules in the red and the absorption of Cs_2 molecules through most of the visible spectrum. Figure 5.10 shows the measured absorption cross section of Cs_2 in the range 470-670nm, where it varies between 10^{-22} and $10^{-20}m^2$. By extrapolation just beyond the short wavelength limit of Lapp and Harris's measurements, an absorption cross section of $\sim 3\times10^{-21}m^2$ is estimated for pump wavelengths in the region necessary to produce SERS on the 6s-7s transition, enhanced by the $6s\text{-}7p_{1/2,3/2}$ resonances. At a vapour pressure of 10 torr, the absorption coefficient would be $\sim 0.05\ cm^{-1}$; this indicates that at low incident intensity, the pump-beam transmittance through a 20 cm vapour column would be less than 30%.

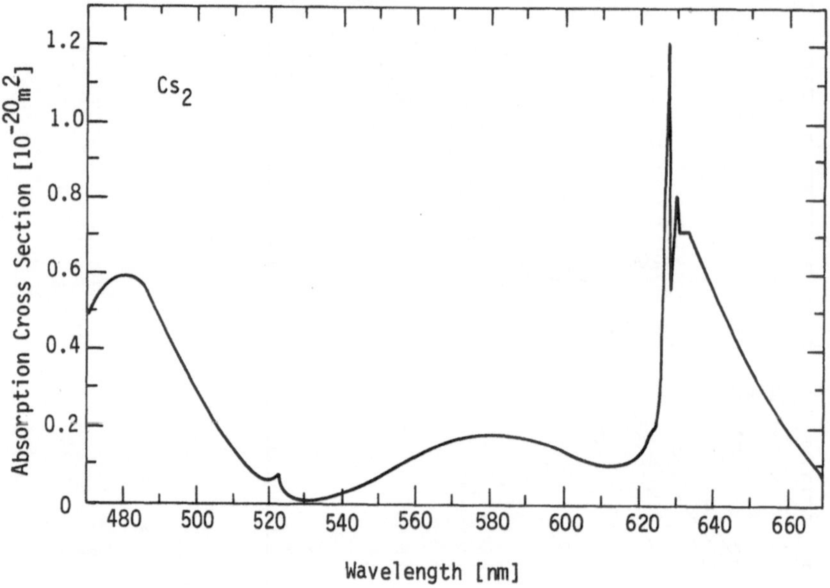

Fig.5.10. Absorption cross section of Cs_2 dimers as a function of wavelength. (After [6.3])

Fortunately, this molecular absorption may be less serious than these figures suggest, because a high pump intensity can saturate (i.e., bleach) the absorption. A requirement for bleaching is that the number of photons needed to excite all of the dimers in the beam path shall be much less than the total number of photons in the incident pulse. For a gaussian laser beam that is focussed to a waist of confocal parameter b_p at the centre of a vapour column of length L, the volume of vapour encountered by the beam (and hence the number of dimers encountered) is a minimum when $L/b_p = \sqrt{3}$. Under the typical experimental conditions of COTTER et al. [4.51], i.e., a 140μJ, 7ns pulse (20kW) from a nitrogen-laser-pumped dye laser and 10 torr vapour pressure, less than 10% of the incident pump energy would be absorbed for this focussing. In fact, COTTER et al. observed that the optimum Stokes output energy and tuning range on the 6s-7s Cs SERS transition was obtained when $L/b_p \sim 1$ (see Sect.5.2). Therefore, despite the fact that, according to (5.13), which ignores dimer absorption, the Raman gain should increase with stronger focussing as $\tan^{-1}(L/b_p)$, an actual increase of L/b_p to ~ 6 resulted in a poorer performance. The peak output energies were somewhat less and there was more-rapid reduction of output when the dye laser was tuned further away from the intermediate resonances. Also, increasing the vapour pressure from 10 to 30 torr had little influence on the Stokes tuning range; it actually reduced the peak output energy to about half. This can be explained by the increased density of dimers with increased pressure, which exacerbates the problem of dimer absorption. Thus, the presence of a significant population of dimers can immediately place some restrictions on the values of focussing and vapour pressure that can be used for efficient SERS with the relatively low-energy pulses from nitrogen-laser-pumped dye lasers. When the higher energies (> 5mJ) available from dye-lasers pumped by more-powerful uv lasers are used, the restrictions are much less severe, at least on account of molecular absorption, because such absorption can then be bleached more readily.

There is still relatively little known about the processes that occur when alkali dimers are excited by intense laser radiation. SOROKIN and LANKARD [5.29,33] observed various infrared-laser emissions from K, Rb and Cs vapours when these were irradiated by a powerful pulsed dye laser. The wavelengths of these infrared emissions corresponded to various atomic transitions. SOROKIN and LANKARD suggested that the mechanism of excitation involved two-photon absorption by the alkali dimers followed by dissociation into various excited atomic states. More recently, GRANNEMAN et al. [5.65] used several different lines from a pulsed Ar-ion laser to measure the photoionisation rates of Cs_2

molecules. A specially designed oven allowed the ratio of monomer and dimer concentrations to be controlled. By observing the power dependence of the rate of molecular ionisation and using time-of-flight analysis of the ionisation products, GRANNEMAN et al. were able to show that the channel

$$Cs_2 + 2\hbar\omega \rightarrow Cs^+ + Cs^* + e^-$$

is the dominant molecular photoionisation mechanism for laser wavelengths in the region of the 6s-7p atomic resonances. They measured photoionisation rates (W/F^2) in the range 10^{-43} - 10^{-42} cm^4s. Thus, by assuming a photon flux of 10^{26} photons $cm^{-2}s^{-1}$ of 2.7eV photons (equivalent to a 15kW dye-laser beam focussed confocally over a 25cm vapour column), then the photoionisation rate per molecule is about 10^9 - $10^{10}s^{-1}$. This implies complete photoionisation of the dimers with an incident laser pulse of a few nanoseconds duration. Of course, this transition rate does not represent the total two-photon absorption rate of Cs_2, because alternative photodissociation channels that produce only neutral excited products [5.69] would be unobserved in the ionisation experiment of GRANNEMAN et al.

Parametric Generation

There are many possibilities for parametric-mixing processes that involve sums and differences of the frequencies of the strongest waves present in the vapour. These are the pump and Stokes waves and also some fixed-frequency emissions due to ASE (Sect.5.2.2). Most of these parametric processes can be ruled out as serious competitors to SERS. For example, the susceptibilities for third-harmonic generation of the pump wave ($3\omega_p$), anti-Stokes generation ($2\omega_p - \omega_s$) and sum-frequency generation ($2\omega_p + \omega_s$) are low in the case of the alkalis, because the generated frequencies are above the ionisation continuum and also the bound-free matrix elements involved are low. Processes in which the frequencies involved are far removed from single and two-photon resonances will also be weak. An example of this category would be generation of the third harmonic of the Stokes frequency ($3\omega_s$).

Some other processes, however, such as sum mixing ($\omega_p + 2\omega_s$) and difference mixing ($\omega_p - 2\omega_s$), can occur sufficiently strongly to be easily observed. In particular, the latter process (which is equivalent to the two-photon resonant four-wave process depicted in Table 2.2a or Fig.6.1 with $\omega_1 = \omega_p$, $\omega_2 = \omega_3 = \omega_s$) is readily observable in SERS experiments with alkali metals involving the Raman transition between the ground state and first excited s level. The parametrically generated frequency $\omega_p - 2\omega_s$ falls in the visible spectrum the colour is blue, yellow, orange, and red for Na, K, Rb, and Cs, respectively. The radiation emerges from the vapour in a collimated beam which, after filtering, shows up well on a white card. This effect was observed by SOROKIN et al. [1.35], while generating a Stokes output at 2.7μm in potassium vapour. When a 1kW dye laser was used for the pump, the parametrically generated yellow output was observed for only a narrow range of pump frequencies; it peaked sharply at the frequency corresponding to optimum phase matching. However, when COTTER et al. [1.37] used ten times more dye-laser power, they observed the yellow output from potassium over most of the 250 cm^{-1} tuning range for which a strong SERS output (> 1kW) was obtained. The maximum power of the parametric output was ~ 200 mW. This change of behaviour with increased power can be attributed to the decreased ratio of phase mismatch Δk to the Raman gain. As discussed in Section 3.4, this leads to phase locking of the waves and subsequent insensitivity to Δk. Such processes will be discussed further in Chap.6. We only note here that KÄRKKÄINEN [5.34] studied the competition between SERS and these four-wave mixing processes, both theoretically and experimentally; he concluded that the effect on the SERS generation efficiency is negligible.

5.4.2 Line Broadening Processes

The measurements described in Sect.5.2.1 showed that for SERS in alkali vapours the bandwidth of the generated Stokes radiation is significantly broader than that of the dye-laser input. Typical measured linewidths for the Stokes output were 0.3 - 1.0 cm^{-1}, obtained with pump-laser linewidths of ~0.1 cm^{-1}; the calculated Doppler-broadened linewidths of the Raman transitions were typically 0.03 - 0.05 cm^{-1}. The theory of AKHMANOV et al. [2.98], discussed in Sect.5.3.3, shows that, for pump fluctuations short compared to T_2, the Stokes bandwidth can broaden up to the pump bandwidth. Broadening in excess of the pump bandwidth can occur for a transform-limited pulse of duration $<T_2$, but this does not correspond to the typical dye-laser pulses used in SERS experiments. In this subsection, we mention some line-broadening mechanisms that might contribute towards the additional broadening observed experimentally. [Two such mechanisms considered earlier were the effects of hyperfine structure (Sect.5.2.1) and gain focussing (Sect.5.3.2).] It seems likely that a number of different processes could play a significant role and they may occur simultaneously. The dynamic interaction between all of these is very involved; to identify experimentally the dominant

broadening mechanism (if any) is not likely to be easy. The line-broadening mechanisms can be separated into those that are intrinsic to the Raman process (such as the ac Stark effect, which must inevitably be present) and those that are caused extraneously (as for example Stark broadening caused by photoelectrons and ions in the vapour). Before discussing these extraneous broadening effects, we first give a rough estimate of the magnitude of the shift in Stokes frequency (and hence broadening) caused by the ac Stark effect. Consider the three-level system depicted on the left-hand side of Fig.5.2a. Following steps analogous to those that led to (4.74), the expression for the shifted Raman transition frequency is

$$\Omega_{fg} + [|\mu_{gi}|^2 I_p - |\mu_{if}|^2 I_s]/2\hbar^2\varepsilon_0 c\Delta \quad , \tag{5.18}$$

where I_p, I_s are the pump and Stokes intensities, respectively, and Δ is the pump detuning from the intermediate level, i.e., $\Delta = \Omega_{ig} - \omega_p$. For example, taking the 6s-7s Raman transition in Cs, (5.18) accounts for a shift of approximately $(9.3 \times 10^{-8} I_p - 3.5 \times 10^{-5} I_s)/\Delta$ cm^{-1}, where the intensities are in W cm^{-2} and Δ is in cm^{-1}. Thus, an incident pump intensity of 100MW cm^{-2} would result in a shift of 0.5 cm^{-1} at 20 cm^{-1} detuning from resonance. On the other hand, because the generated Stokes frequency is close to resonance with a strongly allowed transition, a similar (but opposite) shift would be induced by Stokes radiation of $\sim$ 250kW cm^{-2} (i.e., over two orders of magnitude less intensity). This shift may tend to curtail further growth by shifting the frequency for maximum Raman gain away from the frequency of the already established Stokes wave. (The combined effects of level shifts and a stochastic pump on SRS have not been discussed in the literature, but AGOSTINI et al. [5.70] did so for two-photon resonant three-photon ionisation.)

Broadening by Charged Particles

The density of electrons and ions produced by the multiphoton ionisation of atoms and molecules can be sufficient to cause significant Stark broadening of the neutral atomic levels involved in the Raman transition. Stark broadening under these conditions was discussed earlier in Sect.4.6.4. In the alkalis, photoionisation processes could lead to free electron and ion densities as great as 10^{21} - 10^{22}m^{-3} at vapour pressures in the region of 10 torr; it was estimated in Sect.4.6.4 that the final Raman levels could thus be broadened to as much as a few cm^{-1}. In contrast, the cross sections for direct single-photon ionisation from the final Raman level, due to the pump, are on the order of 10^{-17} cm^2 [5.71]. This gives a broadening (due to lifetime shortening) of ~ 0.01 cm^{-1} or less for a pump intensity $\sim 10^{-8}Wcm^{-2}$.

Resonance Broadening and ASE

As described in Sect.5.2.2, in the alkalis strong amplified spontaneous emission (ASE) is observed on the transition between the final Raman level (ns) and the adjacent lower p level (n-1 p). This drastically shortens the effective lifetime of the final Raman level, thus resulting in broadening of the Raman transition. Any estimate of the magnitude of the effect on SERS linewidth is complicated, however, by the fact that the necessary population inversion for ASE must be provided by the Raman-scattering process itself.

In the alkalis, the cascading ASE processes lead to a growing radiatively trapped population in the lowest p level (see Sect.5.2.1). A large transition moment connects this level with the final Raman level; therefore, this excited p-level population can have the effect of resonantly broadening the Raman transition by collisions. The resonance-broadening calculations in the impact approximation predict a lorentzian lineshape of width $\Delta\omega$, given by [4.68]

$$\Delta\omega = \left[4\pi r_e c^2 k_{JJ'} \left(\frac{2J+1}{2J'+1}\right)^{1/2} \frac{f_{JJ'}}{\Omega_{J'J}}\right] N_J \quad \text{(FWHM)} \quad , \tag{5.19}$$

where N_J is the population in the state J, which broadens the state J' (J, J' are angular-momentum labels), $f_{JJ'}$ is the oscillator strength that connects the two levels and $\Omega_{J'J}$ is the transition frequency. The constant $k_{JJ'}$ is close to unity. For example, when the 4s-5s Raman transition in potassium (Fig.5.3) is used, if 10% of the atoms are excited to the 4p levels at a vapour pressure of 10 torr (N_J = 10% of 1.35×10^{23} m^{-3}), then according to (5.19) the 5s level will be broadened to $\Delta\omega \sim 0.05$ cm^{-1}.

5.4.3 Saturation

Apart from the influence of the various competing processes described earlier, two intrinsic causes of saturation limit the maximum Stokes energy. These are the limited number of pump photons (pump depletion), and the limited number of atoms in the initial Raman level (atom depletion). Pump depletion refers to the situation in which the number of Stokes photons generated becomes comparable to the number of incident pump photons, in which case, the Stokes output is linearly proportional to the incident pump-beam intensity, (Sect.

3.4). Atom depletion can occur if the atoms that have been excited to the final Raman level are unable to return to the initial level in a time short compared to the duration of the pump pulse.

Although a rigorous treatment of atom saturation requires a dynamic approach, it is instructive to consider the steady-state solutions, which are analogous to the single-photon case, Sects.2.6.8, and 4.6. Thus, when level shifts are ignored (for simplicity), the steady-state Stokes-wave growth equation can be written as

$$\frac{dI_s}{dz} = g_R I_p I_s/(1 + I_p I_s/I_{sat}^2) \quad ,$$

where $I_p I_s/I_{sat}^2 = T_1 T_2|\alpha_R E_s^* E_p/2\hbar|^2$. By use of (5.1,3), the saturation parameter can be written as $I_{sat}^2 = \hbar\omega_s/(g_R/N)T_1$. Because the maximum value of the product $I_p I_s$ occurs when the pump-beam is depleted by 50%, in which case $I_p = I_{po}/2$, $I_s = (\omega_s/\omega_p)I_{po}/2$, a criterion for atomic saturation to be unimportant is that

$$(I_p I_s)_{max} = \frac{\omega_s}{4\omega_p} I_{po}^2 << I_{sat}^2 \quad ;$$

i.e.

$$\left(\frac{G_R}{N}\right)\left(\frac{I_{po}}{\hbar\omega_p}\right) << \frac{1}{T_1} \quad . \tag{5.20}$$

This simply states that the maximum rate of removal of photons from the incident beam (cross section G_R/N times incident photon flux), each of which corresponds to the removal of an atom from its initial state, should be less than the rate of return of atoms to the initial state, $1/T_1$. A rough criterion applicable also to pulsed pump radiation is that, to avoid atom saturation, the number of atoms in the beam path should exceed the number of incident pump photons. To see the effect of such atomic saturation on the Stokes intensity, the growth equation may be integrated, which yields (3.24) for $I_s(z)$, but with a new gain coefficient

$$G_R z \rightarrow \left[G_R z - \left(\frac{G_R}{N}\right)\left(\frac{I_s}{\hbar\omega_s}\right)T_1\right] \quad .$$

In the early stages of growth, the reduction of gain due to the $I_s(z)$ term is negligible. As the gain approaches e^{30}, if the criterion (5.20) shows that saturation can occur, the gain reduction will become important before pump depletion sets in. Thus, for the remainder of the cell length, any increase of $G_R z$ beyond ~ 30 is offset by the saturation term; therefore, over a further distance δz,

$$G_R \delta z = \left(\frac{G_R}{N}\right)\left(\frac{I_s}{\hbar\omega_s}\right)T_1$$

or, rearranging,

$$\frac{I_s}{\hbar\omega_s} = \frac{N\delta z}{T_1} \quad . \tag{5.21}$$

This shows that any further increase of the number of Stokes photons requires the return of an equal number of atoms to their initial state. Hence, to extract more Stokes energy, two approaches may be used. The first is to increase the atomic number density N. Alternatively, the available δz could be increased; this is the length of cell that remains after the initial e^{30} gain has occurred. The Stokes intensity is, in principle, nevertheless ultimately limited by pump depletion.

Both atom and pump depletion were observed by CARLSTEN and DUNN [1.36] in the course of their SERS experiments using Ba vapour (see Sect.5.2.1). They found, first, that the maximum Stokes output energy that they were able to generate represented 40% efficiency for converting pump photons to Stokes photons. As shown in Fig.5.11, they observed the Stokes output energy as a function of the dye-laser energy input; they found that, after an initial rise, and provided that the number of available atoms is sufficient (i.e., the vapour pressure is high enough), there is a region in which the Stokes output is linearly dependent on the dye-laser input. In this region, the output is limited by pump depletion. However, for a given vapour pressure, the Stokes output eventually levelled off when the number of atoms in the beam path became comparable to the number of generated Stokes photons. An even clearer demonstration of the effect of atom depletion is shown in Fig.5.12. This shows the maximum Stokes output energy as a function of Ba vapour pressure when a constant dye-laser energy input was used. After an initial rise, the output becomes linearly dependent on pressure; throughout this range, the number of photons generated is limited by the number of atoms available. But, at ~ 0.2 torr, the number of atoms becomes comparable to the number of incident pump photons; therefore, a further increase of Stokes output is prevented because of pump depletion. These results are in accord with the approximate analysis given above.

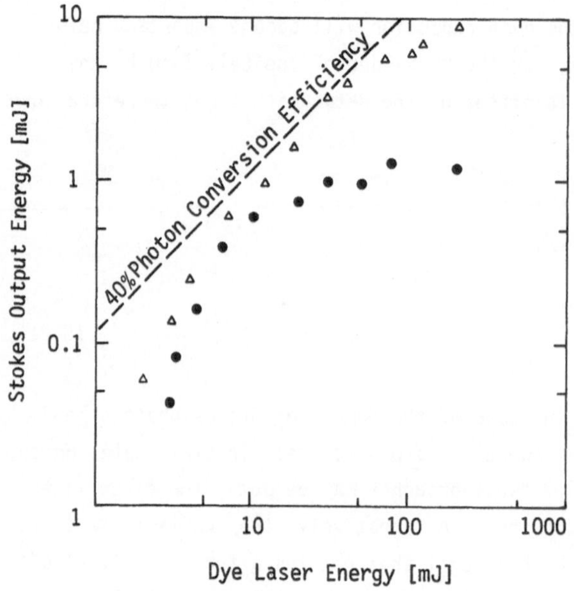

Fig.5.11. Energy of the generated Stokes signal as a function of incident dye-laser energy. For ●, the barium vapour pressure was 0.02 torr, for Δ, 0.2 torr. The dashed line indicates a conversion efficiency of 40% in terms of photon numbers. (After [1.36])

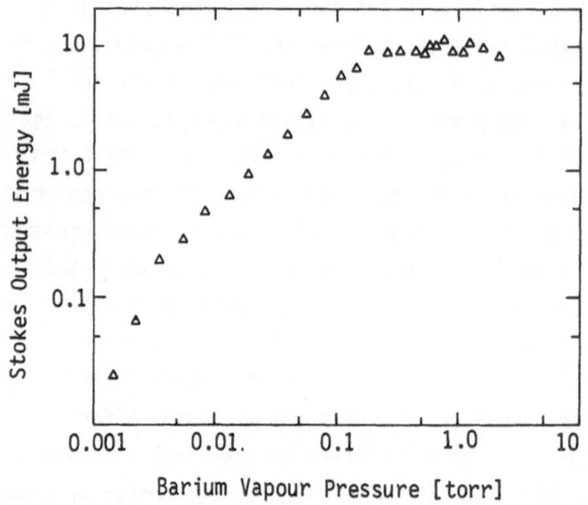

Fig.5.12. Energy of the generated Stokes signal as a function of the barium vapour pressure. Note the linear rise in the atom depletion region up to ~ 0.2 torr, and saturation due to pump depletion at pressures greater than 0.2 torr. (After [1.36])

Saturation due to atom depletion is observed on these intersystem SERS transitions in Ba vapour because the final Raman levels ($^3D_{1,2}$) are metastable; therefore, the recirculation time T_1 is long. However, saturation by atom depletion has also been observed in SERS in alkali vapours [5.72]; in this case, the mechanism is rather more complex. In Sect.5.2, it was pointed out that, in the case of alkali atoms, after excitation to the final Raman level by SERS, they rapidly lose their energy, by a cascade of ASE steps, until the lowest p level is reached, where they become radiatively trapped. These trapped atoms are therefore prevented from replenishing the initial Raman level. (It was noted in Sect.5.2 that a possible advantage of using SERS transitions from the lowest p levels is that atom saturation may be less serious, because the atoms are recirculated. On the other hand, the starting population in the excited initial level will probably be much less than that available in the ground state.)

Since, to avoid atom saturation, it is necessary to have (as a rough guide) more atoms in the beam path than photons in the pump beam, requirements are placed on both the vapour pressure and degree of focussing; the latter affects the volume of vapour intercepted by the beam. In practice, the maximum possible vapour pressure is determined either by technical problems that are associated with high-temperature operation of the vapour cell, or by limitations imposed by the formation of dimers (Sect.5.4.1).

A further important consequence of atom depletion is that it can make the Stokes output energy dependent on the small-signal Raman gain. Thus, as found in the foregoing analysis, once atom saturation has set in, the Stokes energy is proportional to that length of vapour-column over which atom saturation takes place. If the small-signal gain coefficient G_R is reduced, the length of vapour column required to develop the e^{30} growth of the Stokes radiation is increased. There is then a smaller remaining length of vapour within which the Stokes output increases linearly; consequently the Stokes output energy is lowered.

Because G_R depends on the detuning of the pump from resonance, the generated Stokes energy will also depend on the detuning. Figure 5.5 shows the SERS tuning profile in Carlsten and Dunn's experiment, in which barium vapour was used. The output dip at 2.0 torr vapour pressure when the dye laser is tuned close to the intermediate $^3P_1$ level is due to the various resonantly enhanced competing processes. However, outside the region of this dip, the Stokes output energy is observed to decrease steadily as the pump laser is tuned further from resonance. This is typical of the tuning profiles observed

in SERS; see for example Fig.5 of COTTER and HANNA [5.72]. The latter authors
have used the foregoing simple interpretation, supported by a rate-equation
analysis, to obtain quite good agreement between the calculated and observed
behaviour of the Stokes output versus tuning for the 6s-7s transition in
cesium vapour.

5.4.4 Limitations Due to Diffraction

In principle, by tuning the pump frequency towards the Raman transition frequency, it should be possible to generate progressively longer infrared wavelengths. In particular, by choosing a high-lying excited state of an alkali
atom as the final Raman level f, the small energy separations and near unity
oscillator strengths between the Rydberg levels (such as np and ns) would
ensure a strong resonance enhancement for SERS, even when the pump is tuned
near f. However, two main factors tend to raise the threshold for long-wave-
length generation. First, there is the linear dependence of the Raman gain
on the Stokes frequency ω_s, see e.g. (5.2). Second, as the ratio $\kappa = \omega_s/\omega_p$
of Stokes to pump frequencies becomes small, the problem of loss due to
diffraction of the Stokes radiation becomes increasingly severe; it results
in a greatly increased Raman threshold (Sect.5.3.2). To date, the longest
reported wavelength generated by SERS is $\sim 21.5\,\mu m$. This was achieved by
WYNNE and SOROKIN [5.22], using the frequency-doubled output from a nitrogen-laser-pumped rhodamine-dye-laser to pump the 4s-8s Raman transition
in potassium vapour. The uv power available after frequency doubling was
800W; by tuning in the region of the $8p_{3/2}$ level ($32230cm^{-1}$) the Stokes output was observed over a $\sim 2cm^{-1}$ tuning range.

Large Stokes diffraction loss also greatly decreases the effectiveness of
feedback if a Raman resonator is employed. The idea of a Raman resonator,
i.e., placing mirrors that reflect the Stokes wave at each end of the medium,
is particularly effective for vibrational Raman scattering in molecules,
because the Stokes wavelength is only slightly longer than that of the pump,
i.e., $\kappa \to 1$. SCHMIDT and APPT [1.48] used a resonator configuration to generate tunable ir by Raman scattering in high-pressure hydrogen gas. Also using
hydrogen gas, FREY and PRADÈRE [5.73] employed a series of lenses inside the
gas cell to reduce the ir diffraction loss. However, the longest ir wavelengths
generated in these experiments were limited by the effects of diffraction loss.
GRASIUK and ZUBAREV [5.6], RABINOWITZ et al. [1.51] and HARTIG and SCHMIDT
[1.54] had some success using a waveguide structure to reduce the Stokes diffraction loss.

The effects of Stokes diffraction loss can be understood with the help of (5.13). This shows that the net gain is the small difference between two large terms, a Raman gain proportional to $\tilde{P}_p$ and a diffraction-loss term proportional to $\sqrt{\tilde{P}_p}$. A resonator would be useful if by providing an effective increase of cell length L, it allowed a significant reduction of incident pump power, and hence $\tilde{P}_p$, for a given Stokes output. However, because a reduction of $\tilde{P}_p$ reduces the Raman gain more rapidly than it reduces the diffraction loss, it can be seen that only a marginal decrease of the pump power required for threshold is possible when the diffraction loss is large. Experimentally, we found this to be the case. Resonating either or both the pump and Stokes waves (by deliberately aligning the beams perpendicular to the vapour-cell windows or by using external mirrors) produced only marginal benefits.

In principle, the ir diffraction loss in SERS could also be significantly reduced by using stainless-steel capillary tubing for waveguides, although this may introduce practical difficulties due to the presence of liquid metal within the capillary.

5.5 Related Stimulated Processes

5.5.1 Stimulated Hyper-Raman Scattering (SHRS)

Hyper-Raman scattering is the name given to the three-photon analogue of Raman scattering. In this process (see Fig.2.3a), two pump photons (ω_{p1} and ω_{p2}) are annihilated and a single Stokes photon is emitted. The atom or molecule undergoes an upward transition of energy $\hbar\Omega_{fg}$ between states g, f, and $\omega_s \simeq \omega_{p1} + \omega_{p2} - \Omega_{fg}$. For atoms or other media whose states may be classified by parity, the transition is between states of opposite parity (in the electric-dipole approximation). The hyper-Raman susceptibility is of fifth order, and was given in Table 2.3.

Hyper-Raman scattering was first observed as a weak spontaneous process [5.74]; the first observation of the stimulated effect was by YATSIV et al. [1.28] who mention it as one of a number of nonlinear optical effects that they observed in potassium vapour. The system studied by them was, however, somewhat complicated in that the initial level for the hyper-Raman transition was an excited state that was first populated by a simultaneous third-order Raman process. Moreover, the two pump photons for the hyper-Raman process had different frequencies. More-recent studies of SHRS by VREHEN and

HIKSPOORS [2.52,5.75] (cesium), COTTER et al. [1.40] (sodium) and REIF and WALTHER [5.76] (strontium) used transitions from the ground states; the pump photons were provided by a single laser source.

The SHRS gain coefficient G_{HR} has been given in Sect.3.3, (3.25), where it is expressed in terms of the hyper-Raman susceptibility (2.22), Table 2.3. Similar to the SRS gain (5.6), we can write, for parallel linearly polarised light, and a single pump beam,

$$G_{HR} = g_{HR} I_{po}^2 = \frac{N\omega_s I_{po}^2 \mu_{fi}^2 \mu_{ij}^2 \mu_{jg}^2 / 4\varepsilon_o^3 c^3 \hbar^5 \Gamma}{(\Omega_{ig}-2\omega_p)^2 (\Omega_{jg}-\omega_p)^2} , \qquad (5.22)$$

where we assume that the dominant denominators are for single-photon resonance of ω_p with Ω_{jg} and two-photon resonance of $2\omega_p$ with Ω_{ig}.

In the SHRS process studied by VREHEN and HIKSPOORS, cesium atoms initially in the 6s ground state were excited to the $6p_{3/2}$ state by annihilation of two Nd:YAG laser photons at 1.064µm. Owing to the relatively large detuning from the nearest two-photon resonance ($2\omega_p - \Omega_{7s6s} = 254\ cm^{-1}$), a high incident pump intensity (2.5GW cm^{-2}) was required to reach the SHRS threshold. At these high intensities, the optical Stark shift of the initial and final levels was found to produce a significant shift and broadening of the Raman line.

The SHRS transition in sodium shown in Fig.5.13 was used by COTTER et al. [1.40] to generate a tunable infrared Stokes output. In the scheme used by VREHEN and HIKSPOORS, the intermediate level with which the pump frequency was most strongly single-photon resonant also acted as the final level of the process (i.e., j = f in 5.22); therefore, the wavelengths of the pump and scattered radiation were not widely different. However, by producing SHRS to a higher final level, as in Fig.5.13, it is possible to generate Stokes radiation with a considerably longer wavelength than the pump. COTTER et al. used a rhodamine 6G dye-laser (500kW peak power) as the pump; by tuning $2\omega_p$ in the region of the 3s-4d two-photon resonance, they observed a Stokes output of up to 5kW peak power. This output, representing a 2% photon-conversion efficiency, was limited by depletion of the atomic-ground-state population (Sect.5.4.3). The SHRS emission was tunable over a 160 cm^{-1} range and, by exploiting the two-photon resonance, could be observed with pump powers as low as ~ 2kW.

Further enhancement of the hyper-Raman gain is, in principle, possible by use of two pump frequencies ω_{p1} and ω_{p2}, with ω_{p1} tuned to take advantage

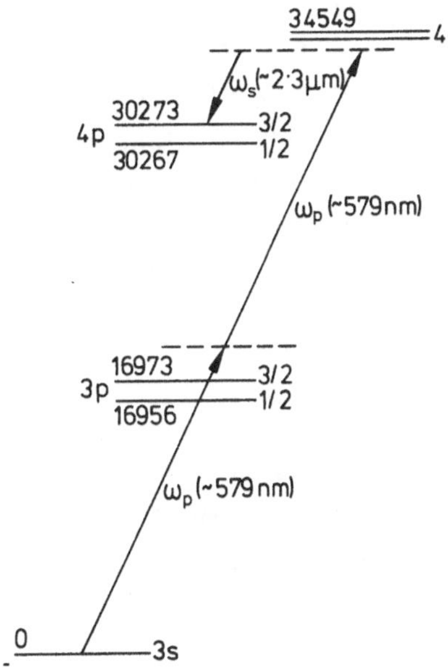

Fig.5.13. Partial energy-level diagram for sodium, showing the 3s-4$p_{3/2}$ hyper-Raman transition used for tunable infrared generation. (After [1.40])

of the single-photon resonance [$\omega_{p1} \simeq \Omega_{jg}$ in (5.22)], while $\omega_{p1} + \omega_{p2} \simeq \Omega_{ig}$. In practice, however, the single-photon resonance may lead to considerable absorption of the pump ω_{p1}, when the transition $g \rightarrow j$ is a principal line, because the large oscillator strength then increases the absorption β (5.16), both directly and by resonance broadening (5.19). Such is the case for the Na level scheme shown in Fig.5.13, where the single-photon absorption caused significant pump attenuation for detunings from the 3s-3p transitions of up to $\sim$ 100 cm^{-1} at 10 torr vapour pressure.

Strontium vapour has been used by REIF and WALTHER [5.76] to generate tunable radiation around 16μm. A rhodamine 6G dye-laser of 20kW power was used as the pump, and produced SHRS between the $(5s)^2\ ^1S_1$ ground state and $5s6p\,^1P_1$ excited state, resonantly enhanced by tuning near to the $5s5d\,^1D_2$ level. The generated power at 16μm was only 20mW, however, owing to the low laser power available; the infrared tuning range was 4 cm^{-1}.

It should be noted that under the experimental conditions appropriate to SHRS, there is also a large gain for the four-wave parametric process in which the two pump photons generate signal (ω_{sig}) and idler (ω_i) radiation, $\omega_{sig} + \omega_i = \omega_{p1} + \omega_{p2}$. This process was illustrated in Table 2.2c, and discussed in Sect.3.4.1 (BLOOM et al. [1.24] used the observation of signal and idler radiation as a simple means of verifying that $\omega_{p1} + \omega_{p2}$ was tuned into exact two-photon resonance). For definiteness, we will take ω_i to be the frequency resonant with the hyper-Raman transition, so that $\omega_{sig} \simeq \omega_{p2} - \Omega_{fg}$.

This shows that the parametrically generated frequency ω_{sig} may be close to the Stokes frequency ω_s generated by SHRS. Indeed, it is possible that a mixed 3rd order (parametric) and 5th order (nonparametric) process, similar to those discussed in Sect.3.4, may occur, which generates $\omega_s \equiv \omega_{sig}$. One difference from those processes discussed in Sect.3.4 is that now ω_i can be subject to a large single-photon absorption, because the transition $g \to f$ is allowed. This may moderate against the mixed process, and so result in ω_{sig} and ω_s being generated with different frequencies, and thus to a certain extent independently. Because in any case ω_s and ω_{sig} may be expected to have nearly equal frequencies, there is clearly some possibility of experimentally confusing the processes. A test that the observed radiation is indeed due to SHRS is the observation of a backward-travelling Stokes wave. This is because the coherence length for the backward-wave four-wave process is very short; this precludes parametric generation.

The relative gains of the hyper-Raman and parametric processes, and their degree of interaction, is likely to be affected by a number of parameters, such as pump linewidth, atomic density, and degree of focussing. Recently, HARTIG [5.35] reported that, under somewhat different experimental conditions than those of COTTER et al. [1.40], but with the same level scheme in sodium, the output was predominantly due to parametric scattering, with some evidence also of a weaker hyper-Raman scattering component.

5.5.2 Two-Photon Emission (TPE) and Anti-Stokes Stimulated Raman Scattering (ASRS)

Successful operation of a two-photon laser has been one of the more elusive prizes of quantum electronics. Such a device was proposed very early in the development of nonlinear optics [5.77,78], but has not yet been realised. The two-photon laser scheme is shown in Fig.5.14a. A population inversion on a two-photon transition $g \to f$ is first achieved, and the energy thus stored in the medium is extracted by triggering with an intense input laser at frequency ω_t. This laser induces stimulated TPE at ω_t and at the complementary frequency ω_g, such that $\omega_t + \omega_g = \Omega_{fg}$. Such a two-photon device would possess, in theory, a number of attractive properties, which have made it the subject of some interest. First, the two-photon amplifier could offer wide-band performance, because the emitted frequencies can be widely varied while still satisfying the relation $\omega_t + \omega_g = \Omega_{fg}$. Unlike SRS, where the maximum Stokes energy is limited by the energy of the pump, in TPE, once triggering has occurred, the process proceeds until the population inversion has been exhausted. Thus, the output energy is not limited by the incident

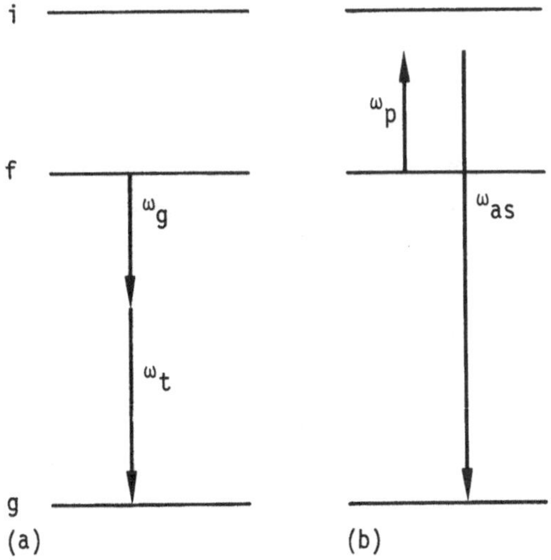

Fig.5.14a and b. Schematic representation of two-photon amplifying schemes: (a) two-photon emission, (b) anti-Stokes Raman scattering

fields, but by the energy stored in the medium. Another interesting feature of the TPE process is that, whereas in stimulated single-photon emission the energy storage is limited by spontaneous emission (ASE), in TPE schemes the energy would be stored, ideally, in a metastable level whose lifetime is not limited by single-photon emission. Thus it should be possible to store more energy per unit volume, which could be extracted when required by triggering. CARMAN [5.79] has suggested that these characteristics of TPE would be advantageous in the very large lasers needed for laser-induced thermonuclear fusion.

Despite the considerable interest over many years in the two-photon laser, the device has defied attempts to realise it [5.79-81]. The lack of success can be attributed to the conflicting requirements of the level scheme. Thus, if the two-photon gain is to be large then, ideally, the intermediate level should lie between the initial and final levels of the two-photon transition. In this way, it is possible to take advantage of considerable single-photon resonance enhancement of the TPE susceptibility χ_{TPE}. However, this also implies that the upper level f of the two-photon transition is not metastable, and that, consequently, the decay to the intermediate level would be extremely rapid (for most electric-dipole transitions). If the intermediate level is above f then χ_{TPE} will be smaller.

On the other hand, the existence of intermediate single-photon resonances above the populated level f will resonantly enhance the Raman susceptibility in such a way as to favour anti-Stokes stimulated Raman scattering (ASRS), in which f acts as the initial level (see Fig.5.14b), in preference to TPE. The principles of ASRS are identical to SERS, with the exception that, because the final level (here g) lies below the initial level, the anti-Stokes frequency ω_{as} is higher than the pump frequency ω_p, i.e., $\omega_{as} = \Omega_{fg} + \omega_p$. ASRS is simpler to achieve than TPE, and some experimental success has been had (this will be described subsequently). The process is of interest as an alternative to TPE for storing energy in an excited level f, but like ordinary Raman scattering, the energy that may be extracted is limited by the pump intensity. Therefore, the maximum energy gain over the incident pump energy is limited to ω_{as}/ω_p.

LOY [2.81] recently succeeded in using the technique of adiabatic rapid passage (ARP) to obtain population inversion on a two-photon transition in NH_3; he was thus able to give the first demonstration of two-photon stimulated gain and emission. This use of the ARP technique had been proposed earlier by GRISCHKOWSKY and LOY [2.79], (Sects.2.6.7,4.6.5). Their suggestion was that the process could be induced by the applied laser fields themselves, by using the optical Stark effect to sweep the two-photon transition frequency relative to the optical field frequencies, while satisfying the adiabatic condition. This, however, requires critical tailoring of the laser pulse shapes. LOY [2.81] used instead an external electric field to Stark-switch the energy levels of the medium (see [5.82] for a review of this technique). This applied Stark field was used to sweep the NH_3 two-photon transition frequency through the sum frequency of two pulsed CO_2 lasers. Provided that the Stark sweep rate was low enough and the laser intensity high enough, a population inversion could be achieved (analogous to single-photon adiabatic rapid passage; Fig.2.5). The measured two-photon gain was, however, rather low (~ 0.2%) because of the need to use low NH_3 gas pressures (~ 10 m torr) to avoid too rapid collision-induced relaxation of the two-photon inversion. The NH_3 system does not, therefore, look promising for a two-photon laser, although it may be that the ARP technique will be of value in the continuing search for a two-photon laser.

Although the first observation of stimulated two-photon emission has been made only recently, both spontaneous [5.83,84] and enhanced [1.29] TPE have been reported. The distinction between these regimes is most easily appreciated by considering the transition rate τ for TPE; we shall later relate this to the susceptibility approach. Let n_g and n_t be the number of photons per mode of given polarisation at frequencies ω_g and ω_t, and let ρ_f and ρ_g

be the occupation probabilities of the levels f and g. Then it is easily shown (see e.g., [2.19]) that, for a single atom,

$$\frac{dn_g}{dt} = \frac{dn_t}{dt} = -\frac{d\rho_f}{dt} \equiv \frac{1}{\tau}$$

$$= \frac{\pi}{2\varepsilon_0^2 V^2} \overline{|\alpha_{fg}(\omega_g;\omega_t)|^2} \omega_g \omega_t \{\rho_f(n_g + 1)(n_t + 1) - \rho_g n_g n_t\} \delta(\Omega_{fg} - \omega_g - \omega_t) \,, \tag{5.23}$$

where V is the quantisation volume of the radiation field. The most important part of (5.23) is that enclosed by braces. The factor $-\rho_g n_g n_t$ arises from two-photon absorption, and causes reduction of the photon numbers n_g, n_t. The factor $\rho_f(n_g + 1)(n_t + 1)$, on the other hand, corresponds to a mixture of stimulated and spontaneous TPE terms. Thus, purely spontaneous emission of the two photons comes from the part ρ_f, whereas purely stimulated TPE comes from $\rho_f n_g n_t$. The enhanced TPE region is where, for example, a triggering field is applied so that n_t is large, but $n_g = 0$, so that amplification occurs through $\rho_f n_t$. To summarise, the regimes are:

(i) $n_g, n_t \sim 0$; spontaneous TPE

(ii) $n_g \sim 0$, $n_t \gg 1$ or $n_t \sim 0$, $n_g \gg 1$; enhanced TPE

(iii) $n_g, n_t \gg 1$; stimulated TPE (or TPA if $\rho_f < \rho_g$) .

To proceed with (5.23), it is necessary to know the photon distributions, for example the number of photons n_g per mode for frequencies around ω_g. Consider the TPE amplifier, in which monochromatic radiation of frequencies ω_g and ω_t are input. The intensities are related to the photon numbers by $I_{g,t} = n_{g,t}\hbar\omega_{g,t}c/V$. We can consider the delta function to be broadened by the transition linewidth Γ, so that $\pi\delta(\Omega_{fg} - \omega_g - \omega_t) \rightarrow \Gamma/[(\Omega_{fg} - \omega_g - \omega_t)^2 + \Gamma^2]$. Thus, for exact two-photon resonance, the stimulated part of (5.23) becomes

$$\frac{\omega_g}{\omega_t}\frac{dI_t}{dz} = \frac{dI_g}{dz} = \frac{\mathcal{N}\hbar\omega_g}{\tau} = -\left\{\frac{N(\rho_g - \rho_f)\omega_g}{2\varepsilon_0^2 c^2 \hbar\Gamma} \overline{|\alpha_{fg}(\omega_g;\omega_t)|^2}\right\} I_t I_g \ . \tag{5.24}$$

The expression in braces is simply the two-photon absorption coefficient k_{TPA} (Sect.3.3) for absorption at frequency ω_g. Because, in a two-photon amplifier, the transition will be inverted, i.e., $\rho_f > \rho_g$, k_{TPA} is in fact negative. When the two-photon gain at frequency ω_g is defined as $g_{TPE} = -k_{TPA}$, (5.24) becomes

$$\frac{\omega_g}{\omega_t}\frac{dI_t}{dz} = \frac{dI_g}{dz} = g_{TPE}I_tI_g \quad . \tag{5.25}$$

Therefore, there will be gain for *both* waves and I_g, I_t will continue to grow so long as $\rho_f > \rho_g$. This is in contrast to SRS or ASRS, where the Stokes and anti-Stokes waves grow at the expense of depletion of the pump. Hence, for sufficiently high intensities, (5.25) must be accompanied by an equation for the change-of-population inversion, and thus gain. However, where little energy has been extracted, we can solve (5.25) directly. Consider the incident intensities to be such that $I_{to} \gg I_{go}$. Then, for a pathlength L,

$$I_g = I_{go}\exp(g_{TPE}I_{to}L) \quad . \tag{5.26}$$

This equation, which describes a stimulated two-photon amplifier, is analogous to the familiar Raman amplifier equations in Chap.3. Pursuing this analogy we can attempt to describe the growth from noise in the two-photon amplifier taking I_{go} to be a typical noise intensity (Sect.3.5). This amounts to using the stimulated regime (iii) in conjunction with a fictitious noise source to model the spontaneous and enhanced two photon emission regimes (i) and (ii). This simple semiclassical treatment of noise fails to give the complete picture, however, the implication from (5.26) is that growth of I_g requires population inversion ($g_{TPE} > 0; \rho_f > \rho_g$). However, by returning to the photon-number rate equations (5.23) we can see that inversion is not essential. Thus, in (5.23), the enhanced and stimulated terms may be regrouped, as $(\rho_f - \rho_g)n_gn_t + \rho_fn_t$, for $n_t \gg n_g$. The term ρ_fn_t is not accounted for in the semiclassical approach. However, if the product n_gn_t is sufficiently small, then it is possible for ρ_fn_t to exceed $(\rho_f - \rho_g)n_gn_t$. Thus, even if there is no population inversion, enhanced TPE can occur. This type of emission is, however, self-quenched as n_g, and therefore $\rho_gn_gn_t$, acting as a TPA term, grows.

Enhanced two-photon emission was demonstrated by YATSIV et al. [1.28] between the 6s and 4s states in potassium, and has also been observed from metastable deuterium atoms [5.85], excited mercury atoms [5.86] and from trivalent praseodymium ions in a crystalline host [5.87]. Recently, HARRIS [5.88] proposed that enhanced two-photon emission and enhanced anti-Stokes scattering (i.e., below the threshold for stimulated ASRS) could be used as a method of generating tunable extreme vacuum-uv radiation. He showed that the brightness of such a source can be made to approach that of a blackbody at the Boltzmann temperature characteristic of the population stored in the upper metastable level. In the experimental realisation of this device [5.89] a continuous glow discharge in He was used to produce a population in the metastable $2s^1S_0$ level at 166278 cm^{-1}. An excited population density ρ_f of about 2.6×10^{-5} was obtained, corresponding to a Boltzmann temperature of 22,700 K. The output of a mode-locked Nd:YAG laser (1.064μm) was then focussed into the discharge to produce enhanced two-photon and anti-Stokes xuv emissions from the excited level. These two outputs appeared as upper and lower sidebands $\omega_{xuv} = \Omega_{fg} \pm \omega_p$; i.e., at 56.9nm and 63.7nm. The brightness of the generated xuv radiation exceeded that of the 58.4nm resonance line by a factor of 140. In more recent work, HARRIS and co-workers [5.90] used a tunable infrared source (a $LiNbO_3$ optical parametric oscillator) to pump the two-photon medium, and thereby obtained a tunable xuv output. This radiation was used to measure the He $1s2s^1S_0$ isotopic shift. Although incoherent, the laser-pumped two-photon source possesses signficant advantages over conventional laboratory xuv sources; higher brightness, tunability, narrower bandwidths, and pulse times of the order of a picosecond.

KOMINE and BYER [5.91] have proposed the use of atomic mercury vapour as a stimulated two-photon amplifying medium (Fig.5.15), by using an inversion between the $6^1P_1^o$ and $6^3P_{0,1,2}^o$ levels. The $6^1P_1^o$ level is not truly metastable, but because it does not decay by single-photon emission to the $6^3P_{0,1,2}^o$ levels, it can be made effectively metastable by radiative trapping with respect to the ground state, at high vapour densities. KOMINE and BYER calculated the stimulated TPE and ASRS gain coefficients for this level scheme with 1.06μm and 10.6μm input wavelengths. There are a number of strong intermediate single-photon resonances, as shown in Fig.5.15, which result in calculated TPE gain coefficients g_{TPE} as large as 0.7 cm GW^{-1} for both 1.06μm and 10.6μm inputs. These calculations were made by assuming a $10^{22}m^{-3}$ inversion density and pressure-broadened transition linewidths. On the other hand, the fact that these intermediate resonances lie above the inverted $6^1P_1^o$ level means that the gains for ASRS are somewhat greater than for TPE. With a 10.6μm input

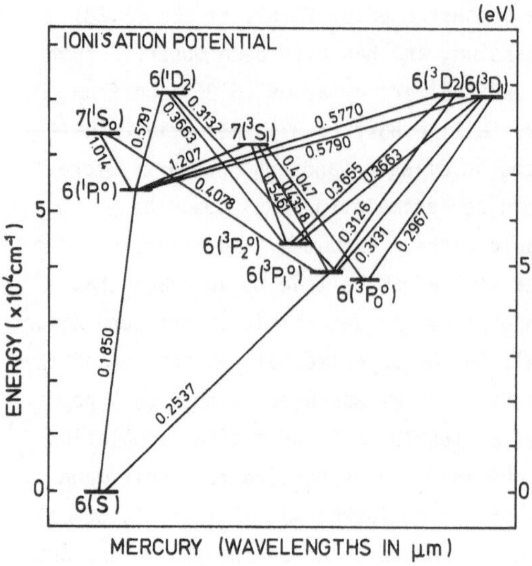

Fig.5.15. Simplified energy-level scheme for atomic mercury, showing the transitions used in forming the sum over intermediate levels for calculating the anti-Stokes and two-photon gain coefficients. The population inversion exists between the $6^1P_1^o$ level and $6^3P^o_{0,1,2}$ levels. (After [5.91])

wavelength, the calculated ASRS gain coefficient is 1.0 $cmGW^{-1}$. With a 1.06μm input, close-lying intermediate levels provide strong resonance enhancement for ASRS, and the gain coefficient in this case is as large as 65 $cmGW^{-1}$. These gains are more than adequate for efficient single-pass anti-Stokes conversion at intensities of less than $1GWcm^{-2}$. Electric discharge with subsequent selective quenching was proposed as a method for producing the required population inversion, although preliminary experiments with optical pumping (either direct optical pumping or excitation of higher-lying levels by multi-photon absorption with subsequent cascade to $6^1P_1^o$) were also suggested.

The threshold condition for sustained operation of standing- and travelling-wave two-photon lasers was analysed by YUEN [5.92]. Also, CARMAN [5.79] made a detailed analysis of the competition between TPE and ASRS in stimulated two-photon decay. He points out that initially, TPE could dominate over ASRS in certain favourable circumstances; there should be strong input radiation I_{to}, a large difference between the frequencies ω_g and ω_t when $\omega_g > \omega_t$, and some resonance enhancement of TPE over ASRS. The latter condition is possible only if the intermediate resonance levels are between the inverted (initial) level and the lower final level; as mentioned previously, this introduces the possibility of premature single-photon decay of the inverted level. This can be minimised if the matrix element between the initial and intermediate states is very much less than that between the intermediate and final states, while the product of the two matrix elements remains as large as possible. Some linear absorption at the anti-Stokes frequency ω_{as} is also beneficial in suppressing ASRS until absorption saturation occurs. A further point is that the frequencies ω_g and ω_t can take part in a parametric four-wave mixing process $2\omega_t + \omega_g = \omega_{as}$ to generate radiation at the anti-Stokes frequency, which can subsequently trigger stimulated ASRS. Thus, a large phase mismatch for this mixing process can also help to suppress ASRS, initially. However, Carman's analysis showed that even if TPE does dominate in the early stages, the parametric coupling between the waves means that ASRS will be triggered ultimately and will dominate over TPE in most practical situations, unless inversion depletion occurs early. It is clear, therefore, that the realisation of two-photon laser emission is hampered by a very large number of conflicting requirements and constraints, although the potential value of such a device inspires continuing research.

One of the first demonstrations of stimulated Raman scattering in an atomic system [1.32] was, in fact, an anti-Stokes process. Ruby-laser light, after being frequency shifted in a liquid Raman converter to within 11 cm^{-1} of the frequency of the potassium resonance line $4s\text{-}4p_{3/2}$ (as described in Sect.5.1) was passed through a heated potassium vapour cell. At the output, a new frequency component was observed, which was shifted 58 cm^{-1} higher than the input frequency. This shift corresponds to the spin-orbit splitting of the 4p states; consequently, it was inferred that the process responsible is anti-Stokes SERS, in which excited atoms initially in the $4p_{3/2}$ state make a transition to the lower $4p_{1/2}$ state. The gain coefficient for the process observed by SOROKIN et al. is resonantly enhanced because the incident frequency is close to resonance with a strongly allowed transition to an intermediate level, in this case the 4s ground state. Figure 5.2b shows the two resonance-enhancement schemes that can be used in ASRS; the experiment described here corresponds to the right-hand part of Fig.5.2b. For the ASRS process observed by SOROKIN et al. to occur, it is necessary to have a population in the initial (upper) $4p_{3/2}$ level greater than that in the final (lower) $4p_{1/2}$ level. In this experiment, the incident frequency ω_p was sufficiently near to resonance with the $4s\text{-}4p_{3/2}$ transition that a significant excited population could be created by direct optical pumping in the wings of the absorption line.

The possible use of ASRS in atomic iodine vapour to produce frequency up-conversion of high-power lasers was suggested by VINOGRADOV and YUKOV [5.93]. The scheme is based on creating an initial population inversion between the $5p_{3/2}$ ground state and metastable $5p_{1/2}$ excited state of atomic iodine. The $5p_{3/2}$ - $5p_{1/2}$ spin-orbit splitting is 7604 cm^{-1}. CARMAN and LOWDERMILK [5.94,95] obtained an inversion by flash photolysis of CF_3I, and then succeeded in observing ASRS by measuring the gain at the anti-Stokes wavelength experienced by a laser probe pulse that had passed through the inverted iodine, coincident with a 1.06μm laser pump pulse. The probe pulse was provided by the broad-band fluorescence produced by a travelling-wave rhodamine B dye-laser, the output of which spanned the anti-Stokes frequency of the 1.06μm pump. CARMAN and LOWDERMILK were thus able to measure a peak ASRS gain of e^7, which agreed closely with theoretical predictions.

6. Raman-Resonant Four-Wave Processes

6.1 Background Material

In Sect.2.5, the two kinds of two-photon resonance were described. In the "two-photon absorption" (TPA) type of resonance, the frequencies ω_1 and ω_2 of two of the incident waves satisfy the condition $\omega_1 + \omega_2 \simeq \Omega_{fg}$, where Ω_{fg} is a two-photon-allowed transition frequency. The TPA resonance enhances the third-order nonlinear susceptibility for the processes of difference-frequency generation, $\omega_4 = \omega_1 + \omega_2 - \omega_3 \simeq \Omega_{fg} - \omega_3$ (Fig.2.2c) and sum-frequency generation $\omega_4 = \omega_1 + \omega_2 + \omega_3 \simeq \Omega_{fg} + \omega_3$ (Fig.2.2d, but note that if ω_3 rather than ω_4 is generated, $\omega_3 = \omega_4 - (\omega_1 + \omega_2) \simeq \omega_4 - \Omega_{fg}$; this is more properly referred to as a form of difference-frequency generation). In Chap.4, we concentrated mainly on sum-frequency processes, of which THG is a particular case.

We now consider four-wave mixing processes that are enhanced by a Raman-type of resonance, i.e., in which the condition $\omega_1 - \omega_2 \simeq \Omega_{fg}$ is satisfied. The Raman resonance also enhances the nonlinear susceptibility for sum- and difference-frequency generation. The sum-frequency process $\omega_4 = \omega_1 - \omega_2 + \omega_3$ (Fig.2.2b) is also known as coherent anti-Stokes Raman scattering (CARS), where, in the most usual arrangements, $\omega_1 = \omega_3$ and ω_2 are input, and the anti-Stokes frequency $\omega_4 = 2\omega_1 - \omega_2$ is generated. CARS has received considerable attention over the past few years, as it greatly extends the techniques of conventional Raman spectroscopy. Because the primary interest in CARS has been of a spectroscopic nature, we shall not pursue it further here. It should, however, be noted that CARS can be used as a means of generating tunable uv radiation. In fact, a recent demonstration of multiple CARS processes in H_2 gas, with the generation of the 8th anti-Stokes frequency of a dye-laser input, underlines its potential in this direction [1.56]. A process that is closely related to CARS, from a theoretical point of view, but which is very different in its practical aspects, is when ω_3 rather than ω_4 is generated, via the process $\omega_3 = \omega_4 - (\omega_1 - \omega_2) \simeq \omega_4 - \Omega_{fg}$ (c.f. the foregoing remarks concerning TPA resonant sum mixing). This is known as "biharmonic pumping" or "coherent Raman mixing", the theory of which was discussed in Sect.3.4. The use of this process for ir generation in molecules will be described in Chap.7.

Thus, we come to Raman-resonant difference mixing, in which $\omega_4 = \omega_1 - \omega_2 - \omega_3 \simeq \Omega_{fg} - \omega_3$ (Figs.2.2a,1.2) is generated. This process was demonstrated by SOROKIN et al. [1.35] as a means of generating widely tunable coherent ir radiation, and is the main subject of this chapter. To date, complete coverage of the range 2-30µm has been achieved [5.34,5.102,6.1-3]. The scheme used originally by SOROKIN et al., in potassium vapour, is depicted in Fig.6.1. (Notice that the labelling of the ω_4, ω_3 arrows is opposite to that used in Table 2.2a and Fig.1.2. This is to facilitate a later discussion of phase-matching techniques, and does not imply that there is a favoured route for resonant denominators in the nonlinear susceptibilities. Indeed, in the alkalis this is far from the case, as will be seen in Sect.6.4.) In their experiment, SOROKIN et al. focussed collinear 1kW beams from two nitrogen-laser-pumped dye-lasers (frequencies ω_1 and ω_3) into a 30 cm column of potassium vapour at 10 torr pressure. The frequency of one of the dye lasers (ω_1) was tuned in the region of the 4s-5p resonance, and so generated radiation at frequency $\omega_2 = \omega_1 - \Omega_{5s-4s}$ by SERS. The radiation at ω_1 and ω_2 then mixed together with the second dye laser, of frequency ω_3, to generate radiation at the difference frequency $\omega_4 = \omega_1 - \omega_2 - \omega_3 = \Omega_{5s-4s} - \omega_3$. By varying the frequency ω_3, a tunable ir output ω_4 was obtained, with ω_4 tuning in the opposite way to ω_3. SOROKIN et al. were thus able to generate a tunable ir output over the wavelength range 2-4µm with a maximum output power of 100mW. A similar experiment using rubidium vapour as the nonlinear medium gave an ir output tunable between 2.9 and 5.4µm.

Before we consider this process further, it is interesting to compare the properties of the TPA and Raman resonances. A convenient technique for achieving Raman resonance, mentioned previously, is to use an intense input wave ω_1 to produce SERS (see Chap.5), thereby generating Stokes radiation at the frequency $\omega_2 = \Omega_{fg} - \omega_1$ inside the medium. As well as eliminating the need for a second laser to provide ω_2, this method automatically ensures that exact Raman resonance is achieved. Thus, the generated frequency ω_4 is given by $\omega_4 = \Omega_{fg} \pm \omega_3$, and is independent of the pump-source frequency ω_1. This means that ω_1 can be tuned to optimise the mixing efficiency, leaving the actual tuning of the generated frequency ω_4 to the source at frequency ω_3. In practice, some resonance enhancement of the Raman gain is required in order to ensure efficient generation at the Stokes frequency ω_2, so restricting the tuning of ω_1, but this is not a severe constraint and in

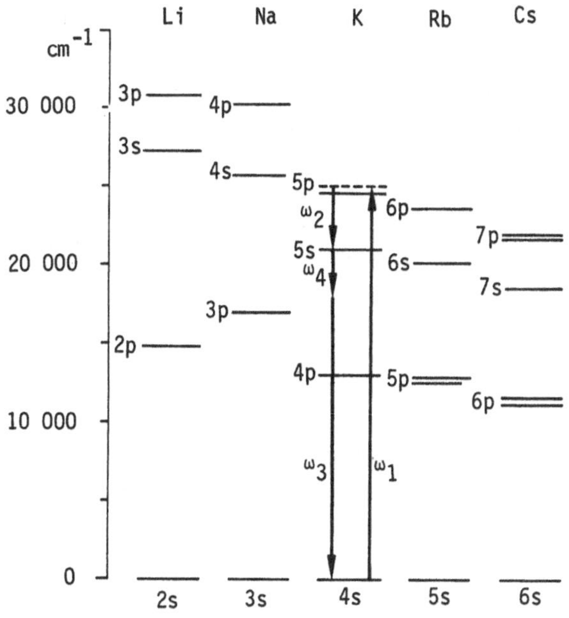

Fig.6.1. Raman-resonant difference mixing in alkali vapours. Only the lowest levels are shown, because these are of most interest for this process. The scheme is illustrated for potassium, but is similar in all of the alkalis; that is, ω_3 is tuned above the n_gp level and ω_1 is tuned near to the (n_g+1)p level, causing SERS on the n_gs - (n_g+1)s Raman transition

any case has the simultaneous benefit of enhancing the four-wave-mixing susceptibility. (These last points are a little over-simplified and will be elaborated upon later.)

If, on the other hand, the TPA type of two-photon resonance were used, then, in general, ω_4 would depend on ω_1 and ω_2, unless care was taken to maintain $\omega_1 + \omega_2$ constant and $\simeq \Omega_{fg}$. Moreover, some single-photon resonance enhancement would again be desirable, if not essential; this would ideally require the use of two tunable lasers to provide ω_1 and ω_2. However, this then loses the simplicity afforded by a single source of $\omega_1 = \omega_2 \simeq \Omega_{fg}/2$.

6.2 Raman-Resonant Difference Mixing

Because the theory follows the lines of that given in Sect.3.4 for biharmonic pumping, we shall not go through the analysis of Raman-resonant difference mixing, but rather will discuss here some of the results that are found. As mentioned in Chap.1, in this mixed parametric/nonparametric process the interplay has to be considered between SRS, involving ω_1 and ω_2, TPA involving ω_3 and ω_4, and the mixing $\omega_4 = \omega_1 - \omega_2 - \omega_3$. An immediate consequence of the coupled wave equations is that the interacting waves obey the relations:

$$\frac{I_1}{\omega_1} + \frac{I_2}{\omega_2} = \frac{I_{1o}}{\omega_1} + \frac{I_{2o}}{\omega_2} \tag{6.1a}$$

$$\frac{I_3}{\omega_3} - \frac{I_4}{\omega_4} = \frac{I_{3o}}{\omega_3} - \frac{I_{4o}}{\omega_4} \quad . \tag{6.1b}$$

(As usual, the subscripts "o" indicate the incident intensities.) The first result is the familiar expression of the fact that, in SRS, each Stokes photon is created as a result of the destruction of one pump photon, or vice versa. Thus, the maximum possible Stokes intensity is given by $I_2 = \omega_2 I_{1o}/\omega_1$, corresponding to complete depletion of the pump (the noise intensity I_{2o} may be neglected). Of particular interest here, however, is the implication of the minus sign in (6.1b). This shows that the *difference* of the photon fluxes of the beams ω_3 and ω_4 is preserved, i.e., I_3 and I_4 increase (TPE) or decrease (TPA) together. Thus, if ω_3 and ω_4 alone were input, then, for normal atomic populations (i.e., all atoms in the ground state), both the intensities I_3 and I_4 would decrease due to TPA. Let us assume that I_{3o} is a strong wave and I_{4o} is weak, so that I_4 is absorbed with a TPA coefficient $K_T \equiv k_T I_{3o}$, whereas I_3 is little changed (Sect.3.3). Naturally, if the atomic population were inverted on the two-photon transition, then the possibility of TPE would arise, with a gain coefficient equal in magnitude to K_T (Sect.5.5.2). There is of course no such inversion, but as we will now see, in the four-wave process the parametric coupling effectively ensures that the waves $I_{3,4}$ can *increase*; this led KÄRKKÄINEN [6.3] to term the process "coherent two-photon emission". (A further analogy is with coherent anti-Stokes Raman scattering.) Unlike true TPE, however, where $I_{3,4}$ can increase without bound unless depletion of the inversion is taken into account (see Sect.5.5.2), in Raman-resonant difference mixing the generated intensity is limited by a balancing of coherent two-photon emission and TPA. Atomic saturation can also limit the process; the foregoing account and that in Chap.3 takes no account of this.

It will be remembered from Sect.3.4.1 that in biharmonic pumping, for perfect phase matching, the parametric coupling caused the Stokes waves to increase exponentially, with a gain coefficient equal to $G_1 + G_2$ [see (3.35)], where G_1 and G_2 are the Raman-gain coefficients of the two Stokes waves due to their respective pumps. In Raman-resonant difference mixing, the Stokes wave I_2 and the generated wave I_4 increase as $\exp(G_R - K_T)z$, where G_R is the Raman gain coefficient for I_2 due to the pump intensity I_{1o}. This is in accord with a concept of the Raman gain of the SERS process being transferred by the parametric coupling to the waves ω_3, ω_4 which also experience TPA. The net gain coefficient is thus $G_R - K_T$; unless $G_R > K_T$, growth will not occur, a conclusion that is also borne out by a large-signal analysis similar to that in Sect.3.4.2.

The gain coefficients are more complicated when there is a phase mismatch, $\Delta k \neq 0$, where

$$\Delta k = k_1 - k_2 - k_3 - k_4 \quad . \tag{6.2}$$

Similarly to (3.32,33) I_2, $I_4 = |A_{2,4} \exp(G_+z/2) + B_{2,4} \exp(G_-z/2)|^2$, where

$$G_\pm = \frac{1}{2}(G_R - K_T) + i\Delta k \pm \left\{\left[\frac{1}{2}(G_R + K_T) + i\Delta k\right]^2 - G_R K_T\right\}^{\frac{1}{2}} \quad . \tag{6.3}$$

If G_R and $K_T \ll |\Delta k|$, then $G_\pm \simeq$ ($i\Delta k \pm i\Delta k$ + a small real term), with the result that I_2 and I_4 become increasing oscillatory functions of distance, i.e., the process is predominantly parametric in nature, with a coherence length of $|\pi/\Delta k|$ (see Sect.3.2). On the other hand, if $G_R \gg |\Delta k|$, K_T, then it is not difficult to show that the conversion efficiency is given by

$$\frac{I_4}{I_{3o}} = \left(\frac{\omega_4}{\omega_2}\right)^2 \left|\frac{\chi_T}{\chi_R}\right| \left[1 + \left(\frac{2\Delta k}{G_R}\right)^2\right]^{-1} \frac{I_2}{I_{1o}} \quad , \tag{6.4}$$

[cf. (3.36,37c)] where the TPA and SRS susceptibilities are, respectively,

$$\chi_T \equiv \chi^{(3)}(-\omega_3;-\omega_4\omega_4\omega_3) = \frac{iN}{6\hbar^3\varepsilon_o\Gamma} |\alpha_{fg}(\omega_3;\omega_4)|^2 \quad , \tag{6.5a}$$

$$\chi_R \equiv \chi^{(3)}(-\omega_2;\omega_1 -\omega_1\omega_2) = \frac{-iN}{6\hbar^3\varepsilon_o\Gamma} |\alpha_{fg}(-\omega_2;\omega_1)|^2 \quad . \tag{6.5b}$$

[We will ignore any geometrical properties here; see the discussion following (3.31).] This result was noted by KÄRKKÄINEN [6.3]; note, however, that his formula contains a typographical error. The parametric susceptibility may be written

$$\chi_P \equiv \chi^{(3)}(-\omega_4;-\omega_3\omega_1-\omega_2) = i(\chi_T\chi_R)^{\frac{1}{2}} \quad , \tag{6.6}$$

and so

$$\left|\frac{\chi_T}{\chi_R}\right| = \left|\frac{\chi_P}{\chi_R}\right|^2 = \left|\frac{\alpha_{fg}(\omega_3;\omega_4)}{\alpha_{fg}(-\omega_2;\omega_1)}\right|^2 \quad , \tag{6.7}$$

demonstrating that the ratio of susceptibilities in (6.4) is independent of both the atomic number density and the two-photon transition linewidth. For all of the experiments discussed in this chapter, where the ratio

$$R = \left|\frac{\chi_T}{\chi_R}\frac{\omega_3\omega_4}{\omega_1\omega_2}\right|^{\frac{1}{2}}$$

is small, and $G_R >> K_T$, it can be shown that (6.4) is approximately correct, even in the large-signal region, for $\Delta k = 0$ (see Sect.3.4.2).

Equation (6.4) is deceptively simple in appearance, because the factors in it are closely related, making less obvious the optimum conditions for maximum conversion efficiency. It must also be remembered that (6.4) is, strictly, a small-signal result. At this point, we will merely indicate the considerations involved, leaving fuller discussion to Sect.6.4.

The first point to be made is that if the Raman gain is sufficiently great, so that $\Delta k/G_R$ is small, there is no need to achieve exact phase-matching, because then (6.4) applies, making the conversion efficiency insensitive to the precise value of Δk. This point was not appreciated in the earliest experiments. Nevertheless, the efficiency is maximised when $\Delta k = 0$. If, on the other hand, the ratio $|\Delta k/G_R| > 1$, then (6.4) does not apply; we must refer to (6.3). Ultimately, if $|\Delta k/G_R| >> 1$ then the concept of coherence length becomes important again, making the efficiency very sensitive to Δk, and much less than (6.4) would give.

The conversion ratio I_2/I_{1o} from pump to Stokes is usually found by experimental measurement, because then the effects of competing processes, atomic saturation and so on, which are difficult to quantify, may be partially allowed for. However, in using this ratio I_2/I_{1o}, we must not lose sight of the fact that it implies a Raman gain sufficient to reach threshold, and so large G_R; this is seen to be more important when it is recalled that for small Δk, the effective gain coefficient is only $G_R - K_T$. It is seen that a further requirement is that K_T should be small.

However, these points appear to conflict with the requirement that $|\chi_T/\chi_R|$ be large. The appearance of this ratio in (6.4) is understood more readily if its alternative form (6.7) is used, $|\chi_P/\chi_R|^2$. Increasing χ_P increases the parametric coupling and, therefore, the transfer of the Raman gain to the generated wave, and so improves the conversion efficiency. Also, the generated intensity at any point is mainly due to the contribution from the previous length G_R^{-1} [more properly, $(G_R - K_T)^{-1}$], this being the distance over which the waves grow appreciably. Increasing χ_R (and, therefore, G_R) thus means that the parametric coupling has a shorter region in which to act, which reduces the conversion efficiency.

A suitable procedure (see Fig.6.2) would appear to be to maximise the ratio $|\chi_T/\chi_R|$ (by an appropriate choice of optical frequencies), whilst using a high enough pump intensity I_{1o} to reach SERS threshold, despite the low χ_R. I_{3o} should then be increased to the point at which the TPA coefficient approaches the Raman gain. I_4 should thus increase with χ_T and I_{3o} until this point, when it will saturate and then decrease. This is confirmed by a theoretical large-signal analysis [3.8]. In practice, one is not likely to have complete control over all of these factors, as will be seen when results for the alkali atoms are discussed in Sect.6.4; for example, in the alkalis the ratio $|\chi_T/\chi_R|$ lies in the range 0.01 - 0.1 for frequencies of interest.

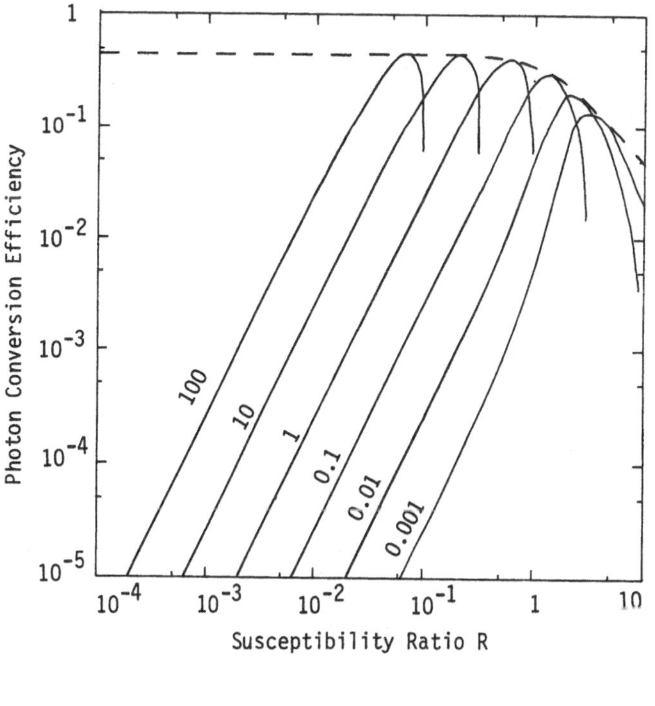

Fig.6.2. Raman-resonant difference mixing. Photon-conversion efficiency, $(\omega_1/\omega_4)I_4/I_{1o}$, versus R (see text), for perfect phase-matching. The parameter on each curve is $(\omega_1/\omega_3)I_{3o}/I_{1o}$. The dotted line represents the locus of maximum efficiency

6.3 Phase Matching

As was discussed in the previous section, for sufficiently large values of the Raman gain, such that $|\Delta k/G_R| << 1$, the precise value of Δk is not important, although it remains the case that $\Delta k = 0$ maximises the conversion efficiency (in the plane-wave limit). However, when K_T approaches G_R, as appears desirable in an optimised process, the dependence on Δk is more pronounced. We now look at the question of how Δk may be adjusted in practice.

In general, the expression (6.2) for the phase mismatch must be replaced by a vector relation, from which it follows that phase matching could be achieved by choosing appropriate angles between the wave vectors $\underline{k}_i$. However, this reduces the spatial overlap of the waves, and hence the mixing efficiency. Ideally, therefore, a method of satisfying $\underline{\Delta k} = \underline{0}$ for collinear waves is sought, in which case (6.2) reduces to

$$c\Delta k = \omega_1\Delta n_1 - \omega_2\Delta n_2 - \omega_3\Delta n_3 - \omega_4\Delta n_4 = 0 \quad , \tag{6.8}$$

where $\Delta n_1 = n_1 - 1 \simeq \chi^{(1)}(-\omega_1;\omega_1)/2$, etc. (To simplify writing, the linear refractive index has been taken to be real.) Formulae for Δn have been given in (2.33) and Sect.4.4.1, both in terms of oscillator strengths and dipole-moment matrix elements. Because each of the indices Δn in (6.8) is proportional to the atomic number density, it follows that for exact phase matching ($\Delta k = 0$), the required combination of frequencies is independent of vapour pressure. This holds quite generally for all parametric processes, but of course is not true if mixtures of gases of differing partial pressures are used.

In the experiments described in Sect.6.1, SOROKIN et al. [1.35] took advantage of the fact that the generated frequency is independent of ω_1. By tuning ω_1 in the region of the 4s-5p resonances, they were able to use the refractive-index dispersion associated with these resonances to achieve phase matching. Thus, because ω_1 is in a strongly dispersive region, Δn_1 is rapidly varying, and the large changes of $\omega_1\Delta n_1$ as ω_1 is varied may be sufficient to fulfil (6.8) over a range of generated frequencies ω_4.

The technique of phase-matching nonlinear mixing processes by adjusting the interacting frequencies in a region of anomalous dispersion has been touched on in Sect.4.4.3; in particular, the recent elaboration of this idea by BJORKLUND et al. [4.54,55] was described. They considered all possible TPA and Raman-resonant four-wave mixing processes, assuming a three-level atom, i.e., one with a ground state g connected to an intermediate state by a single-photon transition and a final (two-photon resonant) level f. They showed that, in this case, for Raman-resonant difference mixing phase matching is always possible by means of a suitable choice of the frequency ω_1. In practice, however, substantial contributions to Δk may come from many states, e.g., when ω_1 and ω_4 are each single-photon resonant with different states. Then it may not be possible to phase match over all of the tuning range for ω_3 by adjusting ω_1 alone.

An example of the dependence of the phase mismatch Δk on the laser frequency ω_1 is shown in Fig.6.3. The depicted curves have been calculated for cesium vapour (at 10 torr, the actual value of Δk being proportional to number density), with ω_1 tuned in the vicinity of the 6s-7p doublet, but the general shape of the curves is common to all of the alkali metals. It may be helpful to refer to Fig.6.1 in the following discussion. This shows the lowest levels of all the alkalis, with the Raman-resonant difference mixing scheme drawn for potassium; the analogous levels in the other alkalis could equally well have been used.

In Fig.6.3, it can be seen that for a given value of the generated frequency ω_4, two values of the pump frequency ω_1 will give exact phase matching ($\Delta k = 0$); one of these lies above the principal doublet 6s-7p, and the other is between the doublet lines. Phase matching with ω_1 tuned below the doublet is not possible. This can be understood by first noting that ω_2 and ω_4 are, in this case, in the infrared, so that the contributions to Δk (6.8) from the products $\omega_2 \Delta\eta_2$ and $\omega_4 \Delta\eta_4$ are small and may be neglected. This gives the approximate phase-matching condition,

$$\omega_1 \Delta\eta_1 \simeq \omega_3 \Delta\eta_3 \quad . \tag{6.9}$$

To generate wavelengths longer than 1.5μm ($\omega_4 < 6700\ \mathrm{cm}^{-1}$) using the 6s-7s Raman resonance in caesium, ω_3 is greater than the frequency of the principal resonance transition 6s-6p, and so $\Delta\eta_3$ is negative. Thus from (6.9), phase matching is only possible if $\Delta\eta_1$ is also negative; this can be the case only if ω_1 is tuned above or between the 6s-7p doublet lines.

As will be shown later, it is found that the efficiency for generation of ω_4 is low when ω_1 is tuned between the doublet lines; therefore, normally, the best choice of ω_1 is the higher of the two possible values for perfect phase matching. However, for the latter choice, Fig.6.3 shows that the longer the wavelength of the generated ir (ω_4) the further must ω_1 be tuned from the 6s - 7p resonances. This is more clearly depicted in Fig.6.4,

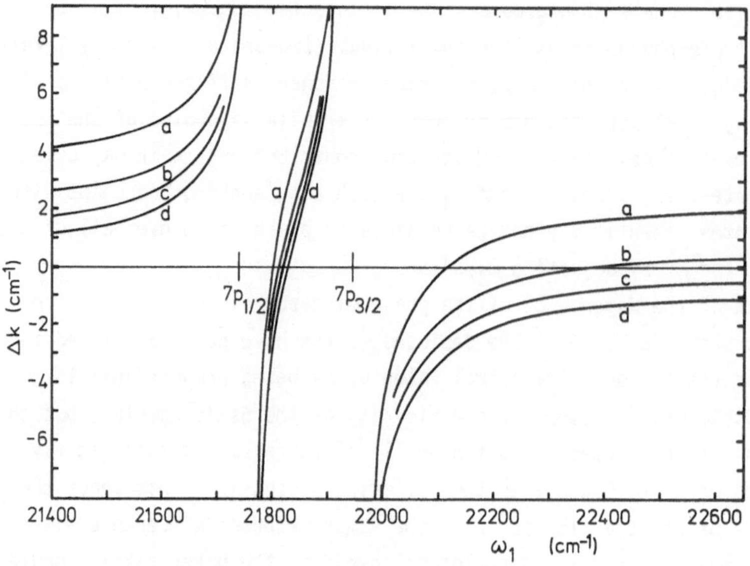

Fig.6.3. Phase mismatch Δk calculated as a function of dye-laser frequency ω_1, for various values of the generated ir frequency ω_4 (a: 4000 cm^{-1}; b: 3000 cm^{-1}; c: 2000 cm^{-1}; d: 500 cm^{-1}). These calculations were made using (6.8,2.110) (taking data from [2.116]) for the case of four-wave difference mixing in Cs vapour (10 torr pressure) using the 6s-7s Raman resonance

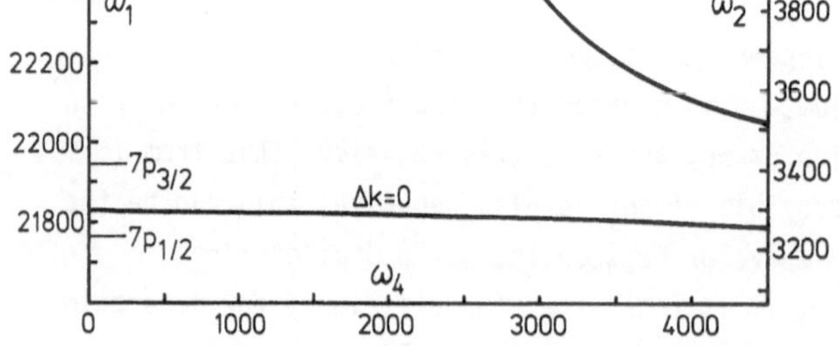

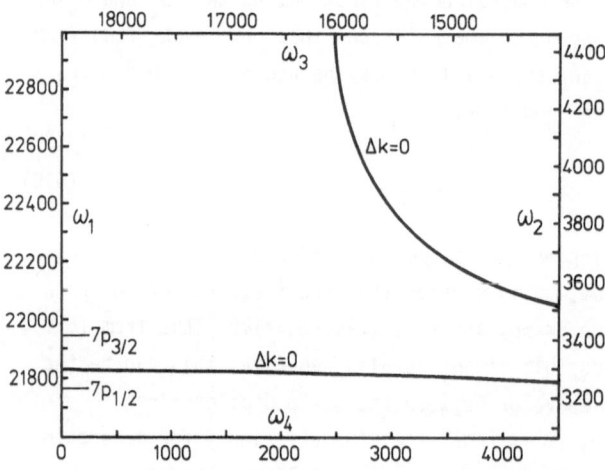

Fig.6.4. Calculated phase matching curves for four-wave difference mixing in Cs vapour (6s-7s Raman resonance). All frequencies are in cm^{-1}

which shows the calculated phase matching curves as functions of the frequencies of all four waves (these curves are of course independent of the atomic number density). It can be seen that, for ω_4 less than ~ 2500cm^{-1} (wavelengths longer than ~ 4µm), it is necessary to tune ω_1 well away from resonance in order to maintain phase matching. In fact, the point is quickly reached at which the dye-laser frequency ω_1 is so far from resonance that the Raman gain becomes too small for the generation of a sufficiently strong Stokes wave ω_2 by SERS; consequently, an upper limit is placed on the frequency ω_1 that can be used. In their early experiments using pure potassium vapour as the nonlinear medium, SOROKIN et al. [1.35] tuned ω_1 above the 4s-5p resonance doublet in order to maintain phase matching. They found that they were unable to generate a detectable infrared power for wavelengths longer than 4µm.

In later experiments, WYNNE et al. [6.1] demonstrated a different phase-matching method that overcame this shortcoming by allowing the dye-laser ω_1 to be held fixed at the frequency that gives the most efficient generation of ω_2. They adjusted the dispersion of the vapour by mixing it with the vapour of a second (or "foreign") alkali metal in a controlled ratio. (Similar techniques, using a mixture of an inert gas and alkali vapour or a mixture of two metal vapours, have been described in Sect.4.4 in the case of vuv generation.) Because the nonlinear process relies on resonance of ω_1 with a transition in the original vapour, and because the energy levels of the various alkalis differ considerably, it is generally true that the nonlinear optical properties of the original vapour are negligibly affected by the foreign vapour. For the same reason, the refractive index at ω_1 is not greatly altered. On the other hand (see Fig.6.1), as ω_4 is tuned into the infrared, the frequency ω_3 will lie above the principal resonance lines of any of the alkalis, and since these lines are strong, the foreign vapour can make a large negative contribution to the refractive-index change Δn_3 (the similar contributions to $\Delta n_{2,4}$ are not so important to the calculation of Δk, as discussed previously). In this way, the relation (6.9) [or (6.8)] can be satisfied even when the detuning of ω_1 from resonance is quite small (i.e., where $\omega_1 \Delta n_1$ is large and negative). It is clear that the further ω_4 is tuned into the infrared, the higher must the principal transition of the foreign vapour be, if phase matching is to be achieved with reasonable mixture ratios. This indicates that the foreign alkali should be lighter than the original. Actually, as can be seen from Fig.6.1, the anomalously high levels of sodium make this an ideal candidate for adding to any of the other vapours.

An example of this is shown in Fig.6.5, which demonstrates the effect on the index of refraction of potassium vapour when sodium vapour at an equal number density is added (10^{17} atoms cm^{-3} each). It can be seen from this that in the region 17000 - 21000 cm^{-1}, i.e., for the range of ω_3 frequencies lying above the 3s-3p sodium D line, the refractive index is more negative than for the potassium vapour alone, as is obvious from inspection of Fig.6.1. Thus, by suitable choice of the ratio Na:K, phase matching can be achieved

In order to establish a homogeneous vapour mixture of sodium and potassium with independent control of the sodium and potassium pressures, WYNNE et al. used a concentric heat-pipe oven (see Sect.4.4.2). The outer pipe was charged with pure sodium and the inner pipe held the mixture of potassium and sodium vapours. With this arrangement, they were able to generate radiation that was tunable over the complete range from 2-30μm. A number of different dyes were needed for ω_3 to cover this range. Rather than vary the Na:K ratio continuously as ω_3 was varied, the procedure adopted was to fix a new Na:K ratio for each dye, and to tolerate the remaining deviations from perfect phase matching. By using dye-laser powers of 1-2kW, WYNNE et al. [6.1,5.102] were able to obtain powers of up to 100mW at 2μm, but falling to around 100μW at 25μm. The longest wavelength that could be detected (30μm) was limited by the response of the Cu:Ge detector used for these experiments.

Before leaving this discussion of phase matching techniques, it should be recalled (see Sect.4.2) that the effect of focussing is to modify the plane-wave phase matching condition $\Delta k = 0$. Depending on the dispersion of the medium, the phase shift produced by focussing can be used to increase the mixing efficiency by compensating for Δk. BJORKLUND [3.1] examined the effects of focussing on the phase matching of a variety of four-wave processes including the parametric process $\omega_4 = \omega_1 - \omega_2 - \omega_3$. However, the theory was developed for the case in which the frequencies ω_1, ω_2 and ω_3 are all applied to the medium as input beams, whereas in the process being discussed here, ω_2 is generated inside the medium by SERS. As we have seen, this considerably modifies the phase matching and coupling behaviour of the waves. Further restrictive assumptions in Bjorklund's treatment which are not justified in this case are that the input beams have identical waist locations and identical confocal-beam parameters. Despite the fact that Bjorklund's analysis does not provide results that can be used quantitatively here, there remains the important qualitative conclusion that the optimum value of Δk may be nonzero for tightly focussed beams.

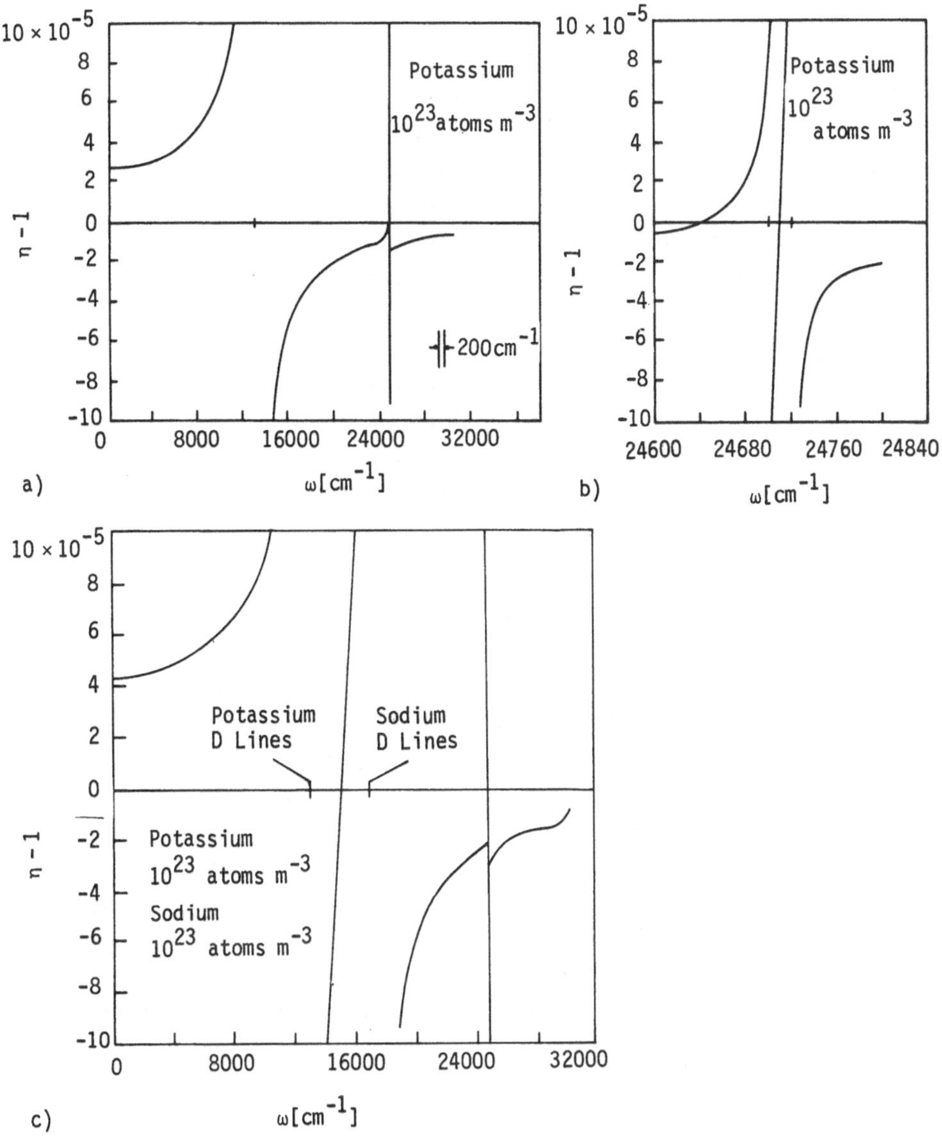

Fig.6.5. Index of refraction vs frequency for pure K vapour (a). Curve (b) shows an expanded view of the 200 cm^{-1} region around the 5p resonances. Curve (c) shows the index of refraction for a mixture of equal parts of sodium and potassium. All of the curves are drawn for the case of no damping. (After [6.1])

6.4 Raman-Resonant Difference-Frequency Generation in Alkali Vapours

It can be seen from Fig.6.3 that on the 6s-7s Raman resonance in cesium, the phase mismatch will be at most 3-4cm^{-1} (at 10 torr) over a wide range of the generated frequency ω_4 when the pump laser ω_1 is tuned to 22050 cm^{-1}, a frequency that gives efficient SERS generation of ω_2. This suggests that it should be possible to generate radiation out to long ir wavelengths without resorting to deliberate attempts to phase match. Experimental evidence for this was given by KÄRKKÄINEN [6.3] who succeeded in generating an ir output that was continuously tunable from 3.5-20μm. This was achieved by use of a simple heat-pipe oven that contained pure caesium vapour. KÄRKKÄINEN used dye-laser powers (6kW and 17kW for ω_1 and ω_3, respectively) that were an order of magnitude greater than those used by WYNNE et al. [6.1]; the measured peak output power of 2mW at 20μm was also rather more than an order of magnitude greater.

These results, in themselves, are not sufficient to test the theoretical predictions of Sect.6.2. However, KÄRKKÄINEN also investigated how the efficiency of the mixing process depends on the pump frequency ω_1. Some results that he obtained using the 4s-5s Raman resonance in potassium vapour are shown in Figs.6.6,7. The calculated values for the phase mismatch Δk, similar to those for cesium shown in Fig.6.3, indicate that, over the range of pump frequencies in Figs.6.6,7, the coherence length varies between fractions of a millimetre and infinity (Δk = tens of cm^{-1} to 0). Yet, the ir output power was always observed to be a smooth function of ω_1, with one or, at most, two maxima. This should be compared with the behaviour observed by YOUNG et al. [1.16] for third-harmonic generation (Fig.4.21), in which the generated output varied periodically with the number of coherence lengths. Despite a careful search, these "Maker's fringes" were not observed in Kärkkäinen's four-wave-mixing experiments. This does provide some evidence to support the idea that, in this resonant process, the phase matching dependence is fundamentally altered from that of a purely parametric process. Some dependence on Δk is, however, evident in Fig.6.7; the shift of the maxima further away from the resonance lines when longer ir wavelengths are generated is consistent with the shift of the phase matching point.

The optimum choice of ω_1 is influenced by a number of factors, the most obvious being the need to optimise the SERS generation efficiency and to at least minimise the phase mismatch. As we have seen the latter constraint need not be a severe one, in theory or practice. In Sect.6.2, it was shown to be desirable to have a large ratio $|\chi_T/\chi_R|$. This implies that ω_1 should

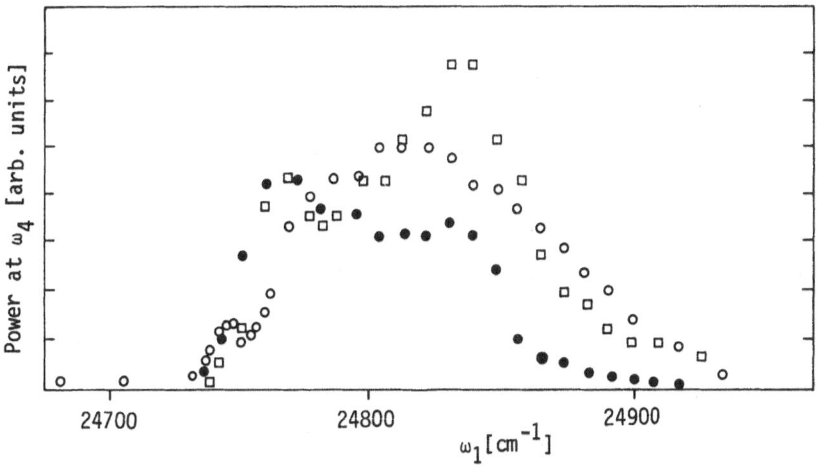

Fig.6.6. Generated ir signal (at λ = 2.4μm) as a function of pump-laser frequency ω_1 at different potassium vapour pressures (open circles: 4.5 torr, squares: 13 torr, filled circles: 20 torr). (After [6.3])

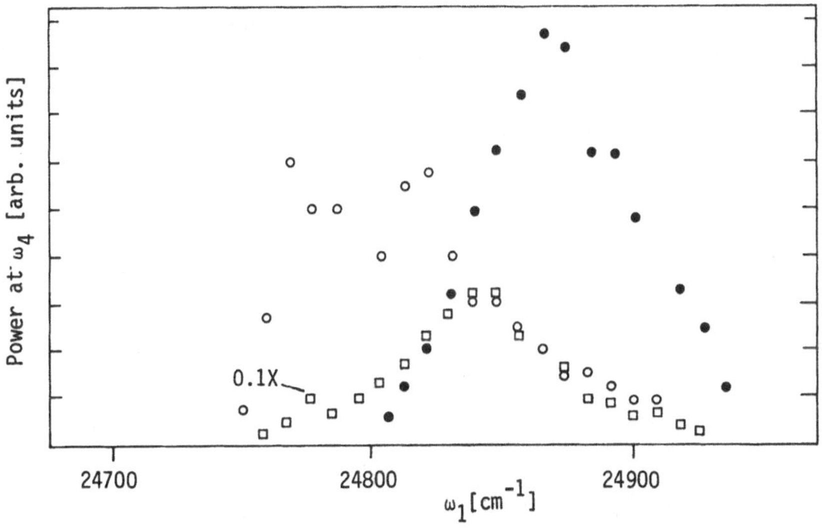

Fig.6.7. Generated ir signal as a function of pump-laser frequency ω_1 at potassium pressure of 10 torr for different output wavelengths (open circles: 1.92μm, squares: 2.27μm, filled circles: 3.5μm). (After [6.3])

be detuned from resonance, as far as is consistent with the requirement of having enough gain to reach Raman threshold.

The calculation of this ratio is straightforward; it reduces to the calculation of the transition polarisabilities $\alpha_{fg}(\omega_3;\omega_4)$ and $\alpha_{fg}(-\omega_2;\omega_1)$, as shown in (6.4,5). However, because χ_R has already been given in Fig.5.7 for the 6s-7s Raman transition in cesium, we have calculated χ_T for the same transition, Fig.6.8, from which the ratio may be deduced. It is seen that χ_T is a relatively slowly varying function of ω_3 or ω_4, since, in the frequency range of interest, there are no strong resonances. It is instructive to consider this point in more detail. We have

$$\chi_T = \frac{iN}{6\hbar^3\varepsilon_0\Gamma}\left|\sum_i \mu_{fi}\mu_{ig}\left(\frac{1}{\Omega_{ig}-\omega_3}+\frac{1}{\Omega_{ig}-\omega_4}\right)\right|^2 \quad , \tag{6.10}$$

where g = 6s, f = 7s and i = 6p,7p... (The spin-orbit splitting may be neglected; as usual, the light is taken to be parallel linearly polarised.) It can be seen from Fig.6.1 that for the range of ω_3, ω_4 shown in Fig.6.8, the denominators $\Omega_{6p}-\omega_3$ and $\Omega_{6p}-\omega_4$ are of opposite sign and not greatly different. There is thus a degree of cancellation, which nullifies the strength of this principal-transition intermediate state (i.e., a large μ_{6s6p}) to the extent that the contribution from the 7p intermediate state is of equal importance. This illustrates that the ordering of the ω_3 and ω_4 arrows in Fig.6.1 has no significance here, in the sense of a dominant route for resonant denominators, in direct contrast to the resonant Raman susceptibility χ_R. It is, moreover, the reason why χ_T is comparatively small. From Figs.5.7,6.8, we deduce that $|\chi_T/\chi_R|$ lies in the range 0.01 - 0.1, a figure that is expected to apply to the other alkalis. This appears to be a serious limitation on the generation efficiency for these transitions in the alkalis.

Another reason why it may not be suitable to work with ω_1 too close to the intermediate resonances is that there may be appreciable attenuation of the pump-laser beam and thus reduction of the Raman-scattering efficiency, due to single-photon absorption and other resonantly enhanced competing processes (see Sect.5.4). Also, it may be necessary to avoid tuning ω_1 in the region of alkali molecular-absorption bands, although, as explained in Sect.5.4, this difficulty can possibly be overcome by using sufficiently powerful dye lasers that are capable of bleaching this absorption.

A situation where a number of these factors conflict is when ω_1 is tuned between the intermediate resonance doublet. By choosing the appropriate frequency between the doublet, Δk can be made zero (cf. Fig.6.4), and yet, when SOROKIN et al. [1.35] tried to observe an ir output generated in this way

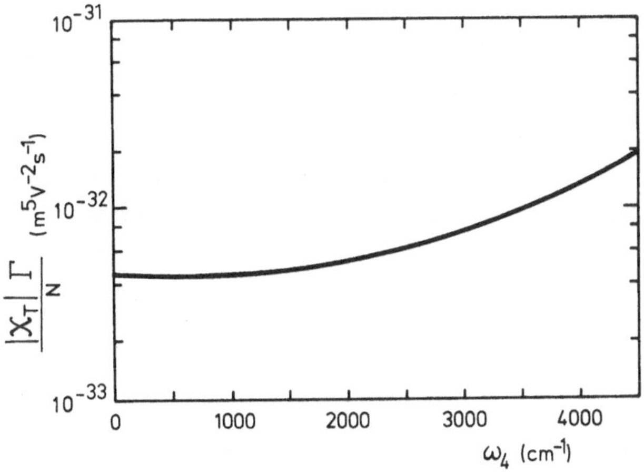

Fig.6.8. Calculated magnitude of the two-photon absorption susceptibility χ_T (normalised with respect to number density and transition linewidth) for cesium vapour with $\omega_3 + \omega_4 = \omega_{6s7s}$

using potassium vapour they were unsuccessful. KÄRKKÄINEN [5.34] later observed an output, although it was very weak. The spin-orbit splitting of the 5p levels in potassium is relatively small (19 cm^{-1}); hence there is considerable absorption of the pump beam ω_1 throughout this region. Cesium appeared more promising, from this point of view, because the splitting is much larger (188 cm^{-1}); indeed, from Fig.6.4 it can be seen that phase matching for the generation of a very wide range of ir wavelengths should be obtained for pump frequencies ω_1 that lie close to a maximum of the SERS efficiency. However, KÄRKKÄINEN found that the generated ir power was significantly less here than with ω_1 tuned above the resonance doublet; this result follows from the fact that the ratio $|\chi_T/\chi_R|$ is less favourable. A complicating factor with regard to the latter statement is the polarisation dependence of the generated waves. This requires a study of the angular-correlation factor (cf. 3.31), which we do not give here.

To date, the longest reported ir wavelength generated using this four-wave-mixing technique is 30μm [5.102]. Only a very small increase of the frequency ω_3 would be required to generate radiation out to much longer ir and far-ir wavelengths. However, the generation efficiency is expected to decrease rapidly, partly because the generated ir power decreases with its frequency ω_4 [cf. (6.3)] and partly because of the increasing losses due to diffraction (cf. the question of ir diffraction loss discussed in Sect.5.3.2). The maximum ir power that has been reported is 9.5W; it was obtained by TYLER et al. [6.2] at a wavelength of 2.2μm using the 4s-5s Raman transition in potassium vapour. For this maximum ir output, the dye-laser powers were 2.6kW and 40kW for ω_1 and ω_3, respectively. The two dye lasers used in the experiment were pumped by a single nitrogen laser in such a way that the uv power apportioned to each dye laser could be varied. By experimenting with different power ratios, TYLER et al. observed an interesting saturation effect. They found that, when the power of laser ω_1 was increased above a certain optimum value, no further increase of ir output was obtained but, at the same time, strong amplified spontaneous emission (ASE) commenced at 1.25μm, the wavelength corresponding to a 5s → 4p transition. For the conditions of their experiment, the optimum value for the power at ω_1 was 2.6kW. On the other hand, with the power at ω_1 set at this optimum level, the power of the laser ω_3 could be increased up to the maximum available (~ 40kW) and the ir output was observed to increase linearly without saturating. TYLER et al. explained this effect by calculating the nonlinear polarisation, using time-dependent perturbation theory to include the effect of the ASE. This calculation showed that, at the critical intensity for ω_1 at which ASE commences, the polarisation at frequency ω_4 is unaffected by further increases of power at ω_1 or ω_2, but remains linearly dependent on the power at ω_3. The significance of these observations by TYLER et al. remains unclear, however. In a similar experiment, KÄRKKÄINEN [6.3] found that as he increased the power of his dye laser ω_1 up to the maximum available (~ 15kW) the ir output power increased correspondingly, without any sign of saturation. This is consistent with the theory presented in Sect.6.2. Another effect that could cause the ir output to saturate is that of atom depletion, as described in Chaps.4,5.

In this section, we have discussed the major features of the Raman-resonant difference-mixing process. However, it can be seen that a detailed understanding of the saturation behaviour is still some way off.

We conclude by briefly assessing this process as a method of tunable ir generation for spectroscopic applications, although for the reason just given an assessment must be based on the few experimental results that have so far been obtained. The greatest attraction of this technique is the very wide range of wavelengths that has been covered (2-30μm). This range can be matched only by the tuning coverage of semiconductor diode lasers and three-wave difference mixing in nonlinear crystals [1.8]. However, in these cases, several different diodes or nonlinear crystals are required to cover completely the medium-ir range, whereas by four-wave difference mixing in a single vapour, cesium, continuous coverage of the range 3.5-20μm has been obtained. Moreover, these diodes and crystals and their ancillary equipment are generally very costly, compared to using an atomic vapour as the active medium. However, a major disadvantage of this mixing technique for spectroscopy is the rather low-power output generated so far, particularly at longer wavelengths. Obviously, any technique that relies on converting visible photons to infrared photons must suffer this penalty. In fact, these power levels are often several orders of magnitude less than various accompanying fixed-frequency ASE outputs; this poses further difficulties. Also, the need to align two pulsed-dye-laser beams so that they remain overlapping and collinear over several-centimetre columns of atomic vapour demands a degree of skill and patience on the part of the experimenter which is not required in the case of SERS, for example. A further important aspect that has received very little attention is the question of the linewidth of the ir output. So far as we know, this has not yet been measured. In the earliest investigations of this process it was assumed that if the linewidths of the dye lasers were greater than the Doppler width of the Raman transition involved, then the generated linewidth would be determined by that of the dye lasers. Because the linewidths of the dye lasers used in the experiments have typically been 0.1 cm^{-1} compared with Doppler widths of ~ 0.03-0.05 cm^{-1}, it was thought that the ir output should also be ~ 0.1 cm^{-1} wide. However, COTTER et al. [1.39,40] found that the spectral width of the Stokes radiation ω_2 is always considerably broader than that of the pump ω_1 (see Sect.5.4.2). It would, therefore, not be surprising if a comparable broadening of the generated ir ω_4 occurred. This would impose a further limitation on spectroscopie applications.

7. Nonlinear Optical Processes in Free Molecules

7.1 Background Material

In reviewing the experimental results, we have so far concentrated on atomic vapours as the nonlinear medium. The important role played by two-photon resonance enhancement has been emphasised; in atoms such as the alkalis or alkaline earths, the two-photon resonance can be conveniently achieved by use of dye lasers. By contrast, the interesting feature of molecules as nonlinear media is that, owing to their vibrational and rotational motion, they have resonances in the infrared region. As in the case of atoms, these resonances can be used in the form of a Raman-type resonance, with incident frequencies ω_1, ω_2 such that $\omega_1 - \omega_2 = \Omega_{fg}$, or in the form of TPA resonance, with $\omega_1 + \omega_2 = \Omega_{fg}$, where, as usual, g and f are the initial and final states of the two-photon transition.

Nonlinear effects due to the Raman type of resonance are of considerable antiquity, with stimulated Raman scattering being first observed in liquid nitrobenzene by WOODBURY and NG [1.26]. Gaseous hydrogen (H_2) soon became a popular medium for detailed studies of the behaviour of stimulated Raman scattering, because it was free from a number of competing processes, such as stimulated Brillouin scattering and self focussing, which were prevalent in many organic liquids. Reviews of this work have been given by BLOEMBERGEN [5.2], KAISER and MAIER [5.3] and MAIER [2.102]. Besides the attraction of being free from other nonlinear processes, hydrogen gas also offers a large vibrational Raman shift (4155 cm^{-1} for the Q branch of $v = 0 \rightarrow v = 1$). Thus, over the years, a number of experiments have examined the possibility of using this shift as a practical and efficient means of down-shifting the frequency of a laser. Using a high-power tunable pump, such as a dye laser, extensive ranges of the infrared spectrum have been covered by generating first, second and third Stokes radiation [1.47-50,54]. Compared with SERS in atoms, the threshold for SRS in H_2 is much higher, because the pump laser is a long way from resonance with the intermediate levels (which are the excited electronic states). On the other hand, for the same photon-conversion efficiency, the energy efficiency of Raman scattering in molecules is better than in atoms, because the molecular Stokes shift (i.e., wasted energy) is less. An added impetus to these attempts at generating tunable infrared radiation has been the much-publicised need for an efficient laser source around 16μm for uranium-isotope separation [7.1]. FREY et al. [7.3] have shown that using a powerful pump beam of wavelength around 3.4μm, the required 16μm output can be generated efficiently by stimulated Raman scattering in liquid N_2. In an interesting proposal, BYER [7.2] suggested an approach that involves stimulated pure-rotational Raman scattering in hydrogen gas, using a CO_2 laser pump. Two problems that confront schemes to generate such long wavelengths by Raman scattering are the inverse relation between Raman gain and Stokes wavelength (see Sects.3.3,5.3) and the large diffraction loss suffered by the Stokes wave (see Sect.5.3.2). One way to overcome the problem of insufficient gain is to use a biharmonic-pumping scheme (Sect.3.4) rather than stimulated Raman scattering alone. Adapting the Byer proposal in this way, LOY et al. [1.63] and SOROKIN et al. [1.64] succeeded in generating ~16μm radiation, although at a disappointingly low efficiency. More recently, BYER and TRUTNA [1.65] used biharmonic pumping to generate some "seed" radiation at 16μm, which was subsequently amplified by SRS; RABINOWITZ et al. [1.52] succeeded in generating significant 16μm output, by SRS from spontaneous Raman-scattered noise. Both schemes used multi-pass arrangements. Further experimental details on this are given in Sect.7.5. Although the foregoing remarks may appear to suggest that it would be out of the question to get to yet longer wavelengths by Raman scattering, in fact an interesting use of resonance enhancement has permitted generation of Stokes radiation at far-infrared wavelengths, [1.71,7.3,4]. This scheme will be discussed further in Sect.7.6.

The first suggestion for using resonance enhancement of the nonlinear susceptibility in molecules appears to have been made by UEDA and SHIMODA [2.17]. The process they considered was third-harmonic generation. There is a gap of several years before any report of experimental investigations of this scheme [7.5,6]. This hiatus is a little surprising, because an appropriate high-power laser source was available in the CO_2 laser (~ 10μm), with sufficient tunability via step tuning to achieve close two-photon resonance in a number of different molecules. Furthermore, third-harmonic generation of CO_2 laser radiation was a subject of some interest, and nonlinear crystals were being investigated for this purpose. In fact, the best results obtained in crystals appear to be a harmonic-generation efficiency of ~ 10^{-6} in a crystal of $CdGeAs_2$ [1.46]. Compared with this figure, however, the efficiences obtained (10^{-11} to 10^{-12}) by use of the scheme proposed by UEDA and SHIMODA have turned out most disappointing, [7.5,6]. The reason for this low efficiency appears to be the very low saturation intensity. This will become clear when the level scheme involved in this process is discussed in Sect.7.2.

An alternative scheme for achieving a TPA resonance in molecules was proposed by SHE and BILLMAN [7.7,8]. Their approach avoids the problem of saturation inherent in that of UEDA and SHIMODA. A detailed discussion of the She and Billman scheme will be given in Sect.7.2; it suffices to indicate here that their proposal has been realised in practice with harmonic-generation efficiencies that seem extremely promising. In fact, BRUECK and KILDAL [1.44,45] reported third-harmonic conversion of a CO_2 laser beam with efficiencies of the order of 2-4% using liquid CO as the nonlinear medium. Also, McNAIR and KLEIN [7.9] and KILDAL and BRUECK [7.10] reported results on two-photon resonant 4-wave mixing in liquid CO. Using CO_2 lasers of modest power, McNAIR obtained a conversion efficiency of $\sim 10^{-5}$ and claims that an efficiency of 10^{-1} (i.e., 10%) should be possible with appropriate scaling. With laser intensities close to the breakdown threshold, KILDAL and BRUECK obtained a conversion efficiency of $\sim$ 0.1% for difference-frequency generation. For comparison, it should be pointed out that LEE et al. [7.11] achieved a conversion efficiency as high as 1% in a similar four-wave mixing experiment in a crystal of germanium, although under these conditions the crystal suffered damage. Clearly then, liquid CO presents a very interesting nonlinear medium, because (i) it appears to be capable of efficient nonlinear generation from CO_2 laser radiation, (ii) is readily scalable, and (iii) does not suffer from irreversible optical damage. In Sect.7.3, we consider the CO medium in some detail; this is preceded, in Sect.7.2, by a general discussion of the two-photon resonant schemes of UEDA and SHIMODA, and SHE and BILLMAN. Although the Ueda and Shimoda approach now appears to be of less interest, we examine in Sect.7.4, as a simple illustrative example of this scheme, the use of HCl gas as the nonlinear medium [7.12].

7.2 Resonance-Enhanced Harmonic Generation in Molecules

Figures 7.1a,b illustrate the two different approaches to third-harmonic generation, suggested by UEDA and SHIMODA [2.17] and SHE and BILLMAN [7.7,8], respectively. In the process shown in Fig.7.1a, the molecule has an allowed

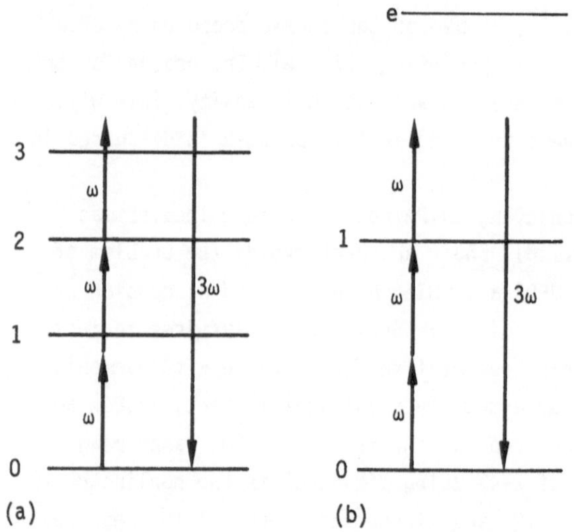

Fig.7.1. Schemes for resonantly enhanced third-harmonic generation in molecules. Levels 0,1,2,3 are rovibrational levels of the lowest electronic state. Level e is an excited electronic state

fundamental vibrational transition (0 → 1) close to the pump frequency ω. The intermediate levels 2 and 3 are the next levels in the vibrational ladder, all of these levels being in the ground electronic level. The rotational level structure is not explicitly indicated in either of these figures; for the moment, each of the levels indicated should be thought of as representing a particular rovibrational level.

A calculation of third-harmonic susceptibility for process a starts from the expressions (2.18) or (4.1). The dominant contribution comes from the first of the four terms in (2.18), because all three frequency denominators are small,

$$\chi_{THG}^{(a)} \simeq \frac{N}{\hbar^3 \varepsilon_0} \frac{\mu_{03}\mu_{32}\mu_{21}\mu_{10}}{(\Omega_{10}-\omega-i\Gamma_{10})(\Omega_{20}-2\omega-i\Gamma_{20})(\Omega_{30}-3\omega-i\Gamma_{30})} \quad , \tag{7.1}$$

where N is the density of molecules in level 0 (we assume that levels 1, 2, 3 are empty); the μ_{ij} are dipole matrix-elements between the rovibrational states i,j; Ω_{10}, Ω_{20} and Ω_{30} are, respectively, the fundamental, first and second overtone frequencies of the vibrational transition and the Γ are the linewidths (HWHM) of these transitions.

Some general conclusions can be made concerning the susceptibility expression (7.1). The matrix elements μ_{21} and μ_{32} are proportional to μ_{10}; in fact, $\mu_{21} = \sqrt{2}\mu_{10}$, $\mu_{32} = \sqrt{3}\mu_{10}$. These relations apply to the anharmonic as well as the harmonic oscillator [2.17]. For a harmonic oscillator, the matrix element μ_{03} would be zero. However, when the anharmonicity is taken into account μ_{03} is found to be proportional to μ_{10} also (in the sense that it is proportional to $\partial\mu/\partial x$, the change of electronic-dipole moment with internuclear spacing), but it is much smaller than μ_{10}, by two or three orders of magnitude typically. Because the absorption coefficient for the fundamental ω is proportional to $|\mu_{10}|^2$, and $|\chi_{THG}^{(a)}|^2 \propto |\mu_{10}|^8$, it follows that a large μ_{10}, favourable to harmonic generation, will also lead to strong absorption of the fundamental frequency.

All three denominators in (7.1) can be made small, thus giving a large resonance enhancement. The greatest enhancement would be possible if the vibration were harmonic. Anharmonicity, by making higher vibrational levels closer together, reduces the enhancement but, on the other hand, increases μ_{03}. The net effect of these conflicting requirements leads, as shown by UEDA and SHIMODA, to a large susceptibility when the anharmonicity is small; but this large susceptibility is obtained at the expense of a lower saturation intensity, due to single-photon absorption of the laser beam. The low saturation intensity is, in practice, a severe limitation on the harmonic-conversion efficiency. KANG et al. [7.6] estimate for their experiment, in which 10.6μm CO_2 laser radiation was tripled in SF_6 gas, that saturation [defined by the condition $W^{(1)}\tau = 1/2$ as in (4.56)] would occur at a fundamental intensity of $\sim 10^{11}$ Wm^{-2}. The intensity they actually used was some 500 times greater than this and the best conversion efficiency they observed was 4×10^{-11}. Similar results for SF_6 were reported by KILDAL and DEUTSCH [7.5].

The scheme shown in Fig.7.1b avoids this problem of absorption, because the incident frequency is chosen to be two-photon resonant with the frequency of the transition $0 \rightarrow 1$, where 1 is the first vibrational level. For the case of a homonuclear diatomic molecule, whose vibration is not infrared active, the intermediate levels for this two-photon transition must be provided by levels in excited electronic states (labelled e in Fig.7.1b). The vibration is, however, Raman active and the selection rules for the two-photon transition are just those for a Raman transition. For heteronuclear molecules, schemes a and b can simultaneously contribute to the nonlinear susceptibility; the relative magnitudes of these contributions will be discussed below.

The fact that all incident and generated frequencies are well away from the intermediate levels allows considerable simplification of the susceptibility expression for scheme b. Thus, starting with (2.18), it is seen that the first and third denominators in each of the four terms can be replaced by Ω, an average electronic transition frequency.

The sum over intermediate states can then be factorised and the final expression for the susceptibility is

$$\chi_{THG}^{(b)} \simeq \frac{2N\left|\sum_{e} \mu_{1e}\mu_{eo}\right|^2}{\hbar^3 \varepsilon_o \Omega^2 (\Omega_{10} - 2\omega - i\Gamma_{10})} , \tag{7.2}$$

where the sum is over all excited vibronic levels and $0 \rightarrow 1$ is the resonant transition. The factor of 2 appears in (7.2) because the first two terms in (2.18) contribute equally to the susceptibility. In comparing $\chi_{THG}^{(b)}$, (7.2), with $\chi_{THG}^{(a)}$, (7.1), we note that both have an exact two-photon resonance and that although $\chi_{THG}^{(b)}$ has the advantage that the electronic-dipole matrix elements are greater than the vibrational matrix elements, it has the disadvantage of the large denominator Ω. KILDAL and DEUTSCH [7.5] made an order-of-magnitude comparison of $\chi_{THG}^{(a)}$ and $\chi_{THG}^{(b)}$ and estimated a typical ratio $\chi_{THG}^{(a)}/\chi_{THG}^{(b)} \simeq 3000$, for the same ground-state density N. However, because scheme b does not suffer from fundamental absorption, much higher densities can be used and the value of $\chi_{THG}^{(b)}$ can, in practice, be made much larger than $\chi_{THG}^{(a)}$. Thus, KILDAL and BRUECK [4.9] used liquid CO in scheme b; whereas, when gaseous SF_6 was used in scheme a, KILDAL and DEUTSCH [7.5] found that the harmonic conversion decreased for pressures greater than a few torr. Scheme b also offers further advantages. Apart from the need for a close two-photon resonance $\omega_1 + \omega_2 \simeq \Omega_{10}$, the susceptibility is insensitive to the third frequency ω_3 in processes such as $\omega_1 + \omega_2 \pm \omega_3 \rightarrow \omega_4$; therefore, the conversion efficiencies for harmonic generation and sum- and difference-frequency generation would all be comparable. In addition, scheme b can be applied to both heteronuclear and homonuclear diatomic molecules. Although, for simplicity, we give diatomic molecules greatest prominence, both schemes can be applied to polyatomic molecules; SHE and BILLMAN mention CH_3F as a possible candidate for up-conversion of CO_2 laser radiation.

We now consider in more detail the calculation of susceptibility in each scheme, starting with scheme a. UEDA and SHIMODA examined the problem of third-harmonic generation in a diatomic molecule, both classically and quantum mechanically. The classical treatment requires the assumption of an anharmonic potential from the outset. They assumed a Morse potential function; i.e., the interatomic potential U(r) was taken to be

$$U(r) = D_e[1 - e^{-\beta(r-r_e)}]^2 , \tag{7.3}$$

where r is the internuclear separation, r_e its equilibrium value, D_e is the dissociation potential and β is a constant. By expanding the Morse function in ascending powers of x, where $x \equiv r - r_e$, the potential is expressed in terms of successive orders of anharmonic correction,

$$U(x) = \frac{1}{2} kx^2 - \frac{1}{3} fx^3 + \frac{1}{4} gx^4 + \dots \quad , \tag{7.4}$$

where k, f, g, etc. are given in terms of the constants D_e and β. The forced vibration of an oscillator that has this potential, subject to a driving field $Ee^{i\omega t}$, is then solved by assuming x to have the form

$$x = x_1 e^{i\omega t} + x_2 e^{i2\omega t} + x_3 e^{i3\omega t} + \dots \quad . \tag{7.5}$$

The third-harmonic amplitude x_3 is thus expressed in terms of k, f and g; these, in turn, are related, via D_e and β, to the conventional spectroscopic parameters ω_e, $\omega_e x_e$ and $\omega_e y_e$, which are extensively tabulated (e.g., [5.67]).

Finally, the amplitude x_3 must be related to the third-harmonic susceptibility. This is done by noting that $x_3 \mu_0 / r_e$ is the dipole moment of a single molecule, oscillating at the third-harmonic frequency (μ_0 being the permanent dipole). Then, x_3 is averaged over all molecular orientations and multiplied by the molecular density to find the macroscopic polarisation. This classical averaging introduces a factor 1/5. The third-harmonic susceptibility is thus obtained in terms of tabulated spectroscopic parameters. The weakness in this classical calculation lies in the procedure for taking the orientation average, because no account is taken of rotational quantisation; i.e., all molecules contribute to the susceptibility. The result is an overestimate.

A quantum-mechanical calculation starts from the susceptibility expression (7.1). If we consider the particular rotational level J as the initial level in the susceptibility expression (level 0 in the notation of this section) then, assuming just P- and R-branch transitions, i.e., $J \rightarrow J \pm 1$, implies the six possible choices of levels 1, 2 and 3 set out in Table 7.1.

In practice, for a light diatomic molecule, if the incident frequency is chosen to produce a strong resonance enhancement along one of these routes for a particular initial J value, the frequency denominators will be considerably larger along any other route or for a different J; therefore, only the one route will contribute significantly. This will be seen quite clearly when we consider a particular example, for the case of HCl, in Sect.7.4.

Table 7.1

Route	State:	0 →	1 →	2 →	3 →	0
1		J	J+1	J+2	J+1	J
2		J	J+1	J	J+1	J
3		J	J+1	J	J-1	J
4		J	J-1	J-2	J-1	J
5		J	J-1	J	J-1	J
6		J	J-1	J	J+1	J

Thus, the reason why the classical calculation yields the overestimate can be seen. The number of molecules that contribute to the susceptibility is not the total number but just those in the particular initial rotational level. After the particular sequence of intermediate levels has been selected, it still remains to carry out an orientation average by summation of the susceptibility over the quantum number M for the initial level, i.e., from -J to +J (see Sect.2.4). This completes the quantum-mechanical calculation of the susceptibility. Again, the final result is expressed in terms of vibrational matrix elements, for which some tabulations are available.

When more complicated molecules are considered, e.g., SF_6, the dense vibrational-rotational structure implies that several routes and several initial states that have different J values, all contribute to the susceptibility. A rough estimate of susceptibility can be found by neglecting this rotational structure and using (7.1) with the linewidths Γ put equal to the observed absorption linewidths. KANG et al. [7.6] have compared this simple estimate for SF_6 with a lengthy calculation that took into account the rotational structure (there are 19 possible routes through the intermediate levels, because P, Q, and R branches are allowed). These two calculations are compared in Fig.7.2. The detailed calculation shows some very sharp resonances with peak susceptibilities that are several orders of magnitude greater than the prediction of the simple estimate. In their experiments on SF_6, KANG et al. found that the observed third-harmonic power was many orders of magnitude less than that predicted by use of the exact susceptibility from Fig.7.2. This was ascribed to saturation, as explained in the foregoing, and the experiment was, therefore, not able to test the theoretical calculation of susceptibility.

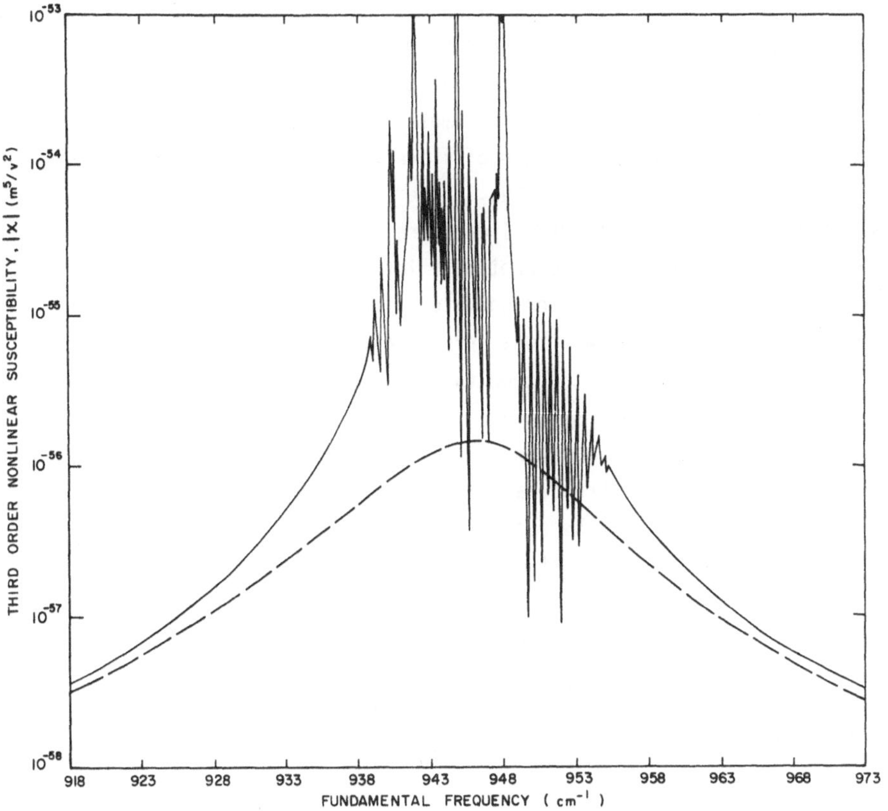

Fig.7.2. Calculated third-harmonic susceptibility (per molecule) of SF_6. The dashed curve corresponds to the simplified calculation referred to in the text. (After [7.6], although note that the units of the susceptibility should in fact be Jm^4V^{-4}, rather than m^5V^{-2}, because their definition of susceptibility is as in the footnote to (A.1) of the Appendix)

To calculate the susceptibility for scheme b, a rather different approach is adopted, because, in general, there are not sufficiently accurate data available on matrix elements between vibronic states. These matrix elements can have either sign; in the sum over intermediate states, there will be a significant degree of cancellation. So, unless the matrix elements are accurately known, the summation could be greatly in error; attempts to calculate polarisabilites by summation over intermediate states have not usually led to accurate results [7.13]. However, for molecular hydrogen there are adequate data; PAN et al. [7.14,15] have shown that a calculation of Raman susceptibility that uses this data (from [7.16]) gives good agreement with experimental measurements of the Raman-scattering cross section.

Although the same approach is not likely to give acceptable accuracy for other molecules, it will be seen below that any two-photon resonant susceptibility, such as for THG, difference-frequency generation, two-photon absorption etc., can be related in a simple way to the Raman susceptibility, provided that the incident frequencies are far from resonance with the intermediate vibronic states. Thus, there is in effect a universal susceptibility. Because experimental data are usually available on Raman-scattering cross sections, the susceptibilities of interest can, therefore, be calculated accurately, even for polyatomic molecules. We now derive these relations between the Raman cross section and the various susceptibilities.

We start by collecting together the various susceptibility formulae, simplified as usual to the case of parallel linearly polarised light. In the two-photon resonance, g is taken as the initial state (Fig.7.3), and f is the two-photon resonant rovibrational level (in the ground electronic state). Thus, from (2.18), χ_{THG} (the superscript (b) used earlier in this section is now omitted) may be approximated by retaining only the first two terms; the second denominator in each of those terms becomes $\Omega_{fg} - 2\omega - i\Gamma$ and may be removed from the summation. The result is

$$\chi_{THG} = \frac{N}{\hbar^3 \varepsilon_0 (\Omega_{fg} - 2\omega - i\Gamma)} \left[\sum_a \mu_{ga}\mu_{af} \left(\frac{1}{\Omega_{ag} - 3\omega} + \frac{1}{\Omega_{ag} + \omega} \right) \right] \left(\sum_c \frac{\mu_{fc}\mu_{cg}}{\Omega_{cg} - \omega} \right) \quad . \tag{7.6}$$

Similarly, the Raman susceptibility becomes

$$\chi_R = \frac{N}{6\hbar^3 \varepsilon_0 (\Omega_{fg} + \omega_s - \omega_p + i\Gamma)} \left| \sum_a \mu_{fa}\mu_{ag} \left(\frac{1}{\Omega_{ag} - \omega_p} + \frac{1}{\Omega_{ag} + \omega_s} \right) \right|^2 \tag{7.7}$$

[see e.g. (5.1b)], and for two-photon absorption of a single frequency ω (see Sect.3.3)

$$\chi_{TPA} = \frac{N}{6\varepsilon_0 \hbar^3 (\Omega_{fg} - 2\omega - i\Gamma)} \left| \sum_a \frac{2\mu_{fa}\mu_{ag}}{\Omega_{ag} - \omega} \right|^2 \quad . \tag{7.8}$$

Reasons for the similarity in form of these formulae have been given in Sect.2.7; a general expression is given in (2.21) and Table 2.3, which shows how susceptibilities for other two-photon resonant processes, such as $\omega_1 + \omega_2 + \omega_3 \rightarrow \omega_4$ where $\omega_1 + \omega_2 \simeq \Omega_{fg}$, may be written. For the present pur-

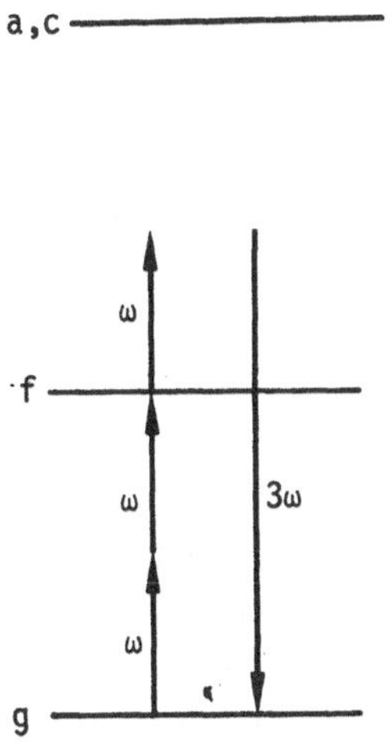

Fig.7.3. Two-photon resonant third-harmonic generation. The levels indicated are rovibrational levels; g and f belong, respectively, to the ground and first excited vibrational levels of the ground electronic state. a and c are in excited electronic states

pose the most important common factor in these expressions is the transition polarisability, which is here simplified to

$$\alpha_{fg}(\omega_1;\omega_2) = \frac{1}{\hbar}\sum_a \mu_{fa}\mu_{ag}\left(\frac{1}{\Omega_{ag}-\omega_1} + \frac{1}{\Omega_{ag}-\omega_2}\right) , \tag{7.9a}$$

in terms of which (7.6-8), etc., may be written. For infrared radiation, $\omega \ll \Omega_{ag}$; therefore, Ω_{ag} may be replaced by an average value, Ω say, and then removed from the summation. For example,

$$\alpha_{fg}(\omega_1;\omega_2) \simeq \frac{1}{\hbar}\left(\frac{1}{\Omega-\omega_1} + \frac{1}{\Omega-\omega_2}\right)\left(\sum_a \mu_{fa}\mu_{ag}\right) \simeq \frac{2}{\hbar\Omega}\left(\sum_a \mu_{fa}\mu_{ag}\right) , \tag{7.9b}$$

which shows α_{fg} to be approximately independent of the applied field frequencies. Equation (7.9b) shows that it is an elementary matter to relate any two α_{fg} that have differing frequency arguments. An ab initio absolute calculation of susceptibility, therefore, requires only evaluation of the sum over states in (7.9b). Clearly, if α_{fg} is to be evaluated with any reasonable accuracy, then the matrix elements must be known with considerable accuracy, because they can have either sign and, therefore, lead to cancellations within the sum. The approximation involved in replacing the frequency denominators by averages must also be reconsidered when accurate calculations are required, because this will affect the degree of cancellation. PAN et al. used the matrix elements of FORD and BROWNE [7.16] for H_2, to calculate the sums over states in (7.6), but it is not clear whether they used the exact denominators.

However, if the approximation with respect to frequency denominators embodied in (7.9b) is accepted, the susceptibilities are all simply related to each other and, in particular, to χ_R. Furthermore, we can deduce the Raman polarisability $\alpha_R = \alpha_{fg}(-\omega_s;\omega_p)$ from experimentally determined Raman-scattering cross sections, using the relations (2.84,85)

$$|\alpha_R|^2 = \frac{(4\pi\varepsilon_0)^2 c^4}{\omega_p \omega_s^3}\left(\frac{d\sigma_R}{d\Omega}\right)_{||,\omega_p} . \tag{7.10}$$

The || notation in (7.10) emphasises that the cross section refers to the situation in which the pump and Stokes waves are linearly polarised in the same direction, because this is the geometry assumed for (7.6-9); moreover, it must be that value of cross section appropriate to a pump frequency ω_p. This does not mean that α_R itself is frequency sensitive, because the frequency factors on the RHS of (7.10) are approximately cancelled by the frequency dependence of the cross section; indeed, to a good approximation, α_R is given by the RHS of (7.9b). Hence, the summation in (7.9b) is essentially given (to within a sign) by the square root of the RHS of (7.10), which in turn is found experimentally. For simplicity, we will rewrite (7.6-8) in terms of α_R rather than the combination on the RHS of (7.10). Thus

$$\chi_{THG} = \frac{N|\alpha_R|^2}{\hbar^3\varepsilon_0(\Omega_{fg}-2\omega-i\Gamma)}\left(\frac{1}{\Omega-\omega_p}+\frac{1}{\Omega+\omega_s}\right)^{-2}\left(\frac{1}{\Omega-3\omega}+\frac{1}{\Omega+\omega}\right)\left(\frac{1}{\Omega-\omega}\right)$$

$$= \frac{2N|\alpha_R|^2}{\hbar^3\varepsilon_0(\Omega_{fg}-2\omega-i\Gamma)}\,\frac{(\Omega-\omega_p)^2(\Omega+\omega_s)^2}{(\Omega-3\omega)(\Omega+\omega)(2\Omega-\omega_p+\omega_s)^2} \tag{7.11a}$$

$$= \frac{32\pi^2\varepsilon_0 c^4 N(\Omega-\omega_p)^2(\Omega+\omega_s)^2}{\hbar\omega_p\omega_s^3(\Omega_{fg}-2\omega-i\Gamma)(\Omega+\omega)(\Omega-3\omega)(2\Omega-\omega_p+\omega_s)^2}\left(\frac{d\sigma_R}{d\Omega}\right)_{||,\omega_p} . \tag{7.11b}$$

Similarly,

$$\chi_{TPA} = \frac{2N|\alpha_R|^2}{3\hbar^3\varepsilon_0(\Omega_{fg}-2\omega-i\Gamma)}\,\frac{(\Omega-\omega_p)^2(\Omega+\omega_s)^2}{(\Omega-\omega)^2(2\Omega+\omega_s-\omega_p)^2} . \tag{7.12}$$

It is interesting to observe that

$$\chi_{THG}/\chi_{TPA} = \frac{3(\Omega-\omega)^2}{(\Omega-3\omega)(\Omega+\omega)} .$$

Also, there is the exact relation

$$\chi_R = \frac{N|\alpha_R|^2}{6\hbar^3\varepsilon_0(\Omega_{fg}+\omega_s-\omega_p+i\Gamma)} . \tag{7.13}$$

In some cases, experimental data are available on the Raman gain coefficient, g_R, and (7.11) can be expressed alternatively, with the help of (3.19) and (7.13), as

$$\chi_{THG} = \frac{4(\Omega-\omega_p)^2(\Omega+\omega_s)^2\Gamma\varepsilon_0 c^2\eta_s\eta_p(g_R/\omega_s)}{(\Omega+\omega)(\Omega-3\omega)(2\Omega+\omega_s-\omega_p)^2(\Omega_{fg}-2\omega+i\Gamma)} , \tag{7.14}$$

where, it should be recalled, g_R is the peak gain coefficient, i.e., the gain for exact Raman resonance, $\Omega_{fg} + \omega_s - \omega_p = 0$.

For the special case of exact two-photon resonance and $\omega, \omega_p, \omega_s << \Omega$, (7.11-14) may be further simplified, to

$$\chi_{THG} = 3\chi_{TPA} = -3\chi_R = -i\varepsilon_0 c^2\eta_s\eta_p g_R/\omega_s = \frac{iN|\alpha_R|^2}{2\hbar^3\varepsilon_0\Gamma} . \tag{7.15}$$

An important feature of the exact two-photon resonance is that the susceptibilities are purely imaginary. A quite different situation arises when all of the frequencies are well below any transition frequency, because then all of the susceptibilities become equal and purely real. An example of this was discussed in Sect.4.1.

Using the result (4.17), for the THG efficiency under plane-wave conditions, we find

$$\frac{P_3}{P_1} = \frac{(\Delta^2+\Gamma^2)}{\eta_1^3\eta_3\Gamma^2}\left[\frac{3(\Omega-\omega)^2}{2(\Omega-3\omega)(\Omega+\omega)}\right]^2(k_{TPA}^{\omega}I_\omega L)^2\left[\mathrm{sinc}\left(\frac{\Delta kL}{2}\right)\right]^2 . \tag{7.16}$$

The quantity $k_{TPA}^{\omega} \equiv k_{TPA}\Gamma^2/(\Delta^2 + \Gamma^2)$ is the two-photon absorption coefficient for a detuning $\Delta = \Omega_{fg} - 2\omega$ from resonance, and k_{TPA} is the peak value, as defined in Sect.3.3. Thus, a high efficiency P_3/P_1 implies that the product $k_{TPA}^{\omega}I_\omega L$ should not be small; in Sect.3.3, it is seen that this further implies significant two-photon absorption. This two-photon absorption does not lead to breaking of phase matching (see Sect.4.3), because the refractive index is independent of the degree of vibrational excitation in the harmonic-oscillator approximation [1.44,45]. (This is readily seen when it is recalled that the vibrational matrix elements obey the relations given earlier, viz. $\mu_{21} = \sqrt{2}\,\mu_{10}$, $\mu_{32} = \sqrt{3}\,\mu_{10}$, etc.) From (7.16), it can be seen that the limitation of THG efficiency by two-photon absorption becomes less severe when the detuning from two-photon resonance is increased (as pointed out by KILDAL [7.17]). This is because the quantity $k^{\omega}_{TPA} I_{\omega} L$ determines the fraction of the fundamental power absorped (see Sect.3.3). For a given fractional absorption of the fundamental (i.e., by adjusting $I_{\omega}L$ to maintain a constant $k^{\omega}_{TPA} I_{\omega} L$ as ω is varied), the THG efficiency is, therefore, determined by the factor $(\Delta^2 + \Gamma^2)$ in the numerator of (7.16).

In their calculations, SHE and BILLMAN [7.7,8], and PAN et al. [7.14] examined in some detail the possibility of efficient third-harmonic generation in H_2. The initial level g was taken to be the J = 1 level of the ground vibrational state in the ground electronic state $(X^1\Sigma^+_g)$, and the two-photon level f was taken to be the J = 1 level of the first excited vibrational level in this electronic state. The transition g → f corresponds, therefore, to the Q branch of the vibrational Raman transition $[Q_{01}(1)]$. The choice of J = 1 as the initial level (rather than J = 0) is made because of the difference of the statistical weights of the ortho (J = 1) and para (J = 0) modification. At room temperature, the J = 1 level has 65% of the total population, whereas J = 0 has only 13%; the susceptibility is, therefore, greatest with J = 1 as the initial level. PAN et al. compared their theoretically calculated Raman susceptibility with those calculated from experimental values of stimulated Raman gain g_R and spontaneous Raman-scattering cross section. From the pressure dependence of the Raman linewidth and the dependence of the Raman cross section on pump wavelength, the experimental data were adjusted to give values appropriate to 1 atmosphere pressure and λ_p = 694.3nm. The theoretical result, given as susceptibility per molecule (see Appendix), was $\chi_R = 1.8 \times 10^{-47} m^5 V^{-2}$, whereas the experimental data gave $\chi_R = 2.1 \times 10^{-47} m^5 V^{-2}$ (from data on g_R) and $\chi_R = 1.7 \times 10^{-47} m^5 V^{-2}$ (from data on Raman-scattering cross sections). The good agreement between these figures gives credence to the calculated value of χ_R and therefore to the calculated harmonic-conversion efficiency. They show that under plane-wave conditions, two-photon resonant THG should have an efficiency of 10% in 1 atmosphere of H_2 for a fundamental intensity of 150MWcm^{-2}. Assuming loose focussing, and 4-atmosphere pressure, the calculation of PAN et al. predicts 10% conversion efficiency, for a total fundamental power of $\sim$ 6MW. The maximum fundamental intensity corresponds to $\sim$ 500MWcm^{-2} and is expected to be well below the threshold for breakdown. In a later correction PAN et al. [7.15] revised these estimates; input powers greater by an order of magnitude are indicated.

Although these calculations indicate the possibility of efficient THG in molecular systems, the disadvantage of H_2 as a medium is that no powerful laser source exists at the appropriate frequency for two-photon resonance [the $Q_{01}(1)$ transition is 4155 cm^{-1}, which would require a source around 2077 cm^{-1}]. Fortunately, the molecule CO provides a two-photon transition that is suited to the CO_2 laser; with this combination, the possibility of high conversion efficiency has become a reality. The results obtained by use of CO will now be examined.

7.3 CO as a Nonlinear Medium

In Table 7.2 are listed the Q-branch transition frequencies ($v = 0 \rightarrow v = 1$, $\Delta J = 0$) of $C^{12}\,O^{16}$ calculated from measured spectroscopic constants [7.18] together with some doubled CO_2 laser frequencies in the $00^01 \rightarrow 02^00$ band at 9.6µm [7.19]. It can be seen that twice the frequency of the R(8) line is within 0.036 cm^{-1} of the CO Q(11) line, and twice the R(10) line frequency is within 0.5 cm^{-1} of the CO Q(0) line.

Using a grating-controlled CO_2 TEA laser, KILDAL and DEUTSCH [7.5] observed third-harmonic generation of the R(8) and R(10) lines. The CO gas was contained in a 1m long cell, at pressures up to one atmosphere. With incident powers of 2-3MW, focussed to a 1.8mm spot at the centre of the cell, THG conversion efficiencies of $\sim 0.4 \times 10^{-12}$ were seen for both the R(8) and R(10) lines. A conversion efficiency of 7.4×10^{-9} was predicted theoretically. Part of this discrepancy was ascribed to the multimode nature of the CO_2 laser beam. In further experiments, KILDAL [7.17] used a TEM_{00} mode CO_2 TEA laser and found good agreement between theory and experiment. Higher powers were also obtained by working at higher CO pressure and using a phase-matching buffer gas.

The susceptibility calculation uses (7.11), generalised to include a sum over initial levels that have different rotational quantum numbers J. This is necessary because the various Q-branch transitions are close in frequency and several may contribute significantly to the susceptibility. Equation (7.11) now becomes,

Table 7.2

CO (0-1) Q branch		$CO_2(00^01-02^00)$	
J	$\nu_{Q(J)}$ [cm^{-1}]	Line	$2\nu_{CO_2}$ [cm^{-1})]
0	2143.272	R(10)	2143.768
1	2143.237		
2	2143.166		
3	2143.061		
4	2142.921		
5	2142.746		
6	2142.536		
7	2142.291		
8	2142.011		
9	2141.696		
10	2141.346		
11	2140.961	R(8)	2140.925
12	2140.541		
13	2140.086		
14	2139.596		
15	2139.071		
16	2138.511		
17	2137.916	R(6)	2138.028
18	2137.285		
19	2136.620		
20	2135.920		
21	2135.185	R(4)	2135.078
22	2134.415		
23	2133.610		
24	2132.770		
25	2131.895		

$$\chi_{THG} = \frac{32\pi^2 c^4 \varepsilon_o (\Omega-\omega_p)^2 (\Omega+\omega_s)^2}{\hbar \omega_p \omega_s^3 (\Omega+\omega)(\Omega-3\omega)(2\Omega-\omega_p+\omega_s)^2} \sum_J \frac{\left[\left(\frac{d\sigma R}{d\Omega}\right)_{||,\omega_p}\right]_J N_J}{\Omega_{Q(J)}-2\omega-i\Gamma_{Q(J)}} , \quad (7.17)$$

where N_J is the population of the v = 0 level of the rotational quantum number J. N_J is given by $N_J = (\mathcal{N}/Z)(2J+1)\exp[-\hbar BJ(J+1)/k_BT]$, where $B/2\pi c = 1.93\ \mathrm{cm}^{-1}$ is the rotational constant, $\mathcal{N}$ is the total molecular density and Z is the partition function. (For a homonuclear diatomic molecule, nuclear-spin statistics must also be allowed for.) In fact, $\left[(d\sigma_R/d\Omega)_{||,\omega_p}\right]_J$ depends only slightly on J; as an approximation the J dependence can be ignored, so that $\left[(d\sigma_R/d\Omega)_{||,\omega_p}\right]_J$ can be replaced by the measured total Q-branch cross section $(d\sigma_R/d\Omega)_{||,\omega_p}$, i.e., that in which the individual Q-branch lines are not resolved. Using (7.17), KILDAL [7.17] calculated that, at atmospheric pressure and room temperature, $\chi_{THG} = 1.9 \times 10^{-24} m^2V^{-2}$ for the R(8) line of the CO_2 laser. It is interesting to compare this with the result based on calculations of PAN et al. for H_2. Re-expressing KILDAL'S result as a susceptibility per molecule (see Appendix) gives $7.7 \times 10^{-50} m^5V^{-2}$, whereas the result for H_2 is $5.8 \times 10^{-48} m^5V^{-2}$. The main reason why the susceptibility for H_2 is greater, is that, at the assumed atmospheric pressure, the two-photon linewidth for H_2 is about an order of magnitude less than that for CO, i.e., $0.013\ \mathrm{cm}^{-1}$ for H_2 [7.20], compared to $0.14\ \mathrm{cm}^{-1}$ for CO [7.21].

From the third-harmonic susceptibility, the expected third-harmonic power can be calculated by use of (4.15). However, in order to evaluate the integral I in (4.15), it is necessary to know the coherence length. The dominant contribution to the refractive indices at ω and 3ω is due to the fundamental infrared-active vibrational transition. Because ω is less than and 3ω is greater than this transition frequency, the medium is negatively dispersive. Approximate values for the refractive indices at ω and 3ω can be obtained by ignoring the rotational structure and using (2.33) with Ω_{ng} put equal to 2ω. A classical averaging over all molecular orientations introduces a factor 1/3, thus giving

$$\delta n \equiv n_3 - n_1 \simeq - \frac{16\mathcal{N}|\mu_{10}|^2}{45\varepsilon_0\hbar\omega} \ ,$$

where μ_{10} is the vibrational dipole matrix element between the $v = 0$ and $v = 1$ states. Using the above expression and the value $\mu_{10} = 0.1D$[1], KILDAL found $\delta n = -5.2 \times 10^{-6}$ per atmosphere, and hence a coherence length L_c of ~ 30 cm at one atmosphere. With this value of L_c, and the value of χ_{THG} given above, KILDAL found good agreement between theory and experiment; the predicted and measured third-harmonic powers agreed within a factor of two.

---

[1] D = Debye = $1/3 \times 10^{-29}$ Cm.

By adding SF_6 gas, which is positively dispersive, KILDAL found that the phase-matching behaviour was in excellent agreement with predictions based on the foregoing value of δn. Figures 7.4a,b show, respectively, the observed and predicted third-harmonic signal versus pressure of the $CO:SF_6$ mixture. In these experiments, the R(10) line of the CO_2 laser was used. Despite the fact that the R(10) line is further from two-photon resonance than the R(8) line, the susceptibility is comparable in each case. This is because twice the R(10) frequency lies just above the highest CO Q-branch frequency (see Table 7.2); therefore, all of the denominators in the sum over J add [see (7.17)], whereas for the R(8) line there is some cancellation due to different signs of the denominators. The best third-harmonic conversion efficiency obtained was 1.9×10^{-8}, with 1.2MW of laser power [R(10)line] incident on a 17.9 cm cell filled with a phase-matched $CO:SF_6$ mixture at 11 atmospheres. KILDAL considered how this performance might be scaled up to give 10% conversion efficiency. The need to avoid breakdown in the gas and the cell windows, led to the rather inconvenient requirement of a cell having a length of $\sim$ 35m. However, it has since been shown that a high conversion efficiency is possible using liquid rather than gaseous CO, and the cell dimensions need then be only a few centimetres [4.9,7.10].

The use of liquid CO has several advantages over gaseous CO. Above a few atmospheres pressure, the susceptibility begins to saturate, due to pressure broadening of the two-photon resonant transition; i.e., the ratio N/Γ in (7.6) tends to a constant value with increasing N. However, at still higher pressures, the discrete vibrational-rotational lines overlap; at liquid densities, the envelope of the Q-branch collapses to a narrow line, due to motional-narrowing effects [7.22,23]. This narrowing was previously observed in Raman scattering in liquid N_2 and O_2 [7.24] and in H_2 (see for example [7.25]). The narrow two-photon resonance gives considerable enhancement to the nonlinear susceptibility. A further advantage of liquid CO is that it shows a high threshold for laser-induced breakdown, in excess of 400J cm^{-2}, which is much higher than the typical value of $\sim$ 10J cm^{-2} for solids.

The cell that contains the liquid CO can be of quite simple design, consisting of a metal-walled inner cell, provided with an infrared-transmitting window at each end. This cell has a surrounding jacket that contains liquid N_2 and the whole assembly is surrounded by a vacuum chamber that is also provided with windows. This arrangement appears to give excellent optical homogeneity over a considerable length of cell [1.44,45,7.9]. Various

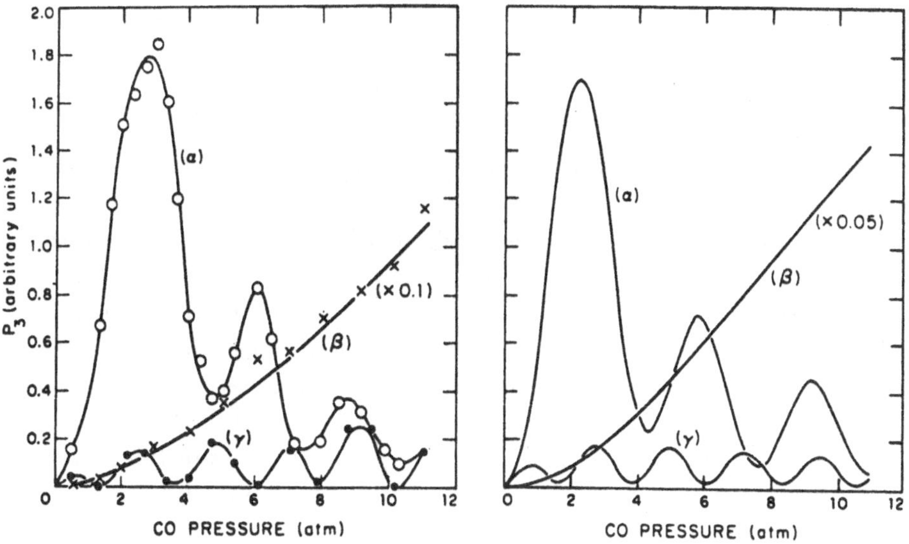

(a) (b)

Fig.7.4a and b. Pressure dependence of the third-harmonic signal from a 17.9 cm cell filled with a $CO:SF_6$ mixture of (α) 1:0, (β) 1:0.0048, and (γ) 1:0.012, and pumped with the CO_2 R(10) line at 9.33μm focussed to a confocal parameter of 32.8 cm at the cell exit window. (a) Measured signal. (b) Calculated signal. (After [7.17])

materials have been used for the cold windows of the inner cell, such as CaF_2, BaF_2 and ZnSe [1.44,45,66,7.9]. The windows can be conveniently sealed using indium wire O-rings. Because the breakdown threshold of the windows is much less than that of the liquid, the input beams are brought to a tight focus within the cell. To obtain an increased interaction length within the cell, FREY et al. [1.66] used mirrors within the liquid (N_2 in their case) to give multiple passes with refocussing. BRUECK and KILDAL [1.45] also used a multi-pass arrangement. To fill the inner cell, gas (which can be purified, if required) is passed into the inner cell, where it is condensed by the cold surrounding jacket. Various cryogenic liquids can be prepared in this way because the jacket of liquid N_2 (boiling point 77K) will condense such liquids as CO (82K), O_2(90K), NO(121K), CH_4(109K), Ar(87K), etc. Mixtures of liquids can be prepared simply by adding the appropriate gas fill; for example KILDAL and BRUECK used $CO:N_2$, $CO:O_2$ and CO:Ar mixtures; SF_6 has been added, to achieve phase matching. The relative concentrations of the constituents can be determined by monitoring their infrared absorption [1.44,45].

The high density of liquid CO implies a short coherence length for third-harmonic generation (KILDAL and BRUECK quote a value of 0.058 cm). Because the medium is negatively dispersive, it is possible, in principle, to use tight focussing (as discussed in Sect.4.2). In fact, the optimum confocal parameter that would be required is too small for practical infrared optics. One approach, which BRUECK and KILDAL [1.44,45] have investigated, is to add a positively dispersive liquid such as SF_6 so as to achieve phase matching. An alternative approach, which they have also reported, is to dilute the liquid CO with solvents, such as the nonpolar liquids O_2, N_2 or Ar; this increases the coherence length. The solvents produce a significant shift of the two-photon resonance. The best conversion efficiency reported by KILDAL and BRUECK (without phase matching) was obtained with the R(6) CO_2 laser line ($00^01 \rightarrow 02^00$ band) in a $CO{:}O_2$ mixture, containing 0.1 mole fraction of CO. The dependence of conversion efficiency on CO concentration is shown in Fig.7.5. (The abscissa acutally shows the peak absorbance of CO on the second vibrational overtone at $\sim$6339 cm^{-1} but this is roughly proportional to the CO concentration.) BRUECK [7.23] has subsequently measured the dependence of the liquid CO Stokes shift and linewidth on solvent concentration and has shown that the reason for the superiority of the $CO{:}O_2$ mixture is twofold. For a particular $CO{:}O_2$ ratio, the Stokes shift comes into exact two-photon resonance with the R(6)CO_2 laser line, and the Raman linewidth at this concentration is less ($\sim$ 0.2 cm^{-1}) than for $CO{:}N_2$. Also, the $CO{:}O_2$ mixture was found to have a breakdown threshold a factor of two greater than for the CO:Ar mixture [4.9].

The experimental set-up used by KILDAL and BRUECK [4.9] involved a $TEM_{00}CO_2$ laser, whose beam was focussed into a dewar with a 5.8 cm liquid path between windows of BaF_2 or ZnSe. The focussed spot size was $\sim$100 μm, which implied a confocal parameter of 0.8 cm. With this arrangement, the best conversion efficiency obtained was 0.3% when 60 mJ of CO_2 energy was incident. The optimum focussing condition for this tight-focussing arrangement is given by $b\Delta k = -4$, rather than the condition $b\Delta k = -2$ given in Sect.4.2. This difference arises as follows. In Sect.4.2, we considered optimisation of $b\Delta k$ while keeping the number density of the nonlinear species (and hence χ_{THG}) fixed. Here, we are considering Δk to be varied by varying N_{CO}, with b fixed at the minimum practical value. Because χ_{THG} and Δk are both proportional to N_{CO} (the solvent is nondispersive) this results in the different optimum condition [3.1]. KILDAL and BRUECK suggest that with a somewhat larger cell and a phase-matched mixture, a THG efficiency of around 10% should be possible.

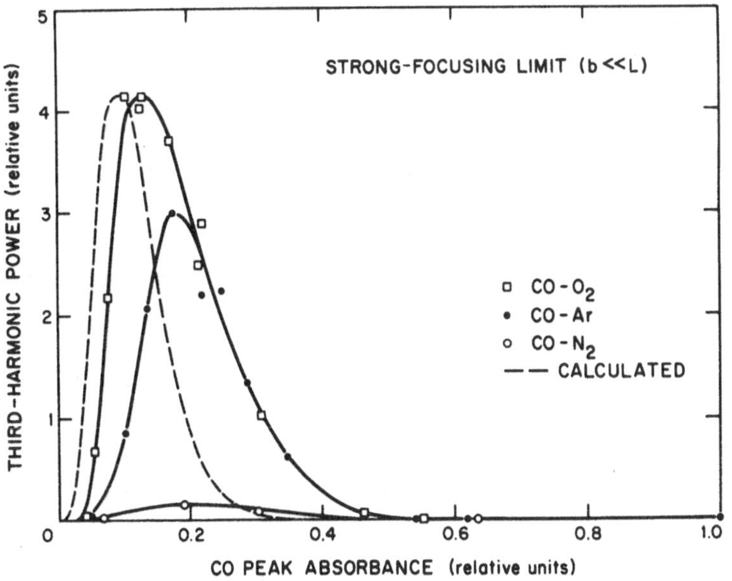

Fig.7.5. The concentration dependence of THG in CO diluted in different solvents at constant laser power (T = 77k, l = 5.8 cm, and b = 0.81 cm). The maximum of the calculated curve has been arbitrarily set equal to the maximum of the $CO:O_2$ experimental curve. (After [4.9])

The calculation of susceptibility follows the same lines as already described for gaseous CO, i.e., using (7.17). However, because the local field at the CO molecule is greater than the macroscopic field by the ratio $(n_\omega^2 + 2)/3$, the right-hand side of (7.16) should be multiplied by a local-field-correction factor, $[(n_\omega^2 + 2)/3]^4$, (see [2.15]). The value of Raman-scattering cross section used in (7.17) is again the total Q branch cross section measured in the gas phase. This cross section consists of a narrow-linewidth, polarised component (associated with the isotropic part of the Raman-polarisability tensor) and a broad, weak, depolarised component (associated with the traceless anisotropic part) which makes a negligible contribution to the measured vibrational cross section. In their calculation of the susceptibility, KILDAL and BRUECK assumed a linewidth of 0.2 cm^{-1} (FWHM = $2\Gamma/2\pi c$) and a two-photon resonance mismatch of 0.4 cm^{-1}. The calculated third-harmonic conversion efficiency for the conditions of their experiment was thus found to be 5×10^{-5}, much less than the actual observed value of 3×10^{-3}. Part of the discrepancy was thought to be due to an overestimate of the resonance mismatch. Also, the laser showed spontaneous mode locking, which would increase the conversion efficiency.

In later experiments, BRUECK and KILDAL [1.44] increased the THG efficiency to 2% by adding SF_6 to phase match the $CO:O_2$ mixture. This increases the coherence length and, by allowing the use of looser focussing (longer confocal parameter b), the problem of breakdown is alleviated. The efficiency was, however, still limited by breakdown; BRUECK and KILDAL [1.44] suggest that, with shorter pulse lengths or longer confocal parameters, the efficiency can be increased yet further; two-photon absorption is expected ultimately to limit the conversion efficiency. As an alternative to longer confocal parameter (which would also require a longer cell and a higher-power laser), BRUECK and KILDAL [1.45] have shown the value of a multi-pass arrangement, achieving a 4% conversion efficiency in this way.

Another experiment in liquid CO, reported by McNAIR and KLEIN [7.9], appears to confirm the above calculated value of susceptibility. In this experiment, two-photon resonant difference-frequency generation was observed, i.e., $\omega_1 + \omega_1 - \omega_2 \rightarrow \omega_4$, where ω_1 was again provided by the R(6) line (1069.05 cm^{-1}) of the CO_2 laser and ω_2 was provided by the R(32) line (983.23 cm^{-1}) of another CO_2 laser. The choice of this particular frequency for ω_2 was dictated by the experimental aim of generating radiation with a wavelength near 8.6μm, because this is of interest for laser isotope separation. A simplifying feature of this difference-frequency process is that the phase mismatch $2\underline{k}_1 - \underline{k}_2 - \underline{k}_4$ is small, because all of the frequencies are similar, and the coherence length, even for pure CO, is quite large, ([7.9] quote L_c = 9.2 cm). The experiment was, therefore, done in pure CO liquid. From the intensity of the generated radiation, they deduced that the susceptibility was $(8.7 \pm 2.0) \times 10^{-22} m^2V^{-2}$, close to the value reported by KILDAL and BRUECK. This suggests that the effect on the susceptibility of the greater density of CO in the pure liquid is offset by the broader linewidth and greater resonance detuning. McNAIR and KLEIN used moderate laser powers but the absence of any observed saturation effects led them to suggest that a suitably scaled version of their experiment could perhaps give 10% conversion efficiency. KILDAL and BRUECK [7.10] did a similar difference-frequency-generation experiment with higher CO_2 laser powers and achieved a conversion efficiency of 0.8%. They found that this efficiency could be further increased (by ~ 25%) using a noncollinear geometry to increase the coherence length.

The use of cryogenic molecular liquids as nonlinear media appears very promising; these early results are likely to stimulate much more research in this area. For example, recent work by KANG et al. [7.26] on THG in gaseous

DCl, indicates a greater susceptibility than for gaseous CO. Liquid DCl may therefore be of interest. Although DBr appears to be less promising as a tripler for CO_2 laser radiation, KUNG et al. [7.27] suggest that it may be of interest with an N_2O laser. There still remain many more media that are relatively unexplored, such as CH_4 and NO, and their isotopic variants. The use of phase-matching liquids and various solvents adds considerable flexibility to their application in nonlinear optics.

7.4 HCl as a Nonlinear Medium

PELLIN and YARDLEY [7.12] have discussed the possibility of using HCl gas as a medium for two-photon resonant up-conversion (generation of $\omega_4 = 2\omega_1+\omega_2$), based on scheme a shown in Fig.7.1. One laser (ω_1) is tuned to the two-photon resonance on a vibration-rotation transition $|v = 0, J\rangle \rightarrow |v = 2, J''\rangle$, having intermediate levels $|v = 1, J'\rangle$ (see Fig.7.6). The second laser (ω_2) is tuned to resonance with the transition $|v = 2, J''\rangle \rightarrow |v = 3, J'''\rangle$. An energy-level route is then sought for which all of the dipole matrix-elements are as large as possible and the frequency denominators in the susceptibility expression as small as possible. Also, the initial level $|v = 0, J\rangle$ should contain a significant population. Clearly, there are several possibilities, depending on the choice of J, J', J'', J''' ; we examine just one of the favourable choices to see what sort of susceptibility results. This value is then used to estimate the conversion efficiency that can be expected.

A calculation of the level populations in the ground vibrational state at 300K yields (Table 7.3)

Table 7.3.

J	Fraction of population
0	0.05
1	0.14
2	0.19
3	0.20
4	0.17

It should also be noted that if HCl is used with the natural isotopic abundances of $^{35}Cl:^{37}Cl \sim 3:1$ then only 3/4 of the total HCl population will

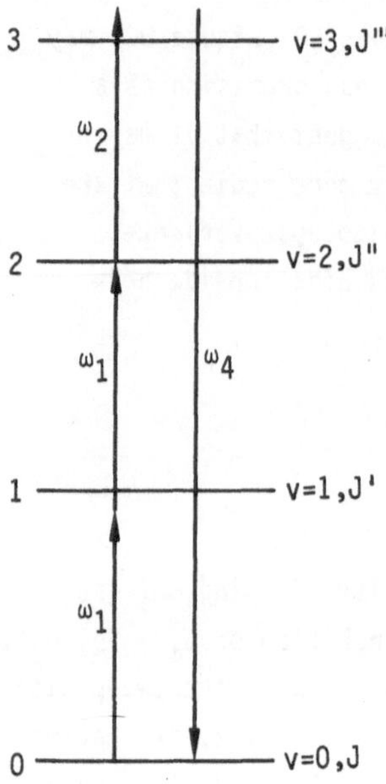

Fig.7.6. Energy-level scheme for two-photon-resonant sum-frequency generation ($\omega_4=2\omega_1+\omega_2$) in HCl

contribute to the susceptibility, because exact two-photon resonance is confined to one isotope. Since, in the case of HCl, only P and R branches are allowed on the single-photon transitions $0 \to 1$, $1 \to 2$, $2 \to 3$, $3 \to 0$, the values of J', J", J''' are restricted to the six possibilities given in Table 7.1. Following PELLIN and YARDLEY, we concentrate on the two-photon transition $|v = 0, J = 3\rangle \to |v = 2, J = 3\rangle$. The intermediate level 1 can be either $|v = 1, J = 2\rangle$ or $|v = 1, J = 4\rangle$ and level 3 can be either $|v = 3, J = 2\rangle$ or $|v = 3, J = 4\rangle$. If ω_2 is assumed to be tunable, the choice of level 3 is immaterial, because ω_2 can be brought into resonance with either of these $2 \to 3$ transitions. For definiteness, we shall assume level 3 to be $|v = 3, J = 4\rangle$. Which of the intermediate levels 1 gives the dominant contribution depends on which is the smaller of the associated frequency denominators $\Omega_{10} - \omega_1$. We have calculated the energy-level spacings using the expression

$$G(v,J) = \omega_e(v + 1/2) - \omega_e x_e(v + 1/2)^2 + \omega_e y_e(v + 1/2)^3 + [B_e - \alpha_e(v + 1/2)]J(J + 1) - [D_e + \beta_e(v + 1/2)]J^2(J + 1)^2 \quad , \tag{7.18}$$

where the spectroscopic parameters are as defined and tabulated by HERZBERG [5.67]. For the two-photon transition $|v = 0, J = 3\rangle \rightarrow |v = 2, J = 3\rangle$ we calculate the frequency to be 5660.80 cm^{-1}; therefore a laser frequency ω_1, corresponding to 2830.40 cm^{-1}, is required. The dominant intermediate level 1 is then found to be $|v = 1, J = 2\rangle$, for which the denominator $\Omega_{10} - \omega_1$ is a calculated -8.96 cm^{-1}, whereas, for $|v = 1, J = 4\rangle$, the value of $\Omega_{10} - \omega_1$ is +132.77 cm^{-1}. This illustrates our earlier remark that there is usually one clearly dominant route. The susceptibility reduces, therefore, to a single term,

$$\chi(-\omega_4;\omega_1,\omega_1,\omega_2) = \frac{N_J}{(2J+1)}\frac{1}{3\hbar^3\varepsilon_0}$$

$$\times \sum_{\substack{M,M'\\ M'',M'''}} \frac{\langle vJM|\mu|v'J'M'\rangle\langle v'J'M'|\mu|v''J''M''\rangle\langle v''J''M''|\mu|v'''J'''M'''\rangle\langle v'''J'''M'''|\mu|vJM\rangle}{(\Omega_{10}-\omega_1-i\Gamma_{10})(\Omega_{20}-2\omega_1-i\Gamma_{20})(\Omega_{30}-2\omega_1-\omega_2-i\Gamma_{30})}, \tag{7.19}$$

where the dipole matrix elements have been written so as to display explicitly their dependences on the quantum numbers v, J, M (M is the z projection of J). Each of the matrix elements in (7.19) can be factorised into a product of rotational and vibrational matrix elements, and the latter can be removed from the sum. N_J is the total population of the initial level $|v = 0, J\rangle$; therefore $N_J/(2J+1)$ is the population of each of its $(2J + 1)$ sublevels. The sum over M provides the orientation average (see Sect.2.8). For general field orientations, the summation is best evaluated by use of spherical tensor techniques [2.21]. However, if all of the fields are assumed to be polarised in the z direction, as in (7.19), the selection rule on M is that M = 0, and the summation is straightforward if tedious [2.17]. Then, for the particular route $J \rightarrow J - 1 \rightarrow J \rightarrow J + 1$,[2], the summation introduces a factor $2J(J + 1)/15(2J + 1)$; thus the susceptibility becomes,

$$\chi(-\omega_4;\omega_1,\omega_1,\omega_2) = \frac{N_J 2J(J+1)\mu_{01}\mu_{12}\mu_{23}\mu_{30}}{45(2J+1)\hbar^3\varepsilon_0(2J+1)(\Omega_{10}-\omega_1-i\Gamma_{10})(\Omega_{20}-2\omega-i\Gamma_{20})(\Omega_{30}-2\omega_1-\omega_2-i\Gamma_{30})} \tag{7.20}$$

[2]This corresponds, in the notation of PELLIN and YARDLEY, to the route Ω_{-+}; their result for the summation reduces, after algebraic simplification, to (7.20). For some of the other routes, however, there are errors in their results [2.21].

where the μ are now dipole matrix elements between vibrational states. These are known [7.28,29] and have the values μ_{01} = +0.067D, μ_{12} = +0.097D, μ_{23} = +0.12D, μ_{03} = $+5.2\times10^{-4}$D. The second and third denominators are simply equal to the linewidth Γ, which we take as 0.1 cm^{-1} at 1 atmosphere pressure [7.30]. This leads to a susceptibility of 3.4×10^{-23}m^{2}V^{-2} at atmospheric pressure. If this is expressed as a susceptibility per molecule, the result is 1.7×10^{-48}m^{5}V^{-2} (i.e., 1.2×10^{-34}cm^6/erg). The susceptibility is, therefore, a little less than in the H_2 example considered earlier, and is between one and two orders of magnitude greater than that of CO gas at one atmosphere. Thus, with incident laser powers of the order of a few MW, a conversion efficiency of ~ 10% is predicted. For third-harmonic generation, the figures are much less promising because ω_2 cannot be freely chosen as in the preceding example and the third denominator is no longer small. Thus, with the third denominator now put equal to 100cm^{-1} (say) instead of 0.1 cm^{-1}, the predicted conversion efficiency is a factor of 10^{-6} less than quoted above. This underlines an important disadvantage of scheme a, viz, that resonant enhancement of two frequency denominators is needed to achieve a reasonably large susceptibility. The process can, therefore, be efficient only for a very restricted choice of input frequencies.

7.5 Biharmonic Pumping in H_2 Gas

The principles of biharmonic pumping (or coherent Raman mixing) have been discussed in Sect.3.4. A strong pump at frequency ω_2 produces a powerful Stokes wave at frequency ω_{2s} by stimulated Raman scattering. A second pump at frequency ω_1, which is not capable, on its own, of producing a Stokes wave ω_{1s} by SRS, can nevertheless generate ω_{1s} by four-wave mixing when the strong pump is also applied. This technique has been used to generate ir radiation in H_2 gas [1.61-65]. In this section, we review the results obtained.

The experiment of BROSNAN et al. [1.62] used as the pump ω_1 the tunable output from a $LiNbO_3$ optical parametric oscillator (OPO) pumped by the 1.06µm output from a Nd:YAG laser. The 1.06µm beam was also used as the powerful pump ω_2 and this readily produced SRS in H_2 at a wavelength around 1.9µm (Stokes shift: 4155 cm^{-1}) with a photon-conversion efficiency of around 40%. The output of the OPO was not sufficiently powerful to produce SRS on its own; however, when combined with the 1.06µm pump an output at

$\omega_{1s} = \omega_1 - (\omega_2 - \omega_{2s}) = \omega_1 - (4155\ cm^{-1})$ was generated. The OPO signal wave could be tuned from 2.12μm (~ 4700 cm^{-1}) to 1.4μm (~ 7000 cm^{-1}) with a corresponding idler-wave tuning range of 2.12μm to 4.4μm (~ 2300 cm^{-1}). Thus, in principle, with ω_1 covering the signal tuning range and part of the idler tuning range, ω_{1s} can be tuned from 0 to 2845 cm^{-1}. This demonstrates clearly the advantage of biharmonic pumping, because the task of producing SRS is given to a powerful fixed-frequency source, of which a number are available, while the lower-power tunable source can be used to produce a yet lower frequency tunable output at its Stokes frequency.

To obtain a rough estimate of the expected output intensity at frequency ω_{1s}, we can use (3.36),

$$\frac{I_{1s}(L)}{I_{2s}(L)} = \left(\frac{\omega_{1s}}{\omega_{2s}}\right)^2 \left(\frac{\chi''_{SRS1}}{\chi''_{SRS2}}\right) \frac{I_{1o}}{I_{2o}} \left(1 + \left(\frac{2\Delta k}{G_1+G_2}\right)^2\right)^{-1} \quad . \tag{7.21}$$

This can be further approximated by noting that $\chi''_{SRS1} \simeq \chi''_{SRS2}$ and that typically (as we shall see below), Δk is small compared to G_2. Thus (7.21), in its usually quoted form, becomes

$$\frac{I_{1s}(L)}{I_{2s}(L)} = \left(\frac{\omega_{1s}}{\omega_{2s}}\right)^2 \frac{I_{1o}}{I_{2o}} \quad . \tag{7.22}$$

BROSNAN et al. calculated the phase mismatch Δk; for the conditions of their experiment (H_2 pressure of 20 atmospheres) Δk falls in the range 0.4 - 0.6cm^{-1} (see Fig.7.7). They also quote a typical value of G_2 as 1.6 cm^{-1} (for 20 atmospheres and $I_2 \sim 200\ MWcm^{-2}$); thus the phase-mismatch term $[2\Delta k/(G_1+G_2)]^2$ should only slightly reduce the conversion efficiency below the theoretical value predicted by (7.22). In practice, it was found that the output power measured at 5μm was about an order of magnitude below the theoretical value. Starting with ~500mJ of 1.06μm energy, in a 10ns pulse, part of this energy was used to pump the $LiNbO_3$ OPO, thus producing ~10mJ of signal output, and the rest was used to produce SRS in H_2, with an efficiency corresponding to $I_{2s}(L)/I_{2o} \simeq 0.2$. The measured energy at 5μm was 3μJ, in a pulse of 5ns duration. Although the over-all energy conversion efficiency is rather small, it should be pointed out that the few μJ is a copious amount of energy for the purposes of infrared spectroscopy. BROSNAN et al. demonstrated the capabilities of this infrared source, especially its wide tuning range, by using it to run an absorption spectrum of polystyrene over the 769-2400 cm^{-1} range.

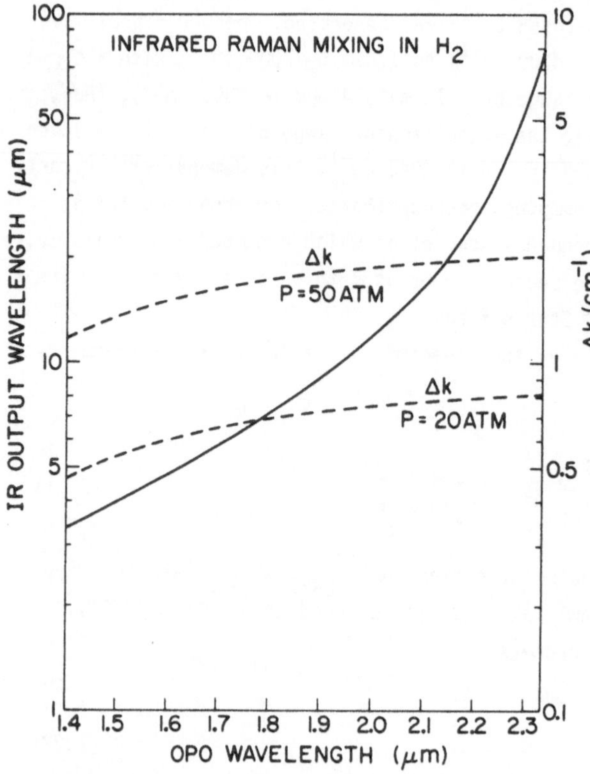

Fig.7.7. Tuning range and phase-matching factor Δk versus tunable input wavelength for infrared generation by coherent Raman mixing in H_2. (After [1.62])

To put this result in perspective, it is worth noting that comparable conversion efficiencies for generation in the 4-25μm region have been obtained by three-wave mixing in crystals, where the difference frequency of the OPO signal and idler waves is generated [7.31-33].

LOY et al. [1.63], SOROKIN et al. [1.64], and BYER and TRUTNA [1.65] have used the pure rotational S(0) Raman transition in hydrogen, which has a Stokes shift of 354.3 cm^{-1}. Stimulated rotational Raman scattering on this transition, using a circularly polarised pump at 694.3nm, was first reported by MINCK et al. [7.34]. Renewed interest in stimulated rotational Raman scattering was generated by a proposal from BYER [7.2] that by use of a powerful CO_2 laser pump it might be possible to generate Stokes radiation with a wavelength around 16μm. For example, using the R(30)$[00^01 \rightarrow 10^00]$ CO_2 laser line at 982.1 cm^{-1} as pump, the Stokes radiation from the S(0) transition in H_2 would be at 627.8 cm^{-1}. This is close to the frequency required for absorption by the Q branch of the ν_3 mode of $^{235}UF_6$. Such a source would, therefore, be of interest for laser-induced uranium-isotope separation, particularly because the Stokes generation would be expected to be efficient, once the SRS threshold is exceeded. Byer made an estimate of the required CO_2 laser-pump intensity and concluded that it should be possible to reach SRS threshold with available CO_2 laser performance. In fact, this estimate is over-optimistic in a number of respects and the first successful generation of 16µm radiation using this energy-level scheme was by the easier route of biharmonic pumping, SOROKIN et al. [1.64] and LOY et al. [1.63]. In their experiments, a ruby laser provided the strong pump (ω_2) to produce SRS. They pointed out, however, that a Nd:YAG laser would be more suitable, both in allowing a higher repetition rate and in reducing ω_{2s} [and hence increasing the conversion efficiency, as implied by (7.22)]. The ruby-laser beam was converted to circular polarisation since then pure rotational SRS, alone, was observed. Pure rotational SRS also occurred with a linearly polarised pump, but the threshold for vibrational SRS was then lower and the rotational Stokes output was always accompanied by a powerful vibrational Stokes output. The ruby laser produced up to 90MW in a 20ns pulse; this was synchronised with a TEA CO_2 laser that produced 2.5MW in a pulse of 60 ns. The beams were combined using a NaCl prism and focussed into a 2m-long cell that contained pure parahydrogen at 4 atmospheres pressure. The cell was surrounded by a liquid N_2 jacket to cool the gas to ~ 100K. With the hydrogen at room temperature the energy generated at 16µm (ω_{1s}) was ~ 10µJ, about an order of magnitude less than the rough prediction based on (7.22) (with $I_{2s}(L)/I_{2o} = 0.3$). With the gas cooled to 100K, a useful increase in output was observed, to ~ 40µJ, although it is not clear how this behaviour should be interpreted within the framework of (7.21). Thus, as with the experiment of BROSNAN et al., it has been shown by LOY and co-workers that biharmonic pumping offers a technique for generating radiation over a very wide range of the infrared, because all that need be done in their experiment is to line-tune the CO_2 laser.

Recently, results reported by BYER and TRUTNA [1.65] and RABINOWITZ et al. [1.52] show considerable progress towards the goal of efficient 16µm generation in para-H_2. Both groups used multipass arrangements to increase the interaction length and hence the gain (BYER and TRUTNA used 25 passes through a 4m-length cell). In the experiment of RABINOWITZ et al. there was enough gain to produce a measurable Stokes output (1 nJ was observed) from noise. In the experiment of BYER and TRUTNA, the gain was somewhat less and they therefore injected a 16μm signal that was subsequently amplified by the Raman process. The 16μm signal was obtained by biharmonic pumping, on the first pass through the H_2 cell, using a Nd:YAG pump in addition to the CO_2 pump. The final 16μm output was an impressive 50mJ, thus illustrating the value of the biharmonic scheme for extending the capability of Raman scattering.

7.6 Far-Infrared (fir) Generation by Stimulated Raman Scattering

In stimulated Raman scattering, when the pump frequency is a long way from resonance with intermediate levels, the Raman gain is inversely proportional to the Stokes wavelength. This imposes a limit on long-wavelength Stokes generation in media such as H_2 or N_2, where the intermediate levels are in excited electronic states. However, in heteronuclear molecules by use of rovibrational levels of the ground electronic state as intermediate levels, low-frequency enhancement of the Raman susceptibility is possible [1.61]. Figure 7.8 illustrates the scheme as suggested by DUCUING et al., [1.61] in which the Q(J) Raman transition is used ($|v = 0, J\rangle \rightarrow |v = 1, J\rangle$), with the pump frequency ω_p close to the R(J - 1) and/or R(J) lines. Notice that the scheme that involves the R(J - 1) resonance is an example of the Raman resonance depicted in Fig.5.2b. The rotational levels $c_1 = |v = 1, J + 1\rangle$ and $c_2 = |v = 0, J - 1\rangle$ are, therefore, the dominant intermediate levels (see Fig.7.8) and the Raman gain G_R, for pump and Stokes waves linearly polarised in the z direction, is given by

$$G_R \equiv g_R I_P = \frac{\pi I_p}{\varepsilon_0^2 c \lambda_s n_s n_p \hbar^3 \Gamma} \cdot \frac{N_J}{(2J+1)}$$

$$\times \sum_{M=-J}^{+J} \left| \frac{\langle b,M|\mu|c_1,M\rangle\langle c_1,M|\mu|a,M\rangle}{\Omega(J)-\omega_p} + \frac{\langle b,M|\mu|c_2,M\rangle\langle c_2,M|\mu|a,M\rangle}{\omega_p-\Omega(J-1)} \right|^2 , \quad (7.23)$$

where the terms with anti-resonant denominators have been dropped. $\Omega(J)$ and $\Omega(J-1)$ are the frequencies of the R(J), R(J-1) transitions, respectively.

It should be noted that when ω_p is in the vicinity of the R(J-1) line (transition $c_2 \rightarrow b$ in Fig.7.8), the Stokes frequency will be close to the

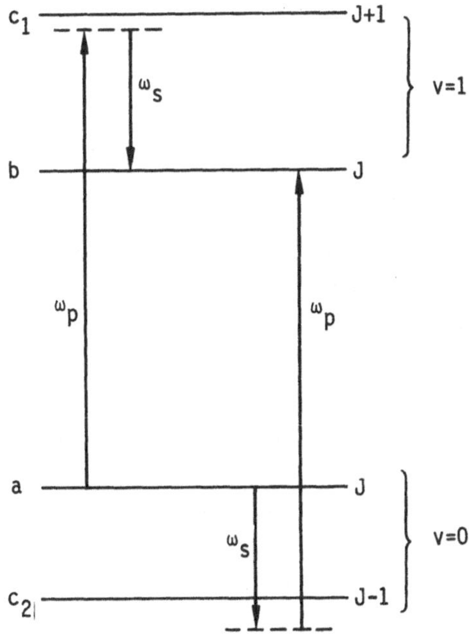

Fig.7.8. Scheme for generation of far-infrared radiation by resonantly enhanced stimulated Raman scattering. Enhancement is achieved with ω_p close to an R(J) transition (shown at the left in the figure) or an R(J-1) transition (at the right in the figure)

J - 1 → J rotational transition in the ground vibrational state. It will, therefore, be subject to strong absorption and hence reduced gain. When ω_p is close to the R(J) line (transition a → c_1 in Fig.7.8), the Stokes frequency will be close to the rotational transition J → J +1 in the excited vibrational state, but the small populations will produce only small absorption. Also this Stokes frequency will not be close to rotational transitions in the ground vibrational state, because vibrational-rotational interaction leads to differing rotational level spacings in the v = 0 and v = 1 vibrational states.

If one of the denominators, say $\Omega(J) - \omega_p$, is much smaller than the other, (7.23) simplifies to

$$G_R = \frac{N_J \pi \mu_{01}^2 \mu_{11}^2 U(J) I_p}{\varepsilon_0^2 c \lambda_s \eta_s \eta_p \hbar^3 \Gamma[\Omega(J)-\omega_p]^2} , \qquad (7.24)$$

where μ_{01} is the vibrational transition dipole moment (v = 0 → v = 1) and μ_{11} is the permanent dipole moment. The factor U(J) comes from the summation over M (i.e., the orientation averaging), which is straightforward here because the assumption of fields polarised in the z direction means that $\Delta M = 0$, i.e., intermediate and final states have the same quantum number

M as the initial state. For the case of a Q(J) transition, the pump being resonant with the R(J) transition, U(J) is given by [1.61],

$$U(J) = \frac{(J+1)(4J^2+8J+5)}{15(2J+1)^2(2J+3)} \quad . \tag{7.25}$$

DUCUING et al. considered, as a numerical example, the case of the Q(3) Raman transition in HF with a pump around 2.4μm, corresponding to a Stokes wavelength around 60μm. They showed that, for a detuning $\omega_p - \Omega(J) \simeq 10\ cm^{-1}$, the gain g_R would be $\sim 5 \times 10^{-4} cmMW^{-1}$. Thus, with a laser intensity of a few hundred $MWcm^{-2}$ and a 1m-length of medium, the single-pass gain should be in excess of e^{30} for Stokes frequencies within a tuning range of several cm^{-1} around 60μm.

FREY et al. [1.67] provided experimental confirmation of these ideas, using HCl rather than HF. (It should also be noted that CHANG and McGEE [7.35] and PETUCHOWSKI et al. [7.36] had shown that SRS plays an important role in fir generation by optical pumping, where a CO_2 laser is commonly used). In their experiment, FREY et al. used a pump wavelength in the 3.3-3.4 μm region. The pump beam, produced by two Raman shifts (in H_s gas) of the output from a ruby-laser-pumped dye laser , had the following characteristics: energy 40mJ, pulse duration 2.5ns, linewidth $<0.1\ cm^{-1}$. The HCl was contained in a 3mm-diameter, 50cm-long, glass light pipe, which provided guidance for the generated far-infrared radiation. Tunable SRS was observed for several Q branch transitions (J = 2 up to J = 7) of $H^{35}Cl$, in the range 60-160 μm. Generated Stokes energies were around 200 μJ corresponding to a peak power of ~80kW and a photon-conversion efficiency of ~12%. For the Q(4) transition, the Stokes tuning range reached a maximum of 2.3 cm^{-1} at 120 torr. A simple calculation of expected tuning range for the experimental conditions ($I_p = 200 MWcm^{-2}$) yielded 2.9 cm^{-1}, which is in very satisfactory agreement.

FREY et al. suggested a number of improvements that should extend the tuning ranges significantly. They also pointed out that HCl was chosen for convenience rather than for optimum Raman gain, and that, for many molecules, including HF, the gain should be as much as an order of magnitude greater. These predictions have since been confirmed; De MARTINO et al. [7.4], using HF, have obtained far-infrared powers and tuning ranges up to three times greater than those obtained in HCl. Therefore, it appears that this technique of resonantly enhanced SRS could provide a broad coverage of the far-infrared spectrum.

7.7 Conclusions

Although the study of nonlinear optical effects in free molecules is not new - the earliest SRS experiments in H_2 were carried out in 1963 [7.37] - the field is now experiencing a new wave of activity. Much of this stems from the idea, first considered for atoms, of exploiting two-photon-resonant processes. Another stimulus has been the increasing availability of high-power tunable infrared sources, plus the desire to extend infrared tuning ranges yet further by various nonlinear processes. Compared with nonlinear effects in atoms, the corresponding effects in molecules, involving the use of infrared rather than visible lasers, require more energetic pump lasers. This is partly because the matrix elements are typically smaller in molecules and partly because, for a given dipole moment, a dipole radiates less power the lower its frequency. Despite these extra difficulties posed by molecules, there are significant advantages, such as their great variety, and the possibility of using high densities. Certainly, the recent experiments of BRUECK and KILDAL [1.44,45], FREY et al. [1.67], and De MARTINO et al. [7.4] suggest that there is considerable scope for infrared and far-infrared applications. Finally, to correct any impression that nonlinear effects in molecules are confined to the infrared, we should mention two more experiments. INNES et al. [4.64] generated vuv radiation in NO gas by four-wave sum mixing using a two-photon resonance between levels in different electronic states, and ABRAMS et al. [7.38,39] observed Stark-induced three-wave mixing in NH_2D, where one of the waves was provided by a 4GHz microwave source.

8. Some Miscellaneous Topics

In this chapter we briefly review a number of areas of recent interest, which also serve to illustrate the wide range of topics encompassed by the field of nonlinear optics.

8.1 Multipole Interactions

The earlier chapters have been limited to the electric-dipole (E1) approximation, which for the vast majority of applications is entirely satisfactory. However, by making use of resonance enhancement of "forbidden" transitions, it is possible for multipole processes to dominate, either because the E1 process is nonresonant and, therefore, weak, or because the E1 process is forbidden.

Although optical second-harmonic generation due to an electric-quadrupole (E2) interaction was observed some time ago (in calcite, [4.1]), and similarly in the microwave spectrum due to a magnetic-dipole (M1) interaction [8.1], nonlinear multipole effects in vapours have only recently received much attention.

For example, LAMBROPOULOS et al. [8.2] pointed out, with reference to two-photon ionisation in the alkalis (lithium in particular), that in regions where the laser frequency is far off resonance with respect to E1 transitions, but close to an E2 transition, the contribution of the latter can be several orders of magnitude greater than the E1 background. However, because in multiphoton ionisation all of the possible final (continuum) states are summed over, there is always a purely E1 route for the process. Thus, the multipole contributions are only significant in the relatively uninteresting non-E1-resonant regions; moreover, they are weak in comparison to resonant E1 routes, as confirmed by calculations for cesium [8.3]. On the other hand, second-order processes are forbidden in the E1 approximation, because of inversion symmetry (or equivalently, parity selection rules). Thus, in the absence of external static symmetry-breaking fields, it is necessary to exploit multipole resonances if second-order processes are to be observed. Such mixing processes in vapours and gases were discussed in detail by HÄNSCH and TOSCHEK [8.4]; BETHUNE et al. [8.5] reported the first observation of second-order sum-frequency generation in free atoms. Figure 8.1 illustrates their scheme, which used sodium. Two lasers, of frequency ω_1 and ω_2 were used to generate the sum frequency $\omega_\sigma = \omega_1 + \omega_2$, where ω_1 and ω_2 were resonant with the 3s - 3p and 3p - 4d E1 transitions, respectively. The matrix element for the transition 4d-3s was necessarily that of a quadrupole interaction, because of parity and angular-momentum selection rules. Before discussing this and other multipole experiments, we first outline the basic theory.

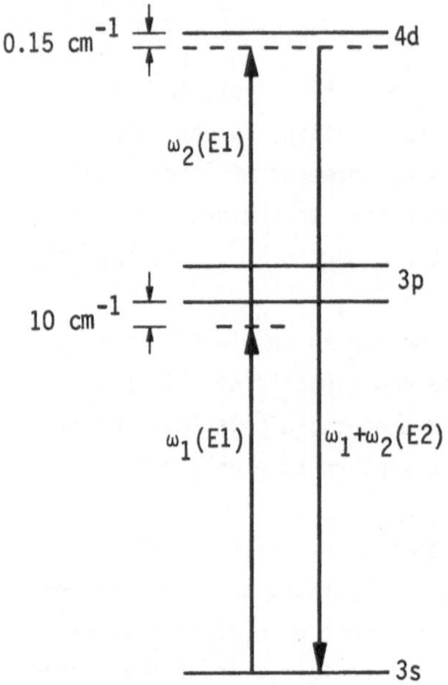

Fig.8.1. Sum-frequency generation in sodium vapour. The process creates an electric quadrupole polarisation at the sum frequency

In general, in a magnetic dielectric both polarisation ($\underline{P}$) and magnetisation ($\underline{M}$) fields act as source terms in Maxwell's equations. In earlier chapters, we have taken $\underline{M} \equiv \underline{0}$ and $\underline{P} \equiv \underline{P}^D$, where, with a change of notation, $\underline{P}^D$ is the electric-dipole polarisation. For the purposes of this chapter, we will be interested in only the lowest-order multipolar contributions for $\underline{P}$ and $\underline{M}$;

$$\underline{P} = \underline{P}^D + \underline{P}^Q \quad , \quad \underline{M} = \underline{M}^D \quad , \tag{8.1}$$

where $\underline{P}^Q$ is the quadrupolar polarisation and $\underline{M}^D$ the dipolar magnetisation. Diamagnetic contributions are ignored, for simplicity. The constitutive relations for Maxwell's equations thus become

$$\underline{D} = \varepsilon_0 \underline{E} + \underline{P} \quad , \quad \underline{B} = \mu_0(\underline{H} - \underline{M}) \quad . \tag{8.2}$$ [1]

As in Sect.2.4, $\underline{P}$ and $\underline{M}$ are computed as expectation values of the polarisation and magnetisation operators $\hat{\underline{P}}$ and $\hat{\underline{M}}$. The perturbed states are evaluated with the multipole-interaction hamiltonian. Thus [2.19,8.7,8], the multipolar operators are

$$\hat{\underline{P}}^D = -\sum_i e\underline{r}_i \equiv \underline{Q} \tag{8.3a}$$

$$\hat{\underline{P}}^Q = -\frac{1}{2} e \sum_i \underline{r}_i\underline{r}_i \cdot \nabla \equiv \underline{Q}_{(2)} \cdot \nabla \tag{8.3b}$$

$$\underline{M}^D = -\mu_B(\underline{L} + 2\underline{S}) \quad , \tag{8.4}$$

and the multipole interactions are

$$V^{E1} = -\hat{\underline{P}}^D \cdot \underline{E}(\underline{0}t) \tag{8.5a}$$

$$V^{E2} = -\hat{\underline{P}}^Q \cdot \underline{E}(\underline{0}t) \tag{8.5b}$$

$$V^{M1} = -\hat{\underline{M}}^D \cdot \underline{B}(\underline{0}t) \quad . \tag{8.6}$$

In (8.3b) $\underline{r}_i$ is the position of electron i with respect to the nucleus (origin), $\underline{Q}_{(2)}$ is the electric-quadrupole tensor (not to be confused with the related spherical-tensor multipole expansions of $\hat{\underline{P}}$ and $\hat{\underline{M}}$, e.g. BRINK and SATCHLER [2.109]), and ∇ operates on the electric fields to its right, at the origin. In (8.4) μ_B (also denoted β) is the Bohr magneton $\mu_B = e\hbar/2m \simeq 9.27 \times 10^{-24} JT^{-1}$, and $\hbar\underline{L}, \hbar\underline{S}$ are the orbital and spin angular-momentum operators. For plane-wave fields, $\underline{E}(\underline{r}t) = \frac{1}{2} \underline{E}_\omega \exp i(\underline{k} \cdot \underline{r} - \omega t)$

---

[1] In much of the literature (e.g., [2.1,15]), these sources are introduced through the current density

$$\underline{J} = \frac{\partial \underline{P}}{\partial t} + \nabla \times \underline{M} \quad ;$$

this is equivalent to (8.2), with $\underline{J} = 0$, [8.6].

+ c.c. as usual, and so

$$V^{E2} = -\frac{1}{2}\, i\underline{k}\cdot\underline{Q}_{(2)}\cdot\underline{E}_{\omega}e^{-i\omega t} + \text{hermitian conjugate} \tag{8.7}$$

$$V^{M1} = -\frac{1}{2}\,(|\underline{k}|c)^{-1}\hat{\underline{M}}^{D}\cdot\underline{k}\times\underline{E}_{\omega}e^{-i\omega t} + \text{hermitian conjugate}\ . \tag{8.8}$$

Use of (8.3-8) in the calculation of the nonlinear susceptibilities is perfectly straightforward, because the interactions are linear in the field. The E1 interactions $\underline{\varepsilon}_1\cdot\underline{Q}\ \ldots\ \underline{\varepsilon}_n\cdot\underline{Q}$ in (2.15) arise from the perturbation solution for the ground state $|g\rangle$, and must be replaced by $\tilde{V}^{E1}+\tilde{V}^{E2}+\tilde{V}^{M1}$, where

$$\tilde{V}^{E1} = \underline{\varepsilon}\cdot\underline{Q} \tag{8.9a}$$

$$\tilde{V}^{E2} = i\underline{k}\cdot\underline{Q}_{(2)}\cdot\underline{\varepsilon} \tag{8.9b}$$

$$\tilde{V}^{M1} = (\mu_B/|\underline{k}|c)(\underline{L}+2\underline{S})\cdot(\underline{k}\times\underline{\varepsilon})\ . \tag{8.9c}$$

[For the M1 case, we can use $\underline{b}=\underline{k}\times\underline{\varepsilon}/|\underline{k}|$, the unit magnetic induction vector, and $E\rightarrow cB$ in (2.9) say, rather than (8.9c).] Similarly, $\underline{\varepsilon}_\sigma^*\cdot\underline{Q}$ is replaced by $\underline{\varepsilon}_\sigma^*\cdot\hat{\underline{P}}$ or $\underline{\varepsilon}_\sigma^*\cdot\hat{\underline{M}}$ depending on whether the polarisation or magnetisation is being calculated. FLYTZANIS [2.1] gave general formulae in terms of the vector potential; BABIKER [2.13] derived a relativistic quantum-electrodynamic nonlinear susceptibility, and POWER and THIRUNAMACHANDRAN [8.9] discussed the multipole polarisability $\alpha_{fg}(\omega_1;\omega_2)$. This formalism is most easily appreciated by reference to an example, for which we return to the experiment of BETHUNE et al. [8.5].

From Fig.8.1 it is clear that the quadrupole-moment density $\underline{P}^Q$ is to be calculated; following the above prescription, we find

$$P_\sigma^Q = \varepsilon_0\chi^{(2)}(-\omega_\sigma;\omega_1\omega_2)E_1E_2\,e^{i(\underline{k}_1+\underline{k}_2)\cdot\underline{r}} \tag{8.10a}$$

$$\chi^{(2)}(-\omega_\sigma;\omega_1\omega_2) = \frac{\mathcal{N}}{2\hbar^2\varepsilon_0}\cdot\frac{\langle 3s|i(\underline{k}_1+\underline{k}_2)\cdot\underline{Q}_{(2)}\cdot\underline{\varepsilon}_\sigma^*|4d\rangle\langle 4d|\underline{\varepsilon}_1\cdot\underline{Q}|3p\rangle\langle 3p|\underline{\varepsilon}_2\cdot\underline{Q}|3s\rangle}{(\Omega_{4d3s}-\omega_1-\omega_2)(\Omega_{3p3s}-\omega_2)}\ . \tag{8.10b}$$

Because $\chi^{(2)}$ contains a field-dependent quantity, $\underline{k}_1 + \underline{k}_2$, it is not a tensor property of the medium alone. This can be remedied by introducing the fourth-rank quadrupolar tensor $\underset{\approx}{\chi}_Q^{(2)}$ defined by

$$i(\underline{k}_1 + \underline{k}_2) \cdot \underset{\approx}{\chi}_Q^{(2)} : \underline{\varepsilon}_1\underline{\varepsilon}_2 = i|\underline{k}_1 + \underline{k}_2|\chi_Q^{(2)} = \chi^{(2)}(-\omega_\sigma;\omega_1\omega_2) \quad . \qquad (8.11)$$

Equation (8.11) can be regarded as but one term in the expansion of spatially dispersive nonlinear susceptibilities as a Taylor series in the wave-vector variables, see FLYTZANIS [2.1].

Taking typical values for $\chi^{(3)}$, BETHUNE et al. [8.5] found that $\underline{P}^Q$ was comparable to third-order dipole polarisations. Thus, despite the weakness of the E2 transition, the use of resonance enhancement ($\Omega_{3p3s} - \omega_2 \simeq 10\ \mathrm{cm}^{-1}$, $\Omega_{4d3s} - \omega_1 - \omega_2 \simeq 0.15\ \mathrm{cm}^{-1}$) resulted in readily observable sum-frequency signals.

A complication in this experiment is that symmetry considerations show that $\chi_Q^{(2)}$ vanishes for collinear propagation. Although an elegant macroscopic argument can be used, e.g., PERSHAN [8.10], a simpler approach for our purposes is to consider the matrix-element selection rules for the process. For an isotropic medium, $\underline{k}_1$ and $\underline{\varepsilon}_1$ etc., are orthogonal. If we assume collinear propagation along the z axis and, further, take the quantisation axis to be the z axis, then the E1 transitions will cause the azimuthal quantum numbers to change by ±1. Thus, from the right in (8.10b), the states go as $|3s\ m = 0\rangle \rightarrow |3p\ m = \pm 1\rangle \rightarrow |4d\ m = 0, \pm 2\rangle$. On the other hand, the E2 matrix element $\sim\langle 3s\ m = 0|(x,y)z|4d\ m = \pm 1\rangle$, and thus angular momentum cannot be conserved in the process: $\chi_Q^{(2)}$ vanishes identically for collinear propagation in an isotropic medium. For this reason BETHUNE et al. used a crossed-beam geometry.

By introducing static transverse magnetic fields, this argument no longer applies, and collinear propagation is possible. Notice that the effects of the static field can be introduced in two ways. Taking (8.10) as an example, the states $|3s\rangle$, $|3p\rangle$, $|4d\rangle$ can be replaced by first-order perturbed wave functions. Then, for example, the magnetic field will couple the $|4d\ m = 0\rangle$ and $|4d\ m = \pm 1\rangle$ sublevels, and thus complete the chain of matrix elements, for collinear propagation. Alternatively, the process can be considered as a four-wave process, in which ω_1, ω_2 and $\omega_3 = 0$ are mixed to produce $\omega_1 + \omega_2$. This is dc-induced sum mixing, and would be described by a $\chi_Q^{(3)}(-\omega_1-\omega_2;\omega_1\omega_2 0)$. Using a suitable (2.15), the result is equivalent to (8.10b) evaluated with perturbed wave functions. Thus, the magnetic mixing of sublevels is viewed as a further transition (at zero frequency) in the process. Because the polarisation $\sim H$, the magnetic field strength, the generated intensity of these parametric processes $\sim H^2$. (In passing, we note that E1 dc-induced processes are well-known, see e.g., Table 2.1 and [8.11]).

A number of collinear-propagation mixing experiments with magnetic fields have been reported. FLUSBERG et al. [8.12] tuned two lasers into Raman resonance with the thallium $6^2P_{1/2}$ - $6^2P_{3/2}$ mixed M1/E2 transition, via the $7^2S_{1/2}$ intermediate state, thereby generating the difference frequency at ~ 7793 cm^{-1}. By increasing the pump laser power, or number density, they found that SRS occurred from noise, so that only one laser was needed; the process is analogous to Raman-resonant difference mixing as described in Chap.6. FLUSBERG et al. [8.13] repeated the experiments of BETHUNE et al. [8.5], but with a magnetic field, and were able to observe some interference effects. MATSUOKA et al. [8.14] observed second-harmonic generation in sodium and calcium vapours, using E2 resonances; similarly, MOSSBERG et al. [8.15] reported second harmonic generation in thallium vapour using M1 and E2 resonances.

By way of contrast, COTTER and YURATICH [5.98] observed stimulated Raman scattering on the 4s-4p opposite-parity transition in potassium vapour. There is no phase-matching requirement for SRS, and the process occurs collinearly, with no need for an external static field. The pump was tuned to the 4s-3d E2 resonance, and thus generated Stokes radiation tunable over a 15 cm^{-1} range around 1.18μm.

Thus multipole processes in atomic vapours are readily observable when use is made of resonance enhancement. Although they are of limited potential as tunable sources, they are of intrinsic interest, and may be helpful for the spectroscopy of forbidden transitions. On the other hand, it is necessary to arrange matters so as to minimise competing effects, particularly the stronger E1 process; in the SRS case this was achieved by working with the lowest transitions.

8.2 Laser-Induced Inelastic Collisions

When the energy defect ΔE between the initial and final states of a system of two colliding atoms is large compared to k_BT then the cross section for inelastic collision is small. GUDZENKO and YAKOVLENKO [8.16] considered a collision process in which a large energy defect is bridged by emission or absorption of a photon; for example a collision in which atom A goes from state 2 to 1 and atom B from 1 to 2, with emission of a photon of energy $\hbar\omega$,

$$A(2) + B(1) \rightleftharpoons A(1) + B(2) + \hbar\omega \quad .$$

The frequency ω is close to $\omega_0 = \Omega_{21}^A + \Omega_{12}^B$, where the Ω are transition frequencies of the isolated atoms. This type of process was independently considered by HARRIS and LIDOW [8.17], who also suggested variations that involve more than one photon. In particular, they considered the four-photon process shown in Fig.8.2. Irradiation by fields at ω_1, ω_2 and ω_3, where $\omega_1 + \omega_2$ is two-photon resonant with the 6s-7s transition in Hg (atom A), produces a large dipole moment in A, which oscillates at the sum frequency $\omega_1 + \omega_2 + \omega_3$. A Kr atom B that passes close to atom A experiences a strong field at this sum frequency, which induces (by dipole-dipole interaction) an oscillating dipole moment at this frequency in atom B. The presence of a fourth frequency, ω_4, then allows atom B to undergo the transition 4p-5p, in a way closely analogous to two-photon absorption. HARRIS and LIDOW analysed this type of process by a perturbation approach and showed that the cross section for four-photon absorption by Kr would be greatly increased if the Hg atoms were also present and subjected to the two-photon resonance radiation. A more detailed analysis has since been given by HARRIS and WHITE [8.18]. This paper and the one by CAHUZAC and TOSCHEK [8.19] are useful sources of further references on this topic. An alternative approach to the analysis, that adopted by GUDZENKO and YAKOVLENKO, is to regard the process as a free-free photon absorption by the A-B quasimolecule. Further details are given in the review by YAKOVLENKO [8.20].

Laser-induced inelastic collisions have now been demonstrated experimentally [8.21,22]. The first reported observation [8.23] was later ascribed to another effect [8.21,24]. The opportunities for competing effects to mask the desired process are many. For example, in addition to the process shown in Fig.8.2, it would be possible for sum-frequency generation to occur in Hg (i.e., photons of frequency $\omega_1 + \omega_2 + \omega_3$ are produced); those photons, together with photons of frequency ω_4 could cause the Kr atom to undergo two-photon absorption. No collision-induced transfer of excitation is involved in this process. This competing effect could be discriminated against (as suggested in [8.17]) by propagating the ω_3 beam counter to the ω_1, ω_2 beams, thus ensuring a very short coherence length for the sum-frequency process.

The energy-level scheme for the successful demonstration by HARRIS et al. [8.21] is shown in Fig.8.3. A mixture of Sr and Ca vapours was used. The 5p level of Sr acted as a storage level, being pumped on the low-frequency side of the 5s-5p transition, by use of a dye laser (4617Å). Excitation from

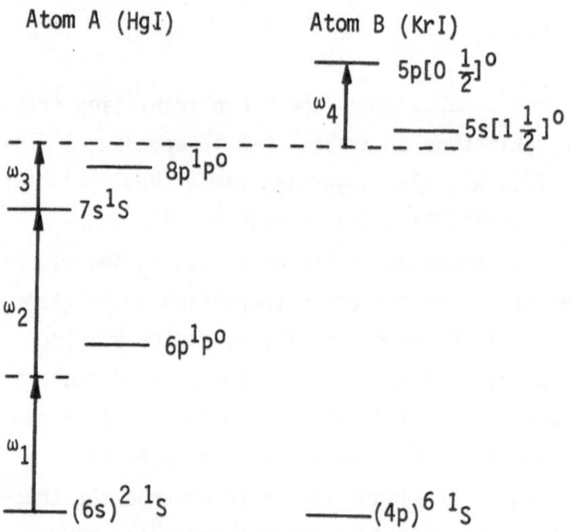

Fig.8.2. Energy-level diagrams for four-photon absorption by a Kr:Hg mixture. Atom A transfers its excitation at the sum frequency $(\omega_1 + \omega_2 + \omega_3)$ to atom B by dipole-dipole interaction. A fourth photon ω_4 excites the Kr atom to the 5p state. (After [8.17])

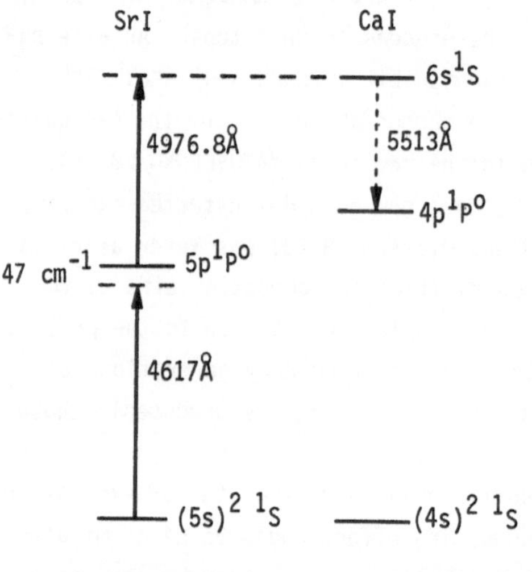

Fig.8.3. Energy-level diagrams for the Sr:Ca laser-induced inelastic-collision experiment. The 4617Å and 4976.8Å were provided by dye lasers, and the 6s population of Ca was monitored by 5513Å fluorescence. (After [8.21])

this level of Sr to the 6s level of Ca was then achieved by a second dye laser (4976.8Å), the 6s-level population being monitored by 5513Å fluorescence as atoms decay to the 4p level of Ca. Further details on the precautions taken to ensure that this was indeed due to laser-induced inelastic collision are to be found in the paper by HARRIS et al. [8.21]; the artifact that misled them in their first experiment is also described there.

HARRIS and his colleagues have suggested a number of possible applications for these inelastic collision processes. These include isotopically selective ionising collisions [8.25], optical-pumping schemes aimed at short-wavelength laser operation [8.17] and up-conversion by a collisional Raman process (see Fig.8.4). There has also been considerable interest in theoretical aspects of the process; references to these papers are given by HARRIS and WHITE [8.18]. In addition to dipole-dipole interactions, VARFOLOMEEV [8.26] has considered a quadrupole-dipole interaction for the process of down-conversion.

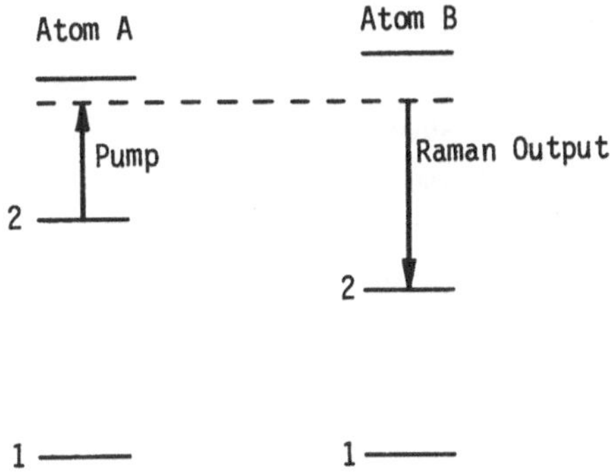

Fig.8.4. A pump applied to atom A initially in state 2 leads to gain at the Raman output frequency, leaving atom B in state 2 and atom A in state 1. Population inversion of state 2 relative to state 1 in atom A is not required. (After [8.21])

8.3 Conjugate Wave-Front Generation by Four-Wave Mixing

Consider the nonlinear polarisation described by a term of the form $P_\omega = \frac{3}{2}\varepsilon_0\chi^{(3)}(-\omega;\omega-\omega\omega)E_\omega(1)E_\omega(2)^*E_\omega(3)$, see Sect.2.3, (2.69). If the three incident fields, all at the same frequency ω, are of the form $E_\omega(1)\sim\exp(i\underline{k}_1\cdot\underline{r})$, $E_\omega(3)\sim\exp(-i\underline{k}_1\cdot\underline{r})$ (i.e., waves 1 and 3 are counterpropagating plane waves) and $E_\omega(2)\sim\exp(i\underline{k}_2\cdot\underline{r})$ then the polarisation P_ω is of the form $\exp(-i\underline{k}_2\cdot\underline{r})$. Thus the polarisation radiates a plane wave, 4, of the form $\exp(-i\underline{k}_2\cdot\underline{r})$, this being the complex conjugate of wave 2, and, therefore, travels in the direction opposite to wave 2. Phase matching for this process is satisfied whatever $\underline{k}_2$, because the general requirement $\underline{k}_4 = \underline{k}_1 - \underline{k}_2 + \underline{k}_3$ is obviously satisfied by $\underline{k}_4 = -\underline{k}_2$, when $\underline{k}_3 = -\underline{k}_1$. Because this arrangement of counter-propagating pump waves 1 and 3 leads to generation of the phase-conjugate (image) wave 4 for each incident (object) plane-wave component, it follows that complicated wave fronts can be phase conjugated. This is equivalent to generation of a time-reversed version of the object wave and means that, for example, phase aberrations produced in the object wave when it passes through an optically imperfect medium can be removed by generating the time-reversed image wave and passing this back through the aberrating medium. The various possible applications that stem from this basic property have aroused much recent interest (e.g. [8.27-33]).

Experimental demonstrations of the effect and its wave-front-correcting behaviour were first demonstrated in CS_2 liquid [8.28,29,34]. Later experiments have been performed in Na vapour with a pulsed laser [8.30] and with CW lasers in both ruby [8.31] and Na vapour [8.32]. By tuning the laser close to a single-photon or two-photon resonance the susceptibilities can be made very large. This is clearly seen in (2.69), where the single-photon and two-photon resonant contributions to $\chi^{(3)}(-\omega;\omega-\omega\omega)$ are shown separately. Thus, for a laser tuned within 1 cm^{-1} of the Na D_1 line, BLOOM et al. [8.30] showed that it is easy to obtain nonlinearities that are several orders of magnitude greater than in CS_2 liquid, despite the density being many orders of magnitude less. Solutions of the coupled equations for growth of the object and image waves show [8.28,29] that both waves experience gain and that for a sufficiently large susceptibility and pump field, self-oscillation can ensue. This behaviour has been observed by BLOOM and BJORKLUND [8.28]. They also point out that, because self-focussing has its origin in the same nonlinearity, the condition under which self-oscillation occurs is also essentially the same as the condition for which whole-beam self-focussing of the pump wave occurs.

The use of two-photon resonance enhancement of $\chi^{(3)}(-\omega;\omega-\omega\omega)$ in a gas leads to very interesting results, because by tuning to exact two-photon resonance the medium is excited to a two-photon coherent state of zero wave momentum. Because this state is independent of atomic position (pointed out in [2.105]), it is not disturbed by thermal motion of the atoms, and the decay of this state is due to homogeneous relaxation processes, such as collisions and spontaneous emission. LIAO et al. [8.35] exploited this behaviour to obtain spectroscopic data with sub-Doppler resolution despite using a wide-bandwidth laser. They subjected the medium (Na vapour) to counter-propagating pump pulses from a dye laser tuned to two-photon resonance with the 3s-4d transition in Na. The same laser also provided a probe pulse (the object wave in the preceding discussion), which could be given a variable delay relative to the pump pulse. By monitoring the intensity of the backward-generated (image) wave as a function of delay, they could observe the decay (T_2) of the two-photon coherent state. In this way, even using a dye-laser linewidth of several GHz, LIAO et al. were able to measure pressure-broadened linewidths (the inverse of the measured T_2) down to several tens of megahertz.

Appendix

Units for Nonlinear Optical Susceptibilities

There are three main sources of numerical error in work with nonlinear susceptibilities: numerical factors that arise from permutation symmetry and the definition of field, choice of macroscopic or microscopic susceptibilities, and units.

The question of numerical factors has been discussed at length in Sect.2.2; our conventions are summarised by the numerical factor K in (2.7,9). In this appendix, K is set equal to unity.

The susceptibilities defined in (2.2) refer to the macroscopic medium, and unless stated otherwise are used throughout this review. The susceptibility per atom is also used in the literature; we shall denote this by $\chi^{(n)mic}$, where

$$\chi^{(n)} = \mathcal{N}\chi^{(n)mic} \quad ; \tag{A.1}$$

$\mathcal{N}$ is the number density of atoms or molecules in the medium.

In this appendix, we concentrate on units. Both SI and cgs/esu units are widely used, but in conformity with present trends we have adopted SI units. The most common definition of susceptibility in SI units, and our choice (Sect.2.1), is of the form (A.2); similarly the cgs/esu definition is (A.3):

$$\left.\begin{array}{ll} P^{(n)} = \varepsilon_0\chi^{(n)}E^n & \mathrm{Cm^{-2}} \\ \chi^{(n)} & (\mathrm{m/V})^{n-1} \\ \chi^{(n)mic} & \mathrm{m^3(m/V)^{n-1}} \end{array}\right\} \mathrm{SI} \qquad (A.2)^{1}$$

[1]An alternative definition of $\chi^{(n)}$ is also found in the literature. This omits $\varepsilon_0(\mathrm{Jm^{-1}V^{-2}})$ from the first of (A.2); because this involves only a simple change of SI units, we will not discuss it further, although the definition of transition hyperpolarisability in Sect.2.7.2 is of this type, as is the susceptibility plotted in Fig.7.2.

$$\left.\begin{array}{lll} P^{(n)} = \chi^{(n)}E^n & \text{esu} & \\ \chi^{(n)} & (\text{cm}^3/\text{erg})^{(n-1)/2} & \\ \chi^{(n)\text{mic}} & \text{cm}^3(\text{cm}^3/\text{erg})^{(n-1)/2} & \end{array}\right\} \text{cgs/esu} \qquad \text{(A.3)}$$

Notice that $\chi^{(1)}$ is dimensionless in both sets of units.

Finally, the relations between the susceptibilities in (A.2) and (A.3) are (c in ms^{-1})

$$\chi^{(n)}(\text{SI})/\chi^{(n)}(\text{esu}) = 4\pi/(10^{-4}c)^{n-1} \qquad \text{(A.4a)}$$

$$\chi^{(n)\text{mic}}(\text{SI})/\chi^{(n)\text{mic}}(\text{esu}) = 4\pi/10^6(10^{-4}c)^{n-1} \quad . \qquad \text{(A.4b)}$$

For convenience, some sample values are given below.

	n = 1	n = 2	n = 3	n = 5
$\chi^{(n)}(\text{SI})/\chi^{(n)}(\text{esu})$	12.6	4.19×10^{-4}	1.40×10^{-8}	1.56×10^{-17}
$\chi^{(n)\text{mic}}(\text{SI})/\chi^{(n)\text{mic}}(\text{esu})$	12.6×10^{-6}	4.19×10^{-10}	1.40×10^{-14}	1.56×10^{-23}

Universal Constants

(used in the text)

Symbol	Value	Name
a_0	$5.29177\cdot10^{-11}$ m	Bohr radius
α	1/137.036	Fine-structure constant (see also, list of major symbols)
c	$2.997925\cdot10^{8}$ ms^{-1}	Velocity of light in vacuo
e	$1.60219\cdot10^{-19}$ C	|Electronic Charge|
ε_0	$8.85419\cdot10^{-12}$ Fm^{-1}	Permittivity of free space
$\hbar = h/2\pi$	$1.05459\cdot10^{-34}$ Js	Dirac constant
k_B	$1.38062\cdot10^{-23}$ JK^{-1}	Boltzmann constant
m	$9.10956\cdot10^{-31}$ kg	Electron Rest mass
μ_B	$9.27410\cdot10^{-24}$ JT^{-1}	Bohr magneton
μ_0	$1.25664\cdot10^{-6}$ Hm^{-1}	Permeability of free space
r_e	$2.81794\cdot10^{-15}$ m	Classical electron radius
R_∞	$109737\cdot31$ cm^{-1}	Rydberg constant
D	$10^{-21}/c$ Cm or 0.39343 au	Debye unit

Useful Relationships

$r_e = \alpha^2 a_0$ (m)

$R_\infty = \alpha^2 mc/4\pi\hbar = \alpha/4\pi a_0$ (m^{-1})

$e^2/4\pi\varepsilon_0 = \alpha\hbar c$ (Jm)

$\mu_B = e\hbar/2m$ (JT^{-1})

$a_0 = \hbar/\alpha mc$ (m)

$\mu_0\varepsilon_0 c^2 = 1$

List of Major Symbols and Acronyms

(see also List of Constants; not all acronyms are used in the text)

Symbol	Unit	Meaning (and, where useful, its first occurrence)
a		Upper state of two-level atom (Sect.2.6)
A_i	Vm^{-1}	$\lvert E_i \rvert$, magnitude of electric field
$\mathscr{A}$	$C^4m^4s^2$	Anti-symmetric part of Raman susceptibility (Sect.5.3)
AF		Adiabatic-Following
ARP		Adiabatic-Rapid-Passage
ASE		Amplified Spontaneous Emission
b	m	Confocal Parameter of gaussian beam (Sect.4.2)
b		Lower state of two-level atom (Sect.2.6)
$\underline{B}$	T	Magnetic induction vector
BHP		Biharmonic Pumping
c.c.		Complex conjugate
CARS		Coherent AntiStokes Raman Scattering/ Spectroscopy
D	Cm	Debye (see list of constants)
$\underline{D}$	Cm^{-2}	Displacement vector
DFG/M		Difference-Frequency Generation/Mixing
$d\sigma_R/d\Omega$	$m^2 sterad^{-1}$	Differential Raman cross section (Sect.5.1)
$(d\sigma_R d\Omega)_{\|\|,\omega_p}$	$m^2 sterad^{-1}$	Differential Raman cross section for parallel linear polarisation, at frequency ω_p (Sect.7.2)
E1		Electric-dipole transition
E2		Electric-quadrupole transition (Sect.8.2)

List of Major Symbols and Acronyms (continued)

Symbol	Unit	Meaning (and, where useful, its first occurrence)
$E(\underline{r}t)$	Vm^{-1}	Total electric field (Sect.2.1)
$E(\underline{r}\omega)$	$Vm^{-1}s$	Fourier transform of total electric field (Sect.2.1)
E_i	Vm^{-1}	Amplitude of monochromatic electric-field component (Sect.2.2)
E_{io}	Vm^{-1}	Incident electric field, $E_i(z = 0)$ (Sect.3.2)
f	m	Position of focus (Sect.4.2)
fir		Far infrared
f_{ij}		Oscillator strength $i \rightarrow j$
FWM, 4WM		Four-wave mixing
g_{HR}	m^3W^{-2}	SHRS gain per (unit pump intensity)^2 (Sect.3.3)
G_{HR}	m^{-1}	SHRS gain (Sect.3.3)
G_p	m^{-1}	Parametric gain (Sect.3.4.1)
g_R, g_1, g_2	mW^{-1}	SRS gain per unit pump intensity (Sect.3.4)
G_R, G_1, G_2	m^{-1}	SRS gain (Sect.3.3)
$G_R(\xi,r), G_{Ro}$	m^{-1}	SRS gain in gain-focussing, and peak value (Sect.5.3.2)
g_{TPE}	mW^{-1}	Stimulated TPE gain per unit pump intensity (Sect.5.5.2)
$G_\pm$	m^{-1}	Characteristic gains, complex (Sect.3.4)
$\underline{H}$	Am^{-1}	Magnetic field vector
ir		Infrared
I_i	Wm^{-2}	Intensity of monochromatic wave (Sect.2.2)
I_{io}	Wm^{-2}	Incident intensity $I_i(z = 0)$ (Sect.3.2)
$I(\Delta k,\xi,\zeta)$		Focussing integral (Sect.4.2)
I_{sat}	Wm^{-2}	Saturation intensity (Sect.4.6.1)
J_{sat}	J	Saturation energy (Sect.4.6.1)

List of Major Symbols and Acronyms (continued)

Symbol	Unit	Meaning (and, where useful, its first occurrence)
k, k_ω	m^{-1}	$\eta_\omega \omega/c$, magnitude of wave-vector (Sect.3.1)
k_{TPA}	mW^{-1}	Two-photon absorption coefficient per unit pump intensity (Sect.3.3)
K_{TPA}	m^{-1}	Two-photon absorption coefficient (Sect.3.3)
K		Spherical-tensor rank (Sect.2.8)
$K(-\omega_\sigma;\omega_1\ldots\omega_n)$		Numerical K-factor, coefficient of $\chi^{(n)}$ in polarisation (Sect.2.2)
L	m	Interaction length in nonlinear medium (Sect.3.2)
L_c	m	$\lvert\pi/\Delta k\rvert$, coherence length (Sect.3.2)
M1		Magnetic-dipole transition (Sect.8.1)
$\hat{\underline{M}}^D$	Am^{-1}	Dipolar magnetisation operator
$\underline{M}$	Am^{-1}	Magnetisation vector
N	m^{-3}	Effective number density of resonant transition (Sect.2.5)
$\mathcal{N}$	m^{-3}	Number density of medium (Sect.2.5)
OKE		Optical Kerr Effect
p		Subscript denoting pump in SRS
p	torr	Vapour pressure (Sect.4.4.1)
P	W	Power
P	Cm^{-2}	Polarisation
$P(\underline{r}t)$	Cm^{-2}	Macroscopic polarisation (Sect.2.1)
$P(\underline{r}\omega)$	$Cm^{-2}s$	Fourier transform of macroscopic polarisation (Sect.2.1)
$P_\sigma^{(n)}$	Cm^{-2}	Monochromatic n^{th} order polarisation at frequency ω_σ (Sect.2.2)
P_{cr}	W	Critical power for self-focussing (Sect.4.6.2)
P_{pth}	W	Threshold pump power in SRS (Sect.5.3.2)
$\tilde{P}_p$	W	$G_{Ro}b_p\kappa$, normalised pump power in SRS (Sect.5.3.2)

List of Major Symbols and Acronyms (continued)

Symbol	Unit	Meaning (and, where useful, its first occurrence)
P^L, P^{NL}	Cm^{-2}	Linear, nonlinear polarisation
$\hat{\underline{P}}^D$	Cm^{-2}	Dipolar polarisation operator
$\hat{\underline{P}}^Q$	Cm^{-2}	Quadrupolar polarisation operator
q		Fano q-parameter in autoionisation (Sect.4.3.2)
Q		Dimensionless vibrational coordinate (Sect.2.7.1)
Q^*_Ω		Slowly-varying envelope of vibrational coordinate (Sect.2.7.2)
$\underline{Q}, Q_{ij}$	Cm	Electric dipole-moment operator, and its matrix element (Sect.2.4)
$\underline{Q}_{(2)}$	Cm^2	Electric quadrupole-moment operator (Sect.8.1)
$\underline{r}$	m	Spatial position
$\underline{r}, \underline{r}'$		Bloch-vector in one- and two-photon vector models (Sect.2.6)
R		Frequency and susceptibility ratio in four-wave mixing (Sect.3.4)
$\mathcal{R}$		Polarisation "flip" ratio (Sect.5.3.1)
s		Subscript denoting Stokes in SRS
sinc(x)		sin(x)/x
$\mathcal{S}$	$C^4m^4s^2$	Symmetric part of Raman susceptibility (Sect.5.3.1)
$\mathcal{S}$		Intrinsic permutation operator (Sect.2.2)
$\mathcal{S}_T$		Overall permutation operator (Sect.2.2)
SFG		Sum-Frequency Generation
SHG		Second-Harmonic Generation
SIT		Self-Induced Transparency
SMBS, SBS		Stimulated Mandelsh'tam-Brillouin Scattering
SRS		Stimulated Raman Scattering

List of Major Symbols and Acronyms (continued)

Symbol	Unit	Meaning (and, where useful, its first occurrence)
SERS		Stimulated Electronic Raman Scattering
SHRS		Stimulated Hyper-Raman Scattering
SVRS		Stimulated Vibrational Raman Scattering
T	K	Absolute Temperature
T_1	s	Longitudinal relaxation time - "recirculation" time of a transition (Sect.2.6.8)
T_2	s	Transverse homogeneous relaxation time - dephasing time $1/\Gamma$ (Sect.2.6.8)
THG		Third-Harmonic Generation
TPA/E/I		Two-Photon Absorption/Emission/Ionisation
u,u'		In-phase component of Bloch-vector (Sect.2.6)
uv		Ultraviolet
v,v'		Quadrature component of Bloch-vector (Sect.2.6)
$\bar{v}$	ms^{-1}	Mean thermal velocity of atom or molecule
vis		Visible
vuv		Vacuum ultraviolet
w,w'		Fractional population inversion, third component of Bloch-vector (Sect.2.6)
w_o,w_{io}	m	Gaussian beam-waist radius (Sect.4.2)
W	Wm^{-3}	Rate of energy density flow from fields to medium (Sect.2.6.3)
$W^{(1)},W^{(2)}$	s^{-1}	One- and two-photon absorption transition rates (Sect.4.6)
z	m	Distance along direction of propagation
z_F	m	Self-focussing distance (Sect.4.6.2)

List of Major Symbols and Acronyms (continued)

Symbol	Unit	Meaning (and, where useful, its first occurrence)
Z		Charge number
Z		Partition function
α		Fine-structure constant (see list of constants)
α		Photon-conversion efficiency (Sect.3.4)
$\alpha, \alpha_{fg}(\omega_1;\omega_2)$	Cm^2V^{-1}	Two-photon matrix element, first order transition hyperpolarisability (Sect.2.5)
$\alpha, \alpha_{gg}(-\omega;\omega)$	Cm^2V^{-1}	Ground-state polarisability (Sect.2.7.2)
$\alpha_R, \alpha_{fg}(-\omega_s;\omega_p)$	Cm^2V^{-1}	Raman polarisability (Sect.2.7.1)
$\alpha^E(Q)$	Cm^2V^{-1}	Electronic polarisability of molecule (Sect.2.7.1)
β	rad	Angular distance (Sect.3.4)
β	m^{-1}	Single-photon absorption coefficient (Sect.5.4.1)
$\beta, \beta', \beta^{(n)}$	s^{-1}	On-resonance Rabi frequencies, one-, two-, n-photon (Sects.2.6,2.7)
$\beta, \beta_{fg}(\omega_1;\omega_2\omega_3)$	Cm^3V^{-2}	Three-photon matrix element, second order transition hyperpolarisability (Sect.2.5)
$\beta_{HR}, \beta_{fg}(-\omega_s;\omega_{p1}\omega_{p2})$	Cm^3V^{-2}	Hyper-Raman polarisability (Sect.2.5)
γ	rad s^{-1}	linewidth
$\gamma_{fg}^{(n)}(-\omega_\sigma';\omega_1\ldots\omega_n)$	$Cm(Vm^{-1})^{-n}$	n^{th} order transition hyperpolarisability (Sect.2.7.3)
Γ	rad s^{-1}	HWHM damping parameter, $1/T_2$ (Sect.2.5)
$\delta(x)$		Dirac delta-function
$\delta_\pm, \delta_\pm'$	rad s^{-1}	One- and two-photon Stark shifts (Sect.2.6.2)
Δ	rad s^{-1}	Detuning from one- and two-photon resonances
Δ'	rad s^{-1}	Detuning from two-photon resonance including quadratic Stark shifts

List of Major Symbols and Acronyms (continued)

Symbol	Unit	Meaning (and, where useful, its first occurrence)
$\Delta E_i/\hbar, \Delta\Omega_i$	rad s^{-1}	Stark shift of state i (Sect.2.6.3)
$\Delta\omega$	rad s^{-1}	Linewidth
Δk	m^{-1}	Phase mismatch (Sect.3.2)
ε	rad s^{-1}	Positive infinitesimal (Sect.2.5)
ε		Normalised detuning from Fano-resonance (Sect.4.3.2)
$\underline{\varepsilon}_i$		Unit polarisation vector of field i (Sect.2.2)
ζ		Normalised distance, see ξ
η_ω		Linear refractive index at frequency ω
η_2	m^2V^{-2}	Coefficient of nonlinear refractive index (Sect.4.6.2)
θ		Tilt angle of adiabatic state and vector model (Sect.2.6.6)
θ	rad	Total phase in four-wave mixing (Sect.3.4.2)
$\Theta^{(K)}$		Angular factor of rank K (Sect.2.8)
κ		ω_S/ω_p, ratio of Stokes to pump frequencies in SRS (Sect.5.3.2)
λ	m	Wavelength
$\underline{\mu}$	Cm	Induced dipole-moment (Sect.2.5)
μ, μ_{ij}	Cm	z-component of dipole-moment operator, and matrix element (Sect.2.5)
μ_{fg}	Cm	Induced transition dipole-moment (Sect.2.7.1)
$\mu^{(n)}_{fg;\sigma'}$	Cm	n^{th} order induced transition dipole moment at frequency ω'_σ (Sect.2.7.2)
ν	Hz	Frequency
ξ, ζ		2(z-f)/b, normalised distance with respect to focus of gaussian beam (Sect.4.2)
$\rho(g)$		Equilibrium diagonal density-matrix element for state g (Sect.2.5)
$\overline{\rho(g)}$		Fraction of population in totality of states of energy $\hbar\Omega_g$ (Sect.2.5)
$\rho(\omega)$	m^{-3}	Density of radiation modes in a cavity (Sect.3.5)

List of Major Symbols and Acronyms (continued)

Symbol	Unit	Meaning (and, where useful, its first occurrence)
σ	m^2	Single-photon absorption cross section (Sect.4.6.1)
σ		Ratio of third-order susceptibilities (Sect.3.4)
$\sigma_{ab}, \sigma'_{fg}$		Off-diagonal density matrix elements in one- and two-photon resonant processes, "material excitation" (Sect.2.6.6)
τ_c	s	Characteristic time of pulse, $\sim$(bandwidth)$^{-1}$ (Sect.2.6)
τ_p	s	Pulse length (Sect.2.6)
υ	ms^{-1}	Characteristic pulse velocity (Sect.3.1)
ϕ_i		Adiabatic state (Sect.2.6.1)
ϕ_i		Field phase (Sect.3.4.2)
ψ		Wave function
$\chi^{(n)}(-\omega_\sigma;\omega_1\cdots\omega_n)$	$(Vm^{-1})^{1-n}$	n^{th} order macroscopic susceptibility (Sect.2.1)
$\chi^{(n)mic}, \chi^{(n)}/\mathcal{N}$	$m^3(Vm^{-1})^{1-n}$	Microscopic susceptibility, "per molecule" (Appendix)
$\tilde{\chi}$	m^2V^{-2}	Skeleton two-photon resonant $\chi^{(3)}$ (Sect.2.8)
χ_{4wm}, χ_p	m^2V^{-2}	$\chi^{(3)}$ for four-wave mixing (Sect.3.4.1)
$\chi_R = \chi'_R + i\chi''_R$	m^2V^{-2}	$\chi^{(3)}(-\omega_s;\omega_p-\omega_p\omega_s)$, SRS (Sect.3.3)
$\chi_{SRS1,2}$	m^2V^{-2}	See χ_R (Sect.3.3)
χ_T, χ_{TPA}	m^2V^{-2}	$\chi^{(3)}(-\omega_1;-\omega_2\omega_2\omega_1)$, TPA (Sect.3.3)
χ_{THG}	m^2V^{-2}	$\chi^{(3)}(-3\omega;\omega\omega\omega)$, THG (Sect.3.2)
$\chi^{(a)}_{THG}$	m^2V^{-2}	Vibrationally enhanced χ_{THG} (Sect.7.2)
$\chi^{(b)}_{THG}$	m^2V^{-2}	Electronically enhanced χ_{THG} (Sect.7.2)
χ_{VUV}	m^2V^{-2}	$\chi(-\omega_{VUV};\omega_1\omega_1\omega_2)$, up-conversion to VUV (Sect.4.3.2)
$\chi^{(2)}_Q$	m^2V^{-1}	Second-order quadrupolar susceptibility (Sect.8.1)
ω	rad s^{-1}	Frequency
ω_p, ω_s	rad s^{-1}	Pump, Stokes frequencies in SRS

List of Major Symbols and Acronyms (continued)

Symbol	Unit	Meaning (and, where useful, its first occurrence)
ω_σ	rad s^{-1}	$\omega_1 + \ldots + \omega_n$ (Sect.2.2)
ω'_σ	rad s^{-1}	$\omega_\sigma - \Omega_{fg}$ (Sect.2.7.2)
Ω	rad s^{-1}	Transition frequency
Ω, Ω'	rad s^{-1}	Off-resonance one- and two-photon Rabi frequencies (Sect.2.6)
Ω_{ij}	rad s^{-1}	Transition frequency $(E_i - E_j)/\hbar$ (Sect.2.5)
$\Omega'_{a,b}$	rad s^{-1}	Stark-shifted eigenfrequencies
Ω_i	rad s^{-1}	Eigenfrequency of state i, $E_i/\hbar$

References

1.1 P.A. Franken, A.E. Hill, C.W. Peters, G. Weinreich: Phys. Rev. Lett. *7*, 118-119 (1961)

1.2 W. Kaiser, C.G.B. Garrett: Phys. Rev. Lett. *7*, 229-231 (1961)

1.3 M. Bass, P.A. Franken, A.E. Hill, C.W. Peters, G. Weinreich: Phys. Rev. Lett. *8*, 18 (1962)

1.4 J.A. Giordmaine: Phys. Rev. Lett. *8*, 19-20 (1962)

1.5 P.D. Maker, R.W. Terhune, M. Nisenoff, C.M. Savage: Phys. Rev. Lett. *8*, 21-22 (1962)

1.6 J. Ducuing, C. Flytzanis: *Optical Properties of Solids*, ed. by F. Abelès (North-Holland, Amsterdam 1972) pp. 859-990

1.7 F. Zernike, J.E. Midwinter: *Applied Nonlinear Optics* (Wiley-Interscience, New York 1973)

1.8 M.J. Colles, C.R. Pidgeon: Rep. Prog. Phys. *38*, 329-460 (1975)

1.9 H. Rabin, C.L. Tang (ed.): *Quantum Electronics*, Vol. I, Pts. A and B (Academic Press, New York 1975)

1.10 R.L. Byer: In *Nonlinear Optics*, Proc. Sixteenth Scottish Universities Summer School in Physics 1975, ed. by P.G. Harper, B.S. Wherrett (Academic Press, London 1977) pp. 47-160

1.11 G.H.C. New, J.F. Ward: Phys. Rev. Lett. *19*, 556-559 (1967)

1.12 J.F. Ward, G.H.C. New: Phys. Rev. *185*, 57-72 (1969)

1.13 W.G. Rado: Appl. Phys. Lett. *11*, 123-125 (1967)

1.14 S.E. Harris, R.B. Miles: Appl. Phys. Lett. *19*. 385-387 (1971)

1.15 R.B. Miles, S.E. Harris: IEEE J. QE-*9*, 470-484 (1973)

1.16 J.F. Young, G.C. Bjorklund, A.H. Kung, R.B. Miles, S.E. Harris: Phys. Rev. Lett. *27*, 1551-1553 (1971)

1.17 D.M. Bloom, G.W. Bekkers, J.F. Young, S.E. Harris: Appl. Phys. Lett. *26*, 687-689 (1975)

1.18 D.M. Bloom, J.F. Young, S.E. Harris: Appl. Phys. Lett. *27*, 390-392 (1975)

1.19 A.H. Kung, J.F. Young, G.C. Bjorklund, S.E. Harris: Phys. Rev. Lett. *29*, 985-988 (1972)

1.20 A.H. Kung, J.F. Young, S.E. Harris: Appl. Phys. Lett. *22*, 301-302 (1973)

1.21 S.E. Harris, J.F. Young, A.H. Kung, D.M. Bloom, G.C. Bjorklund: In *Laser Spectroscopy*, ed. by R.G. Brewer, A. Mooradian (Plenum Press, New York 1973) p. 59

1.22 R.T. Hodgson, P.P. Sorokin, J.J. Wynne: Phys. Rev. Lett. *32*, 343 (1974)

1.23 K.M. Leung, J.F. Ward, B.J. Orr: Phys. Rev. A: *9*, 2440-2448 (1974)

1.24 D.M. Bloom, J.T. Yardley, J.F. Young, S.E. Harris: Appl. Phys. Lett. *24*, 427-428 (1974)

1.25 S.E. Harris: Phys. Rev. Lett. *31*, 341-344 (1973)

1.26 E.J. Woodbury, W.K. Ng: Proc. IRE *50*, 2367 (1962)

1.27 S. Yatsiv, W.G. Wagner, G.S. Picus, F.J. McClung: Phys. Rev. Lett. *15*, 614-618 (1965)

1.28 S. Yatsiv, M. Rokni, S. Barak: IEEE J. QE-*4*, 900-904 (1968)

1.29 S. Yatsiv, M. Rokni, S. Barak: Phys. Rev. Lett. *20*, 1282-1284 (1968)

1.30 S. Barak, M. Rokni, S. Yatsiv: IEEE J. QE-*5*, 448-453 (1968)

1.31 O.J. Lumpkin: IEEE J. QE-*4*, 226-228 (1968)
1.32 P.P. Sorokin, N.S. Shiren, J.R. Lankard, E.C. Hammond, T.G. Kazyaka: Appl. Phys. Lett. *10*, 44-46 (1967)
1.33 M. Rokni, S. Yatsiv: Phys. Lett. *24A*, 277-278 (1967)
1.34 M. Rokni, S. Yatsiv: IEEE J. QE-*3*, 329-330 (1967)
1.35 P.P. Sorokin, J.J. Wynne, J.R. Lankard: Appl. Phys. Lett. *22*, 342-344 (1973)
1.36 J.L. Carlsten, P.C. Dunn.: Opt. Commun. *14*, 8-12 (1975)
1.37 D. Cotter, D.C. Hanna, P.A. Kärkkäinen, R. Wyatt: Opt. Commun. *15*, 143-146 (1975)
1.38 D. Cotter, D.C. Hanna, R. Wyatt: Appl. Phys. *8*, 333-340 (1975)
1.39 D. Cotter, D.C. Hanna, R. Wyatt: Opt. Commun. *16*, 256-258 (1976)
1.40 D. Cotter, D.C. Hanna, W.H.W. Tuttlebee, M.A. Yuratich: Opt. Commun. *22*, 190-194 (1977)
1.41 N. Djeu, R. Burnham: Appl. Phys. Lett. *30*, 473-475 (1977)
1.42 D. Cotter, W. Zapka: Opt. Commun. *26*, 251-255 (1978)
1.43 R. Burnham, N. Djeu: Opt. Lett. *3*, 215-217 (1978)
1.44 S.R.J. Brueck, H. Kildal: Opt. Lett. *2*, 33-35 (1978)
1.45 S.R.J. Brueck, H. Kildal: Appl. Phys. Lett. *33*, 928-930 (1978)
1.46 H. Kildal, R.F. Begley, M.M. Choy, R.L. Byer: J. Opt. Soc. Am. *62*, 1398 (1972)
1.47 W. Schmidt, W. Appt: Z. Naturforsch. *27a*, 1373-1375 (1972)
1.48 W. Schmidt, W. Appt: IEEE J. QE-*10*, 792 (1974)
1.49 R. Frey, F. Pradère: Opt. Commun. *12*, 98-101 (1974)
1.50 M. Bierry, R. Frey, F. Pradère: Rev. Sci. Instr. *48*, 733-737 (1977)
1.51 P. Rabinowitz, A. Kaldor, R. Brickman, W. Schmidt: Appl. Opt. *15*, 2005-2006 (1976)
1.52 P. Rabinowitz, A. Stein, R. Brickman, A. Kaldor: Opt. Lett. *3*, 147-148 (1978)
1.53 A.Z. Grasiuk, I.G. Zubarev: Appl. Phys. *17*, 211-232 (1978)
1.54 W. Hartig, W. Schmidt: Appl. Phys. *18*, 235-241 (1979)
1.55 V. Wilke, W. Schmidt: Appl. Phys. *16*, 151-154 (1978)
1.56 V. Wilke, W. Schmidt: Appl. Phys. *18*, 177-181 (1979)
1.57 T.R. Loree, R.C. Sze, D.L. Barker: Appl. Phys. Lett. *31*, 37-39 (1977)
1.58 T.R. Loree, R.C. Sze, D.L. Barker, P.B. Scott: IEEE J. QE-*15*, 342-368 (1979)
1.59 J.R. Murray, J. Goldhar, D. Eimerl, A. Szöke: Appl. Phys. Lett. *33*, 399-401 (1978)
1.60 J.R. Murray, J. Goldhar, D. Eimerl, A. Szöke: IEEE J. QE-*15*, 337-342 (1979)
1.61 J. Ducuing, R. Frey, F. Pradère: In *Tunable Lasers and Applications*, Proc. Loen Conf. Norway 1976, Springer Series in Optical Sciences, Vol. 3, ed. by A. Mooradian, T. Jaeger, P. Stokseth (Springer, Berlin, Heidelberg, New York 1976) pp. 81-87
1.62 S.J. Brosnan, R.N. Fleming, R.L. Herbst, R.L. Byer: Appl. Phys. Lett. *30*, 330-332 (1977)
1.63 M.M.T. Loy, P.P. Sorokin, J.R. Lankard: Appl. Phys. Lett. *30*, 415-417 (1977)
1.64 P.P. Sorokin, M.M.T. Loy, J.R. Lankard : IEEE J. QE-*13*, 871-875 (1977)
1.65 R.L. Byer, W.R. Trutna: Opt. Lett. *3*, 144-146 (1978)
1.66 R. Frey, F. Pradère, J. Lukasik, J. Ducuing: Opt. Commun. *22*, 355-357 (1977)
1.67 R. Frey, F. Pradère, J. Ducuing: Opt. Commun. *23*, 65-68 (1977)
1.68 F.N.H. Robinson: Bell Syst. Tech. J. *46*, 913-956 (1967)
1.69 J.A. Armstrong, N. Bloembergen, J. Ducuing, P.S. Pershan: Phys. Rev. *127*, 1918-1939 (1962)
1.70 P.N. Butcher: *"Nonlinear Optical Phenomena"* Bulletin 200, Eng. Expt. Station, Ohio State University (1965)
1.71 J. Ducuing: In *Quantum Optics*, Proc. Int. School of Physics, 'Enrico Fermi' Course XLII, ed. by R.J. Glauber (Academic Press, New York 1969) pp. 421-472
1.72 R.P. Feynman, F.L. Vernon, R.W. Hellwarth: J. Appl. Phys. *28*, 49-52 (1957)
2.1 C. Flytzanis: In *Quantum Electronics*, Vol. 1, Pt. A, ed. by H. Rabin, C.L. Tang (Academic Press, New York 1975) pp. 9-207
2.2 Y. Gontier, N.K. Rahman, M. Trahin: Phys. Rev. A: *14*, 2109-2125 (1976)

2.3 D. Grischkowsky, M.M.T. Loy, P.F. Liao: Phys. Rev. A: *12*, 2514-2533 (1975)
2.4 B.J. Orr, J.F. Ward: Mol. Phys. *20*, 513-526 (1971)
2.5 V.S. Butylkin, A.E. Kaplan, Yu.G. Khronopulo: Sov. Phys. JETP *32*, 501-507 (1971)
2.6 V.S. Butylkin, Yu.G. Khronopulo, E.I. Yakubovich: Sov. Phys. JETP *44*, 897-904 (1976)
2.7 M.H. Nayfeh, A.H. Nayfeh: J. Appl. Phys. *47*, 2528-2531 (1976)
2.8 J. Wong, J.C. Garrison, T.H. Einwohner: Phys. Rev. A: *13*, 674-687 (1976)
2.9 M. Takatsuji: Phys. Rev. A: *11*, 619-624 (1975)
2.10 A.M.F. Lau: Phys. Rev. A: *14*, 279-290 (1976)
2.11 W. Heitler: *The Quantum Theory of Radiation* (Clarendon Press, Oxford 1954) Sects.14-16
2.12 J.R. Ackerhalt, K. Rzażewski: Phys. Rev. A: *12*, 2549-2567 (1975)
2.13 M. Babiker: Phys. Rev. A: *11*, 308-315 (1975)
2.14 Y.-R. Shen, N. Bloembergen: Phys. Rev. *137*, A1787-A1805 (1965)
2.15 N. Bloembergen: *Nonlinear Optics* (Benjamin, New York 1965)
2.16 T.P. McLean: In *Interaction of Radiation with Condensed Matter*, Vol. 1, IAEA-SMR-20/1 (Int. Atom. Energy Agency, Vienna 1977)
2.17 Y. Ueda, K. Shimoda: J. Phys. Soc. Jpn. *28*, 196-204 (1970)
2.18 R. Wallace: Mol. Phys. *11*, 457-470 (1966)
2.19 R. Loudon: *The Quantum Theory of Light* (Clarendon Press, Oxford 1973)
2.20 J.F. Ward: Rev. Mod. Phys. *37*, 1-18 (1965)
2.21 M.A. Yuratich: Ph.D. Thesis, University of Southampton (1977)
2.22 P.L. Knight, L. Allen: Phys. Lett. *38A*, 99-100 (1972)
2.23 P.D. Maker, R.W. Terhune: Phys. Rev. *137*, A801-A818 (1965)
2.24 G. Grynberg, F. Biraben, M. Bassini, B. Cagnac: Phys. Rev. Lett. *37*, 283-285 (1976)
2.25 E. Courtens: In *Laser Handbook*, Vol. 2, ed. by F.T. Arecchi, E.O. Schulz-DuBois (North Holland, Amsterdam 1972) pp. 1259-1322
2.26 L. Allen, J.H. Eberly: *Optical Resonance and Two Level Atoms* (Wiley-Interscience, New York 1975)
2.27 N. Tan-No, T. Shirahata, K. Yokoto, H. Inaba: Phys. Rev. A: *12*, 159-168 (1975)
2.28 H. Puell, C.R. Vidal: Phys. Rev. A: *14*, 2225-2239 (1976)
2.29 L. Brillouin: *Wave Propagation and Group Velocity* (Academic Press, New York 1960)
2.30 C.G.B. Garrett, D.E. McCumber: Phys. Rev. A: *1*, 305-313 (1970)
2.31 M.D. Crisp: Phys. Rev. A: *1*, 1604-1611 (1970)
2.32 M.D. Crisp: Phys. Rev. A: *4*, 2104-2108 (1971)
2.33 S.A. Akhmanov, A.P. Sukhorukov, A.S. Chirkin: Sov. Phys. JETP *28*, 748-757 (1969)
2.34 G.W. Series: In *Quantum Optics*, Proc. Tenth Scottish Universities Summer School in Physics 1969, ed. by S.M. Kay, A. Maitland (Academic Press, London 1970) pp. 395-482
2.35 C. Cohen-Tannoudji: J. Phys. (Paris) *32*, C5a, 11-28 (1971)
2.36 C. Cohen-Tannoudji: In *Frontiers in Laser Spectroscopy*, Les Houches 75, Session XXVII, ed. by R. Balian, S. Haroche, S. Liberman (North Holland, Amsterdam 1975)
2.37 S. Feneuille: Rep. Prog. Phys. *40*, 1257-1304 (1977)
2.38 M. Sargent, M.O. Scully, W.E. Lamb: *Laser Physics* (Addison-Wesley, New York 1974)
2.39 R. McWeeny: *Quantum Mechanics: Methods and Basic Applications*, Int. Encylopaedia of Physical Chemistry and Chemical Physics, Vol. 2 (Pergamon Press, Oxford 1973)
2.40 L.I. Schiff: *Quantum Mechanics* (McGraw-Hill, New York 1968) 3rd ed.
2.41 D. Grischkowsky: Phys. Rev. A: *14*, 802-812 (1976)
2.42 A. Flusberg, S.R. Hartmann: Phys. Rev. A: *14*, 813-815 (1976)
2.43 A.M. Bonch-Bruevich, V.A. Khodovoi: Sov. Phys. USP *10*, 637-657 (1968)
2.44 E. Courtens, A. Szöke: Phys. Rev. A: *15*, 1588-1602 (1977)

2.45 P.F. Liao, G.C. Bjorklund: Phys. Rev. Lett. *36*, 584-587 (1976)
2.46 P.F. Liao, G.C. Bjorklund: Phys. Rev. A: *15*, 2009-2018 (1977)
2.47 A. Szöke, E. Courtens: Phys. Rev. Lett. *34*, 1053-1056 (1975)
2.48 J.L. Carlsten, A. Szöke, M.G. Raymer: Phys. Rev. A: *15*, 1029-1045 (1977)
2.49 M.G. Raymer, J.L. Carlsten: Phys. Rev. Lett. *39*, 1326-1329 (1977)
2.50 S.H. Autler, C.H. Townes: Phys. Rev. *100*, 703-722 (1955)
2.51 P.F. Liao, J.E. Bjorkholm: Phys. Rev. Lett. *34*, 1-4 (1975)
2.52 Q.H.F. Vrehen, H.M.J. Hikspoors: Opt. Commun. *21*, 127-131 (1977)
2.53 L.D. Landau, E.M. Lifschitz: *Electrodynamics of Continuous Media*, Course of Theoretical Physics, Vol. 8 (Pergamon Press, Oxford 1960)
2.54 P.F. Liao, J.E. Bjorkholm: Opt. Commun. *16*, 392-395 (1976)
2.55 S.E. Moody, M. Lambropoulos: Phys. Rev. A: *15*, 1497-1501 (1977)
2.56 H.R. Gray, C.R. Stroud: Opt. Commun. *25*, 359-362 (1978)
2.57 A. Schabert, R. Keil, P.E. Toschek: Appl. Phys. *6*, 181-184 (1976)
2.58 M.P. Bondareva, Yu. M. Kirin, S.G. Rautian, V.P. Safonov, B.M. Chernobrod: Opt. Spectrosc. *38*, 219-227 (1975)
2.59 V.S. Letokhov, V.P. Chebotayev: *Nonlinear Laser Spectroscopy*, Springer Series in Optical Sciences, Vol. 4 (Springer, Berlin, Heidelberg, New York 1977)
2.60 A.T. Georges, P. Lambropoulos, J.H. Marburger: Phys. Rev. A: *15*, 300-307 (1977)
2.61 P.L. Knight: Opt. Commun. *22*, 173-177 (1977)
2.62 D. Grischkowsky: Phys. Rev. Lett. *24*, 866-869 (1970)
2.63 D. Grischkowsky, J.A. Armstrong: Phys. Rev. A: *6*, 1566-1570 (1972)
2.64 P.L. Knight, L. Allen: Phys. Lett. A: *38A*, 99-100 (1972)
2.65 I.I. Sobel'man: *An Introduction to the Theory of Atomic Spectra* (Pergamon Press, Oxford 1972)
2.66 P.W. Langhoff, S.T. Epstein, M. Karplus: Rev. Mod. Phys. *44*, 602-644 (1972)
2.67 J.F. Ward, A.V. Smith: Phys. Rev. Lett. *35*, 653-656 (1975)
2.68 R.H. Lehmberg, J. Reintjes, R.C. Eckardt: Phys. Rev. A: *13*, 1095-1103 (1976)
2.69 M. Gell'man, F. Low: Phys. Rev. *84*, 350-354 (1951)
2.70 I.I. Rabi, N.F. Ramsey, J. Schwinger: Rev. Mod. Phys. *26*, 167-171 (1954)
2.71 M.D. Crisp: Phys. Rev. A: *8*, 2128-2135 (1973)
2.72 M.M.T. Loy: Phys. Rev. Lett: *32*, 814-817 (1974)
2.73 H.G. Venkatesh, G.G. Sarkar: J. Phys. A: *9*, 1015-1023 (1976)
2.74 H.G. Venkatesh, J. Ram: J. Phys. A: *9*, 999-1014 (1976)
2.75 Yu.G. Khronopulo: Izv. Vyssh. Uchebn. Zaved. Radiofizika *7*, 674 (in Russian) (1964)
2.76 M. Takatsuji: Physica *51*, 265-272 (1970)
2.77 M. Takatsuji: Phys. Rev. A: *4*, 808-810 (1971)
2.78 J.M. Ziman: *Elements of Advanced Quantum Theory* (Cambridge University Press, Cambridge 1969) Sect. 3.5
2.79 D. Grischkowsky, M.M.T. Loy: Phys. Rev. A: *12*, 1117-1120 (1975)
2.80 J.C. Garrison, T.H. Einwohner, J. Wong: Phys. Rev. A: *14*, 731-737 (1976)
2.81 M.M.T. Loy: Phys. Rev. Lett: *41*, 473-476 (1978)
2.82 R.H. Lehmberg, J. Reintjes: Phys. Rev. A: *12*, 2574-2583 (1975)
2.83 M.H. Nayfeh: Phys. Rev. A: *14*, 1304-1307 (1976)
2.84 M.H. Nayfeh, A.H. Nayfeh: Phys. Rev. A: *15*, 1169-1172 (1977)
2.85 A. Javan, P.L. Kelley: IEEE J. QE-*2*, 470-473 (1966)
2.86 J.R. Ackerhalt, B.W. Shore: Phys. Rev. A: *16*, 277-282 (1977)
2.87 G. Placzek: "The Rayleigh and Raman Scattering" 1934 UCRL Transl. No. 526 (L) (US Dept. of Commerce, Washington, D.C. 1962)
2.88 E. Garmire, F. Pandarese, C.H. Townes: Phys. Rev. Lett: *11*, 160-163 (1963)
2.89 G. Herzberg: *Infrared and Raman Spectra of Polyatomic Molecules*, Molecular Spectra and Molecular Structure II (Van Nostrand, New Jersey 1960)
2.90 C.-S. Wang: Phys. Rev: *182*, 482-494 (1969)
2.91 O. Klein: Z. Phys. *41*, 407-442 (1927)
2.92 A.D. Buckingham: Adv. Chem. Phys: *12*, 107-142 (1967)
2.93 M.P. Bogaard, B.J. Orr: In *Int. Rev. Science*, Physical Chemistry, Series Two Vol. 2, ed. by A.D. Buckingham (Butterworth, London 1975) pp. 149-194

2.94 D.A. Long, L. Stanton: Proc. R. Soc. London Ser. A: *318*,441-457 (1970)
2.95 J.A. Koningstein: *Introduction to the Theory of the Raman Effect* (D. Reidel Publishing Co., Dordrecht 1972)
2.96 M. Maier, W. Kaiser, J.A. Giordmaine: Phys. Rev. *177*, 580-599 (1969)
2.97 S.A. Akhmanov, K.N. Drabovich, A.P. Sukhorukov, A.S. Chirkin: Sov. Phys. JETP *32*, 266-273 (1971)
2.98 S.A. Akhmanov, Yu.E. D'Yakov, L.I. Pavlov: Sov. Phys. JETP *39*, 249-256 (1974)
2.99 R.L. Carman, F. Shimizu, C.-S. Wang, N. Bloembergen: Phys. Rev. A: *2*, 60-72 (1970)
2.100 C.-S. Wang: In *Quantum Electronics*, Vol. 1, Part A, ed. by H. Rabin, C.L. Tang (Academic Press, New York 1975) pp. 419-472
2.101 W. Kaiser, A. Laubereau: In *Nonlinear Optics*, Proc. Sixteenth Scottish Universities Summer School in Physics 1975, ed. by P.G. Harper, B.S. Wherrett (Academic Press, London 1977) pp. 257-305
2.102 M. Maier: Appl. Phys. *11*, 580-599 (1976)
2.103 C.A. Sacchi: Riv. Nuovo Cimento, Serie 2 *2*, 210-237 (1972)
2.104 H. Nakatsuka, J. Okada, M. Matsuoka: J.Phys. Soc. Jpn. *37*, 1406-1412 (1974)
2.105 M. Matsuoka: Opt. Commun. *15*, 84-86 (1975)
2.106 M. Matsuoka, H. Nakatsuka, J. Okada: Phys. Rev. A: *12*, 1062-1065 (1975)
2.107 P. W. Milonni, J.H. Eberly: J. Chem. Phys. *68*, 1602-1613 (1978)
2.108 H. Friedmann, A.D. Wilson-Gordon: Opt. Commun. *24*, 5-10 (1978)
2.109 D.M. Brink, G.R. Satchler: *Angular Momentum* (Clarendon Press, Oxford 1971) 2nd ed.
2.110 B.W. Shore, D.H. Menzel: *Principles of Atomic Spectra* (Wiley-Interscience, New York 1968)
2.111 M.A. Yuratich, D.C. Hanna: J. Phys. B: *9*,729-750 (1976)
2.112 D. Cotter, D.C. Hanna: J. Phys. B: *9*, 2165-2171 (1976)
2.113 J.A. Koningstein, O. Sonnich Mortenson: Phys. Rev. *168*, 75-78 (1968)
2.114 Yu.A. Il'inskii, V.D. Taranukhin: Sov. J. Quantum Electron. *5*, 805-810 (1975)
2.115 G. Grynberg, F. Biraben, E. Giacobino, B. Cagnac: J. Phys. (Paris) *38*, 629-640 (1977)
2.116 H. Eicher: IEEE J. QE-*11*, 121-130 (1975)
2.117 J.N. Elgin, G.H.C. New, K.E. Orkney: Opt. Commun. *18*, 250-254 (1976)
2.118 D. Grischkowsky, R.G. Brewer: Phys. Rev. A: *15*, 1789-1793 (1977)
2.119 N. Bloembergen, Y.-R. Shen: Phys. Rev. Lett. *12*, 504-507 (1964)
3.1 G.C. Bjorklund: IEEE J. QE-*11*, 287-296 (1975)
3.2 R.L. Byer, R.L. Herbst: In *Nonlinear Infrared Generation*, Topics in Applied Physics, Vol. 16, ed. by Y.-R. Shen (Springer, Berlin, Heidelberg, New York 1977) pp. 81-137
3.3 G.V. Venkin, G.M. Krochik, L.L. Kulyuk, I. Maleev, Yu.G. Khronopulo: JETP Lett. *21*, 105-107 (1975)
3.4 J.A. Giordmaine, W. Kaiser: Phys. Rev. *144*, 676-688 (1966)
3.5 V.S. Butylkin, G.M. Krochik, Yu.G. Khronopulo: Sov. Phys. JETP *41*, 247-252 (1975)
3.6 G.M. Krochik, Yu.G. Khronopulo: Sov. J. Quantum Electron. *5*, 917-921 (1976)
3.7 G.V. Venkin, G.M. Krochik, L.L. Kulyuk, I. Maleev, Yu.G. Khronopulo: Sov. Phys. JETP *43*, 873-879 (1976)
3.8 M.A. Yuratich: in preparation
3.9 T.G. Giallorenzi, C.L. Tang: Phys. Rev. *166*, 225-233 (1968)
3.10 D.A. Kleinman: Phys. Rev. *174*, 1027-1041 (1968)
4.1 R.W. Terhune, P.D. Maker, C.M. Savage: Phys. Rev. Lett. *8*, 404-406 (1962)
4.2 P.D. Maker, R.W. Terhune, C.M. Savage: In *Quantum Electronics III*, ed. by P. Grivet, N. Bloembergen (Columia University Press, New York 1964)
4.3 P. Sitz, R. Yaris: *Introduction to the Theory of Atomic Spectra* (Pergamon Press, New York 1968)
4.4 A.D. Buckingham, D.A. Dunmur: Trans. Faraday Soc. *64*, 1776-1783 (1968)
4.5 E.L. Dawes: Phys. Rev. *169*, 41-48 (1968)
4.6 M. Born, E. Wolf: *Principles of Optics* (Pergamon Press, Oxford 1970) pp. 445-448
4.7 H. Kogelnik, T. Li: Opt. *5*, 1150-1567 (1966)

4.8 G.D. Boyd, D.A. Kleinman: J. Appl. Phys. *39*, 3597-3639 (1968)
4.9 H. Kildal, S.R.J. Brueck: Phys. Rev. Lett. *38*, 347-350 (1977)
4.10 I.V. Tomov, M.C. Richardson: IEEE J. *QE-12*, 521-531 (1976)
4.11 K.S. Hsu, A.H. Kung, L.J. Zych, J.F. Young, S.E. Harris: IEEE J. *QE-12*, 60-62 (1976)
4.12 J. Reintjes, C.Y. She, R.C. Eckardt: IEEE J. *QE-14*, 581-596 (1978)
4.13 E.M. Anderson, V.A. Zilitis: Opt. Spectrosc. *16*, 211-214 (1964)
4.14 E.M. Anderson, V.A. Zilitis: Opt. Spectrosc. *16*, 99-101 (1964)
4.15 D.R. Bates, A. Damgaard: Phil. Trans. Roy. Soc. London *A242*, 101-119 (1949)
4.16 H.B. Bebb: Phys. Rev. *149*, 25-32 (1966)
4.17 B. Warner: Mon. Not. R. Astron. Soc. *139*, 115-128 (1968)
4.18 Y. Ohashi, Y. Ishibashi, T. Kobayasi, H. Inaba: Jpn. J. Appl. Phys. *15*, 1817-1818 (1976)
4.19 E.A. Stappaerts, G.W. Bekkers, J.F. Young, S.E. Harris: IEEE J. *QE-12*, 330-333 (1976)
4.20 W.R.S. Garton, G.L. Grasdalen, W.H. Parkinson, E.M. Reeves: J. Phys. B: *1*, 114-119 (1968)
4.21 U. Fano: Phys. Rev. *124*, 1866-1878 (1961)
4.22 J.J. Wynne, P.P. Sorokin: J. Phys. B: *8*, L37-L41 (1975)
4.23 P.P. Sorokin, J.J. Wynne, J.A. Armstrong, R.T. Hodgson: Ann. N. Y. Acad. Sci. *267*, 30-50 (1976)
4.24 J.A. Armstrong, J.J. Wynne: Phys. Rev. Lett. *33*, 1183-1185 (1974)
4.25 J.A. Armstrong, J.J. Wynne: In *Nonlinear Spectroscopy*, Proc. Int. School of Physics 'Enrico Fermi' Course LXIV, ed. by N. Bloembergen (North-Holland, Amsterdam 1977) pp. 152-169
4.26 L. Armstrong, B.L. Beers: Phys. Rev. Lett. *34*, 1290-1291 (1975)
4.27 H. Scheingraber, H. Puell, C.R. Vidal: Phys. Rev. A: *18*, 2585-2591 (1978)
4.28 N.L. Manakov, M.A. Preobrazhenskii, L.P. Rapoport: Opt. Spectrosc. *35*, 14-16 (1973)
4.29 V.V. Slabko, A.K. Popov, V.F. Lukinykh: Appl. Phys. *15*, 239-241 (1978)
4.30 A.K. Popov, V.P. Timofeev: Opt. Commun. *20*, 94-100 (1977)
4.31 V.L. Doitcheva, V.M. Mitev, L.I. Pavlov, K.V. Stamenov: Opt. Quantum Electron. *10*, 131-138 (1978)
4.32 C.S. Wang, L.I. Davis: Phys. Rev. Lett. *35*, 650-653 (1975)
4.33 J. Reintjes, R.C. Eckardt, C.Y. She, N.E. Karangelen, R.C. Elton, R.A. Andrews: Phys. Rev. Lett. *37*, 1540-1543 (1976)
4.34 J. Reintjes, C.Y. She, R.C. Eckardt, N.E. Karangelen, R.A. Andrews, R.C. Elton: Appl. Phys. Lett. *30*, 480-482 (1977)
4.35 R. Mahon, T.J. McIlrath, D.W. Koopman: Appl. Phys. Lett. *33*, 305-307 (1978)
4.36 D. Cotter: Opt. Lett. *4*, 134-136 (1979)
4.37 A.H. Kung: Appl. Phys. Lett. *25*, 653-654 (1974)
4.38 M.H.R. Hutchinson, Ling C.C., D.J. Bradley: Opt. Commun. *18*, 203-204 (1976)
4.39 J. Reintjes, C.Y. She: Opt. Comm. *27*, 469-474 (1978)
4.40 D.I. Metchkov, V.M. Mitev, L.I. Parlov, K.V. Stamenov: Opt. Commun. *21*, 391-394 (1977)
4.41 M.G. Grozeva, D.I. Metchkov, V.M. Mitev, L.I. Pavlov, K.V. Stamenov: Opt. Commun. *23*, 77-79 (1977)
4.42 T.J. McIlrath, T.B. Lucatorto: Phys. Rev. Lett. *38*, 1390-1393 (1977)
4.43 R.E. Honig, D.A. Kramer: RCA Rev. *30*, 285-305 (1969)
4.44 H.E.J. Schins, R.W.M. Van Wijk, B. Dorpema: Z. Metallkd. *62*, 330-336 (1971)
4.45 P.J. Leonard: At. Data Nucl. Data Tables *14*, 21-37 (1974)
4.46 H. Puell, K. Spanner, W. Falkenstein, W. Kaiser, C.R. Vidal: Phys. Rev. A:*14*, 2240-2257 (1976)
4.47 A.I. Ferguson, E.G. Arthurs: Phys. Lett. *58A*, 298-300 (1976)
4.48 J.R. Taylor: Opt. Commun. *18*, 504-508 (1976)
4.49 A.H. Kung, J.F. Young, S.E. Harris: Appl. Phys. Lett. *28*, 239 (erratum) (1976)
4.50 C.R. Vidal, J. Cooper: J. Appl. Phys. *40*, 3370-3374 (1969)
4.51 D. Cotter, D.C. Hanna: Opt. Quantum Electron. *9*, 509-518 (1977)

4.52 C.R. Vidal, F.B. Haller: Rev. Sci. Instr. *42*, 1779-1784 (1971)
4.53 C.R. Vidal, M.M. Hessel: J. Appl. Phys. *43*, 2776-2780 (1972)
4.54 G.C. Bjorklund, J.E. Bjorkholm, P.F. Liao, R.H. Storz: Appl. Phys. Lett. *29*, 729-732 (1976)
4.55 G.C. Bjorklund, J.E. Bjorkholm, R.R. Freeman, P.F. Liao: Appl. Phys. Lett. *31*, 330-332 (1977)
4.56 R.R. Freeman, G.C. Bjorklund, N.P. Economou, P.F. Liao, J.E. Bjorkholm: Appl. Phys. Lett. *33*, 739-742 (1978)
4.57 N.P. Economou, R.R. Freeman, G.C. Bjorklund: Optics Lett. *3*, 209-211 (1978)
4.58 I.N. Drabovich, D.I. Metchkov, V.M. Mitev, L.I. Pavlov, K.V. Stamenov: Opt. Commun. *20*, 350-353 (1977)
4.59 S.C. Wallace, G. Zdasiuk: Appl. Phys. Lett. *28*, 449-451 (1976)
4.60 T.J. McKee, B.P. Stoicheff, S.C. Wallace: Optics Lett. *3*, 207-208 (1978)
4.61 R.R. Freeman, G.C. Bjorklund: Phys. Rev. Lett. *40*, 118-120 (1978)
4.62 P.P. Sorokin, J.A. Armstrong, R.W. Dreyfus, R.T. Hodgson, J.R. Lankard, L.H. Manganaro, J.J. Wynne: In *Laser Spectroscopy*, Proc. 2nd Int. Conf. Megève, ed. by S. Haroche, J.C. Pebay-Peyroula, T.W. Hänsch, S.E. Harris (Springer, Berlin, Heidelberg, New York 1975) pp. 46-54
4.63 C.Y. She, J. Reintjes: Appl. Phys. Lett. *31*, 95-97 (1977)
4.64 K.K. Innes, B.P. Stoicheff, S.C. Wallace: Appl. Phys. Lett. *29*, 715-717 (1976)
4.65 E.A. Stappaerts: IEEE J. *QE-15*, 110-118 (1979)
4.66 S. Ch'en, M. Takeo: Rev. Mod. Phys. 29, 20-73 (1957)
4.67 W.R. Hindmarsh, J.M. Farr: Prog. Quantum Electron. *2*, Pt 3 (1972)
4.68 W.R. Hindmarsh: In *Atoms, Molecules and Lasers*, IAEA-SMR.12/12 (Int. Atom. Energy Agency, Vienna 1974) pp. 133-173
4.69 H.R. Griem: *Spectral Line Broadening by Plasmas* (Academic Press, New York 1974)
4.70 H. Puell, C.R. Vidal: Opt. Comm. *27*, 165-170 (1978)
4.71 L.J. Zych, J.F. Young: IEEE J. *QE-14*, 147-149 (1978)
4.72 S.A. Akhmanov, R.V. Khoklov, A.P. Sukhorukov: In *Laser Handbook*, Vol. 2, ed. by F.T. Arecchi, E.O. Schulz-Dubois (North Holland, Amsterdam) pp. 1151-1228
4.73 Y.R. Shen: Prog. Quantum Electron. *4*, 1-34 (1977)
4.74 J.H. Marburger: Prog. Quantum Electron. *4*, 35-110 (1977)
4.75 R.H. Lehmberg, J. Reintjes, R.C. Eckhardt: Appl. Phys. Lett. *25*, 374-376 (1974)
4.76 R.H. Lehmberg, J. Reintjes, R.C. Eckardt: Phys. Rev. A*13*, 1095-1103 (1976)
4.77 A. Owyoung: Appl. Phys. Lett. *26*, 168-170 (1975)
4.78 S.E. Harris, D.M. Bloom: Appl. Phys. Lett. *24*, 229-230 (1974)
4.79 R.M. Measures: J. Appl. Phys. *48*, 2673-75 (1977)
4.80 P. Lambropoulos, M. Lambropoulos: *Proc. Int. Symp. on Electron & Photon Interaction with Atoms*, ed. by H. Kleinpoppen, M.R.C. McDowell (Plenum Press, New York 1976)
4.81 C. Grey-Morgan: Rep. Prog. Phys. *38*, 621-665 (1975)
4.82 P. Lambropoulos, M.R. Teague: J. Phys. B: *9*, 587-603 (1976)
4.83 E.H.A. Granneman, M.J. van der Wiel: J. Phys. B: *8*, 1617-1626 (1975)
4.84 M.R. Teague, P. Lambropoulos, D. Goodmanson, D.W. Norcross: Phys. Rev. A: *14*, 1057-1064 (1976)
4.85 A. Rachman, G. Laplanche, M. Jaouen: Phys. Lett. *68A*, 433-436 (1978)
4.86 A.T. Georges, P. Lambropoulos, J.H. Marburger: Opt. Commun. *18*, 509-512 (1976)
4.87 T.B. Lucatorto, T.J. McIlrath: Phys. Rev. Lett. *37*, 428-431 (1976)
4.88 G.H.C. New: Opt. Commun. *19*, 177-181 (1976)
4.89 J.N. Elgin, G.H.C. New: J. Phys. B: *11*, 3439-3457 (1978)
5.1 N. Bloembergen, G. Bret, P. Lallemand, A. Pine, P. Simova: IEEE J. QE-*3*, 197-201 (1967)
5.2 N. Bloembergen: Am. J. Phys. *35*, 989-1023 (1967)

5.3 W. Kaiser, M. Maier: In *Laser Handbook*, Vol. 2, ed. F.T. Arecchi, E.O. Schulz-DuBois (North Holland, Amsterdam 1972) pp.1077-1150
5.4 M. Schubert, B. Wilhelmi: Sov. J. Quantum Electron. *4*, 575-588 (1974)
5.5 Y.-R. Shen: In *Light Scattering in Solids*, Topics in Applied Physics, Vol. 8, ed. by M. Cardona (Springer, Berlin, Heidelberg, New York 1975) pp. 275-328
5.6 A.Z. Grasiuk, I.G. Zubarev: In *Tunable Lasers and Applications*, Proc. Loen Conf. Norway 1976, Springer Series in Optical Sciences, Vol. 3, ed. by A. Mooradian, T. Jaeger, P. Stokseth (Springer, Berlin, Heidelberg, New York 1976) pp. 88-95
5.7 S.K. Kurtz, J.A. Giordmaine: Phys. Rev. Lett. *22*, 192-195 (1969)
5.8 M.A. Piestrup, R.N. Fleming, R.H. Pantell: Appl. Phys. Lett. *26*,418-421 (1975)
5.9 S.D. Smith, R.B. Dennis, R.G. Harrison: Prog. Quantum. Electron. *5*, 205-292 (1977)
5.10 W.R. Fenner, H.A. Hyatt, J.M. Kellman, S.P.S. Porto: J. Opt. Soc. Am. *63*, 73-77, (1973)
5.11 T.J. McKee, B.P. Stoicheff, S.C. Wallace: Appl. Phys. Lett. *30*, 278-280 (1977)
5.12 R. Burnham, N. Djeu: Appl. Phys. Lett. *29*, 707-709 (1976)
5.13 J.M. Hoffman, A.K. Hays, G.S. Tisone: Appl. Phys. Lett. *28*, 538-539 (1976)
5.14 M.L. Bhaumik, R.S. Bradford, E.R. Ault: Appl. Phys. Lett. *28*, 23-24 (1976)
5.15 R.T.V. Kung, I. Itzkan: IEEE J. QE-*13*, 73-79 (1977)
5.16 R. Wyatt: PhD Thesis, University of Southampton (1976)
5.17 B. Godard, O. DeWitte: Opt. Commun. *19*, 325-328 (1976)
5.18 C. Rullière, J.P. Morand, O. DeWitte: Opt. Commun. *20*, 339-341 (1977)
5.19 H. Bücher, W. Chow: Appl. Phys. *13*, 267-269 (1977)
5.20 C. Rullière, J. Joussot-Dubien: Opt. Commun. *24*, 38-40 (1978)
5.21 R.L. Byer, R.L. Herbst: Laser Focus *14*, No. 5, 48-57 (1978)
5.22 J.J. Wynne, P.P. Sorokin: J. Phys. B: *8*, L37-L41 (1975)
5.23 A.M.F. Lau, W.K. Bischel, C.K. Rhodes, R.M. Hill: Appl. Phys. Lett. *29*, 245-247 (1976)
5.24 R.T.V. Kung, I. Itzkan: Appl. Phys. Lett. *29*, 780-783 (1976)
5.25 T. Holstein: Phys. Rev. *72*, 1212-1233 (1947)
5.26 T. Holstein: Phys. Rev. *83*, 1159-1168 (1951)
5.27 L. Krause: Adv. Chem. Phys. *28*, 267-316 (1975)
5.28 D.R. Grischkowsky, J.R. Lankard, P.P. Sorokin: IEEE J. *QE-13*, 392-396 (1977)
5.29 P.P. Sorokin, J.R. Lankard: J. Chem. Phys. *55*, 3810-3813 (1971)
5.30 D.R. Grischkowsky, P.P. Sorokin, J.R. Lankard: Opt. Commun. *18*, 205-206 (1976)
5.31 D.S. Bethune, J.R. Lankard, P.P. Sorokin: Opt. Lett. *4*, 103-105 (1979)
5.32 A.B. Peterson, I.W.M. Smith, D.C. Hanna: J. Opt. Soc. Am. *68*, 655 (1978)
5.33 P.P. Sorokin, J.R. Lankard: J. Chem. Phys. *51*, 2929-2931 (1969)
5.34 P.A. Kärkkäinen: PhD Thesis, University of Southampton (1976)
5.35 W. Hartig: Appl. Phys. *15*, 427-432 (1978)
5.36 R.T. Hodgson: Appl. Phys. Lett. *34*, 58-60 (1979)
5.37 J.E. Bjorkholm, P.F. Liao: Phys. Rev. Lett. *33*,128-131 (1974)
5.38 P.M. Stone: Phys. Rev. *127*, 1151-1156 (1962)
5.39 A. Kancerevicius, S. Zilionyte: Litov. Fiz. Sb. *7*, 73-83 (in Russian) (1967)
5.40 E. Koenig: Physica *62*, 393-408 (1971)
5.41 D.W. Norcross: Phys. Rev. A: *7*, 606-616 (1973)
5.42 J.C. Weisheit: Phys. Rev. A: *5*, 1621-1630 (1972)
5.43 M. Fabry: J. Quant. Spectrosc. Radiat. Transfer *16*, 127-135 (1976)
5.44 G.S. Kvater, T.G. Meister: Vestn. Leningr. Univ. *9*, 137 (in Russian) (1952)
5.45 S.M. Gridneva, G.A. Kasabov: *Proc. VII Int. Conf. Phenomena in Ionised gases*, Vol.II, Belgrade Gradevinska Knjiga (1965)
5.46 L. Agnew: Bull. Am. Phys. Soc. *11*, 327 (1966)
5.47 F.G. Fulop, H.H. Stroke: In *Atomic Physics III*, ed. by S.J. Smith, G.K. Walters (Plenum Press, New York 1973)
5.48 A.N. Klyucharev, A.V. Lazerenko: Opt. Spectrosc. *30*, 628-629 (1971)
5.49 R.J. Exton: J. Quant. Spectrosc. Radiat. Transfer *16*, 309-314 (1976)
5.50 G. Pichler: J. Quant. Spectrosc. Radiat. Transfer *16*, 147-151 (1976)

5.51 A. Corney, K. Gardner: J. Phys. D: *11*, 1815-1823 (1978)
5.52 D. Cotter: J. Phys. D: *12*, L9-12 (1979)
5.53 H. Kogelnik: App. Opt. *4*, 1562-1569 (1965)
5.54 S.A. Akhmanov: In *Nonlinear Spectroscopy*, Proc. Int. School of Physics "Enrico Fermi" Course LXIV, ed. by N. Bloembergen (North-Holland, Amsterdam 1977) pp. 255-275
5.55 E.E. Hagenlocker, W.G. Rado: Appl. Phys. Lett. *7*, 236-238 (1965)
5.56 V.A. Mikhaĭlov, V.I. Odintsov, L.F. Rogacheva: JETP Lett. *25*, 138-140 (1977)
5.57 F.A. Korolev, V.A. Mikhaĭlov, V.I. Odintsov: Opt. Spectrosc. *44*, 535-538 (1978)
5.58 G.P. Dzhotyan, Yu.E. D'Yakov, I.G. Zubarev, A.B. Mironov, S.I. Mikhaĭlov: Sov. Phys. JETP *46*, 431-435 (1977)
5.59 N.M. Kroll: J. Appl. Phys. *36*, 34-43 (1965)
5.60 N.M. Kroll, P.L. Kelley: Phys. Rev. A: *4*, 763-776 (1971)
5.61 É.M. Belenov, I.A. Poluéktov: Sov. Phys. JETP *29*, 754-756 (1969)
5.62 I.A. Poluéktov, Yu.M. Popov, V.S. Roĭtberg: JETP Lett. *20*,243-244 (1974)
5.63 J.N. Elgin, T.B. O'Hare: J. Phys. B: *12*, 159-168 (1979)
5.64 B.M. Miles, W.L. Wiese: At. Data *1*, 1-17 (1969)
5.65 E.H.A. Granneman, M. Klewer, K.J. Nygaard, M.J. Van der Wiel: J. Phys. B: *9*, 865-873 (1976)
5.66 W.H. Evans, R. Jacobson, T.R. Munson, D. Wagman: J. Res. Nat. Bur. Stand. *55*, 83-96 (1955)
5.67 G. Herzberg: *Molecular Spectra and Molecular Structure I. Spectra of Diatomic Molecules* (Van Nostrand, New Yersey 1950)
5.68 M. Lapp, L.P. Harris: J. Quant. Spectrosc. Radiat. Transfer *6*, 169-179 (1966)
5.69 M. McClintock, L.C. Balling: J. Quant. Spectrosc. Radiat. Transfer *9*, 1209-1214 (1969)
5.70 P. Agostini, A.T. Georges, S.E. Wheatley, P. Lambropoulos, M.D. Levenson: J. Phys. B: *11*, 1733-1747 (1978)
5.71 A. Burgess, M.J. Seaton: Mon. Not. R. Astron. Soc. *120*, 121-151 (1960)
5.72 D. Cotter, D.C. Hanna: IEEE J. QE-*14*, 184-191 (1978)
5.73 R. Frey, F. Pradère: Infrared Phys. *16*, 117-120 (1976)
5.74 R.W. Terhune, P.D. Maker, C.M. Savage: Phys. Rev. Lett. *14*, 681-684 (1965)
5.75 Q.H.F. Vrehen, H.M. Hikspoors: Opt. Commun. *18*, 113-114 (1976)
5.76 J. Reif, H. Walther: Appl. Phys. *15*, 361-364 (1978)
5.77 P.P. Sorokin, N. Braslau: IBM J. Res. Dev. *8*, 177-181 (1964)
5.78 A.M. Prokhorov: Science *149*, 828-830 (1965)
5.79 R.L. Carman: Phys. Rev. A: *12*, 1048-1061 (1975)
5.80 E.B. Gordon, Yu.L. Moskvin: Sov. Phys. JETP *43*, 901-907 (1976)
5.81 D.S. Bethune, J.R. Lankard, P.P. Sorokin: J. Chem. Phys. *69*, 2076-2081 (1978)
5.82 R.G. Brewer: Physics Today *30*, No 5, 50-59 (1977)
5.83 G. Bret, E. Teller: Astrophys. J. *91*, 215- (1940)
5.84 M. Lepeles, R. Novick, N. Tolk: Phys. Rev. Lett. *15*, 690-693 (1965)
5.85 P. Braünlich, P. Lambropoulos: Phys. Rev. Lett. *20*, 1282-1284 (1968)
5.86 P. Platz: Appl. Phys. Lett. *17*, 537-539 (1970)
5.87 C. Delalande, A. Mysyrowicz: Opt. Commun. *7*, 10-12 (1973)
5.88 S.E. Harris: Appl. Phys. Lett. *31*, 498-500 (1977)
5.89 L.J. Zych, J. Lukasik, J.F. Young, S.E. Harris: Phys. Rev. Lett. *40*, 1493-1496 (1978)
5.90 R.W. Falcone, J.R. Willison, J.F. Young, S.E. Harris: Optics Lett. *3*, 162-163 (1978)
5.91 H. Komine, R.L. Byer: Appl. Phys. Lett. *27*, 300-302 (1975)
5.92 H.P. Yuen: Appl. Phys. Lett. *26*, 505-507 (1975)
5.93 A.V. Vinogradov, E.A. Yukov: JETP Lett. *16*, 447-448 (1972)
5.94 R.L. Carman, W.H. Lowdermilk: IEEE J. QE-*10*, 706 (1974)
5.95 R.L. Carman, W.H. Lowdermilk: Phys. Rev. Lett. *33*, 190-193 (1974)
5.96 P.P. Sorokin, J.R. Lankard: IEEE J. QE-*9*, 227-230 (1973)

5.97 W. Zapka, U. Brackman (to be published, 1979) have recently developed a new laser dye—2,2"-dimethyl-*p*-terphenyl—which extends the spectral coverage of dye lasers down to 311 nm
5.98 D. Cotter, M.A. Yuratich: Opt. Commun. *29*, 307-310 (1979)
5.99 A.Z. Grasiuk, I.G. Zubarev: Appl. Phys. *17*, 211-232 (1978)
5.100 W.R. Trutna, Y.R. Park, R.L. Byer: IEEE J. QE-*15*, 648-655 (1979)
5.101 M.G. Raymer, J. Mostowski, J.L. Carlsten: Phys. Rev. A *19*, 2304-2316 (1979)
5.102 J.J. Wynne, P.P. Sorokin: In *Nonlinear Infrared Generation*, Topics in Applied Physics, Vol.16, ed. by Y.-R. Shen (Springer, Berlin, Heidelberg, New York 1977) pp.159-214
6.1 J.J. Wynne, P.P. Sorokin, J.R. Lankard: In *Laser Spectroscopy*, ed. by R.G. Brewer, A. Mooradian (Plenum Press, New York 1974) pp. 103-111
6.2 I.L. Tyler, R.W. Alexander, R.J. Bell: Appl. Phys. Lett. *27*, 346-347 (1975)
6.3 P.A. Kärkkäinen: Appl. Phys. *13*, 159-163 (1977)
7.1 R.J. Jensen, J.G. Marinuzzi, C.P. Robinson, S.D. Rockwood: Laser Focus *12*, No 5, 51-63 (1976)
7.2 R.L. Byer: IEEE J. QE-*12*, 732-733 (1976)
7.3 R. Frey, F. Pradère, J. Lukasik, J. Ducuing: Opt. Commun. *22*, 355-357 (1977)
7.4 A. De Martino, R. Frey, F. Pradère: Opt. Commun. *27*, 262-266 (1978)
7.5 H. Kildal, T.F. Deutsch: IEEE J. QE-*12*, 429-435 (1976)
7.6 M.H. Kang, K.M. Chung, M.F. Becker: J. Appl. Phys. *47*, 4944-4948 (1976)
7.7 C.Y. She, K.W. Billman: Appl. Phys. Lett. *27*, 76-79 (1975)
7.8 C.Y. She, K.W. Billman: Appl. Phys. Lett. *27*, 636 (erratum) (1975)
7.9 R.E. McNair, M.B. Klein: Appl. Phys. Lett. *31*, 750-752 (1977)
7.10 H. Kildal, S.R. Brueck: Appl. Phys. Lett. *32*, 173-176 (1978)
7.11 N. Lee, R.L. Aggarwal, B. Lax: J. Appl. Phys. *48*, 2470-2476 (1977)
7.12 M.J. Pellin, J.T. Yardley: Appl. Phys. Lett. *29*, 304-307 (1976)
7.13 H.F. Schaefer: *The Electronic Structure of Atoms and Molecules. A Survey of Rigorous Quantum Mechanical Results* (Addison-Wesley, New York 1972)
7.14 C.L. Pan, C.Y. She, W.M. Fairbank, K.W. Billman: IEEE J. QE-*13*, 763-769 (1977)
7.15 C.L. Pan, C.Y. She, W.M. Fairbank, K.W. Billman: IEEE J. QE-*15*, 54 (1979)
7.16 A.L. Ford, J.C. Browne: Phys. Rev. A: *7*, 418-426 (1973)
7.17 H. Kildal: IEEE J. QE-*13*, 109-113 (1977)
7.18 H. Kildal, R.S. Eng, A.H.M. Ross: J. Mol. Spectrosc. *53*, 479-488 (1974)
7.19 F.R. Peterson, D.G. McDonald, J.D. Cupp, B.L. Danielson: In *Proc. Laser Spectrosc. Conf.*, ed. by A. Mooradian, R.G. Brewer (Plenum Press, New York 1974) pp. 555-569
7.20 M.A. Henesian, L. Kulevskij, R.L. Byer, R.L. Herbst: Opt. Commun. *18*, 225-226 (1976)
7.21 G.M. Hoover, D. Williams: J. Opt. Soc. Am. *59*, 28-33 (1969)
7.22 A.D. May, J.C. Stryland, G. Varghese: Can. J. Phys. *48*, 2331-2335 (1970)
7.23 S.R.J. Brueck: Chem. Phys. Lett. *53*, 273-277 (1978)
7.24 W.R.L. Clements, B.P. Stoicheff: Appl. Phys. Lett. *12*, 246-248 (1968)
7.25 J.R. Murray, A. Javan: J. Mol. Spectrosc. *29*, 502-504 (1969)
7.26 M.H. Kang, V.T. Nguyen, T.Y. Chang, T.C. Damen, E.G. Burkhardt: Appl. Phys. Lett. *33*, 303-304 (1978)
7.27 R.T.V. Kung, K.O. Tong, D. Lay: IEEE J. QE-*14*, 871-872 (1978)
7.28 W.S. Benedict, R. Herman, G.E. Moore, S. Silverman: J. Chem. Phys. *26*, 1671-1677 (1957)
7.29 R.A. Toth, R.H. Hunt, E.K. Plyler: J. Mol. Spectrosc. *35*, 110-126 (1970))
7.30 D.H. Rank, B.S. Rao, T.A. Wiggins: J. Mol. Spectrosc. *17*, 122-130 (1965)
7.31 G.C. Bhar, D.C. Hanna, B. Luther-Davies, R.C. Smith: Opt. Commun. *6*, 323-326 (1972)
7.32 D.C. Hanna, B. Luther-Davies, R.C. Smith, R. Wyatt: Appl. Phys. Lett. *25*, 142-144 (1974).
7.33 D. Andreou: Opt. Commun. *23*, 37-43 (1977)
7.34 R.W. Minck, E.E. Hagenlocker, W.G. Rado: Phys. Rev. Lett. *17*, 229-231 (1966)
7.35 T.Y. Chang, J.D. Mc Gee: Appl. Phys. Lett. *29*, 725-727 (1976)

7.36 S.J. Petuchowski, A.T. Rosenberger, T.A. DeTemple: IEEE J. QE-*13*, 476-481 (1977)
7.37 R.W. Minck, R.W. Terhune, W.G. Rado: Appl. Phys. Lett. *3*, 181-184 (1963)
7.38 R.L. Abrams, A. Yariv, P.A. Yeh: IEEE J. QE-*13*, 79-82 (1977)
7.39 R.L. Abrams, C.K. Asawa, T.K. Plant, A.E. Popa: IEEE J. QE-*13*, 82-85 (1977)
8.1 A.V. Voskanyan, D.N. Klyshlo, V.S. Tusmanov: Sov. Phys. JETP *18*, 967-972 (1964)
8.2 P. Lambropoulos, G. Doolen, S.P. Rountree: Phys. Rev. Lett. *34*, 636-639 (1975)
8.3 Y. Flank, G. Laplanche, M. Jaouen, A. Rachman: J. Phys. B. *9*, L409-L412 (1976)
8.4 T. Hänsch, P. Toschek: Z. Phys. *236*, 373-382 (1970)
8.5 D.S. Bethune, R.W. Smith, Y.-R. Shen: Phys. Rev. Lett. *37*, 431-434 (1976)
8.6 F.N.H. Robinson: *Macroscopic Electromagnetism* (Pergamon Press, Oxford 1973)
8.7 E.A. Power, S. Zienau: Proc. R. Soc. London, Ser.A: *251*, 427-454 (1959)
8.8 M. Babiker, E.A. Power, T. Thirunamachandran: Proc. R. Soc. London, Ser. A: *338*, 235-249 (1974)
8.9 E.A. Power, T. Thirunamachandran: J. Phys. B: *8*, L170-L172 (1975)
8.10 P.S. Pershan: Phys. Rev. *130*, 919-929 (1963)
8.11 I.J. Bigio, J.F. Ward: Phys. Rev. A: *9*, 35-39 (1974)
8.12 A. Flusberg, T. Mossberg, S.R. Hartmann: Phys. Rev. Lett. *38*, 59-62 (1977)
8.13 A. Flusberg, T. Mossberg, S.R. Hartmann: Phys. Rev. Lett. *38*, 694-697 (1977)
8.14 M. Matsuoka, H. Nakatsuka, H. Uchiki, M. Mitsunaga: Phys. Rev. Lett. *38*, 894-898 (1977)
8.15 T. Mossberg, A. Flusberg, S.R. Hartmann: Opt. Commun. *25*, 121-124 (1978)
8.16 L.I. Gudzenko, S.I. Yakovlenko: Sov. Phys. JETP *35*, 877-881 (1972)
8.17 S.E. Harris, D.B. Lidow: Phys. Rev. Lett. *33*, 674-676 (1974)
8.18 S.E. Harris, J.C. White: IEEE J. QE-*13*, 972-978 (1977)
8.19 Ph. Cahuzac, P.E. Toschek: Phys. Rev. Lett. *40*, 1087-1090 (1978)
8.20 S.I. Yakovlenko: Sov. J. Quantum Electron. *8*, 151-169 (1978)
8.21 S.E. Harris, R.W. Falcone, W.R. Green, D.B. Lidow, J.C. White, J.F. Young: In *Tunable Lasers and Applications*, Proc. Loen Conf. Norway 1976, Springer Series in Optical Sciences, Vol. 3, ed. by A. Mooradian, T. Jaeger, P. Stokseth (Springer, Berlin, Heidelberg, New York 1976) pp. 193-206
8.22 Ph. Cahuzac, P.E. Toschek: In *Laser Spectroscopy III*, ed. by J.J. Hall, J.L. Carlsten (Springer, Berlin, Heidelberg, New York 1977)
8.23 D.B. Lidow, R.W. Falcone, J.F. Young, S.E. Harris: Phys. Rev. Lett. *36*, 462-464 (1976)
8.24 D.B. Lidow, R.W. Falcone, J.F. Young, S.E. Harris: Phys. Rev. Lett. *37*, 1590 (erratum) (1976)
8.25 S.E. Harris, D.B. Lidow: Appl. Phys. Lett. *26*, 104-105 (1975)
8.26 A.A. Varfolomeev: Sov. Phys. JETP *39*, 985-988 (1974)
8.27 R.W. Hellwarth: J. Opt. Soc. Am. *67*, 1-3 (1977)
8.28 D.M. Bloom, G.C. Bjorklund: Appl. Phys. Lett. *31*, 592-594 (1977)
8.29 A. Yariv, D.M. Pepper: Opt. Lett. *1*, 16 (1977)
8.30 D.M. Bloom, P.F. Liao, N.P. Economou: Opt. Lett. *3*, 58-60 (1978)
8.31 P.F. Liao, D.M. Bloom: Opt. Lett. *3*, 4-6 (1978)
8.32 P.F. Liao, D.M. Bloom, N.P. Economou: Appl. Phys. Lett. *32*, 813-815 (1978)
8.33 D.M. Pepper, J. Au-Yeung, D. Fekete, A. Yariv: Opt. Lett. *3*, 7-9 (1978)
8.34 S. Jensen, R.W. Hellwarth: Appl. Phys. Lett. *32*, 166-168 (1978)
8.35 P.F. Liao, N.P. Economou, R.R. Freeman: Phys. Rev. Lett. *39*, 1473-1476 (1977)

Subject Index

Page numbers in *italics* refer to tables or figures.